FRANKIA SYMBIOSES

Developments in Plant and Soil Sciences
Volume 12

Frankia Symbioses

Edited by

A.D.L. AKKERMANS
Department of Microbiology
Agricultural University of
Wageningen
Wageningen, The Netherlands

D. BAKER
C.F. Kettering Research
Laboratory
Yellow Springs, Ohio, USA

K. HUSS-DANELL
Department of Plant Physiology
University of Umeå
Umeå, Sweden

J.D. TJEPKEMA
Department of Botany
and Plant Pathology
University of Maine
Orono, Maine, USA

First published as *Plant and Soil*, Vol. 78, Nos. 1–2 (1984)

1984 **MARTINUS NIJHOFF/DR W. JUNK PUBLISHERS**
a member of the KLUWER ACADEMIC PUBLISHERS GROUP
THE HAGUE / BOSTON / LANCASTER

Proceedings of the Workshop on *Frankia* Symbioses, held in Noordwijkerhout, The Netherlands on September 1, 1983 and in Wageningen, The Netherlands on September 5 and 6, 1983.

Distributors

for the United States and Canada: Kluwer Boston, Inc., 190 Old Derby Street, Hingham, MA 02043, USA
for all other countries: Kluwer Academic Publishers Group, Distribution Center, P.O.Box 322, 3300 AH Dordrecht, The Netherlands

Library of Congress Catalog Card Number: 84-6110

ISBN-13:978-94-009-6160-9   e-ISBN-13:978-94-009-6158-6
DOI: 10.1007/978-94-009-6158-6

Dedicated to
Professor **A. Quispel**
on the occasion of his retirement

Contents

Preface

Five years have now passed since the first symposium on frankiae was held at Harvard Forest, Petersham, Massachusetts, USA and the inauguration of the term actinorhiza. Many advances have been made during these five years in our understanding of the actinorhizal symbioses. Evidence for this was provided by the papers presented at the Wageningen Workshop on Frankia symbioses, held in Wageningen at the Department of Microbiology of the Agricultural University. Most of these papers are now published in this volume of PLANT AND SOIL. We kindly acknowledge the assistance of Anton Houwers, editor of the journal, in planning, reviewing and publishing these studies.

Although the papers presented at Wageningen described the active research areas, they also illuminated those aspects of these symbioses which remain beyond our understanding. Primary among the areas of our ignorance is the concept of species within the bacterial symbiont, *Frankia*. At present groupings of bacterial strains are based on cell chemistry, physiology, serology, DNA homology and symbiotic capabilities (cross-inoculation). When these classification schemes are merged no clear species framework is obtained. Undoubtedly part of the difficulty is due to a lack of strains for analysis. Currently bacterial strains from only half of the actinorhizal symbioses known to exist, have been isolated and studied in pure culture. We must postpone therefore any comprehensive taxonomic classification until a larger majority of the symbioses are represented.

Another research area wherein our understanding is insufficient is host-symbiont interaction. Biochemical, physiological and regulatory relationships between the bacterium and the plant have not been clearly defined. These interactions ultimately affect the performance of the symbiotic plants and must be studied if we wish to optimize the growth and usefulness of actinorhizae. Similarly how the symbiosis interacts with factors of its environment is little understood.

Our research efforts have been concentrated for the most part on the *Alnus-Frankia* relationship. This is understandable since most of the pure-cultured *Frankia* strains were isolated from *Alnus*. However, this emphasis on the alders should not blind us to the differences and potentials of other actinorhizal symbioses. More research should be devoted to these currently less well-known actinorhizae to complement the work in progress on alders. In particular, efforts should be directed to the application of actinorhizal symbioses for revegetation and land reclamation and for fuelwood production in developing countries. The actinorhizal genera are woody perennials and provide an important long-term contribution to their environments. This character should be exploited not only in contrast to the herbaceous legumes but also to the tree legumes being used in land reclamation projects.

If our research efforts in the next few years proceed as enthusiastically and vigorously as in the past five years we will realize large gains in our knowledge of the actinorhizal symbioses. Hopefully this will lead to utilization of actinorhizal plants on a larger scale in the near future.

We wish to record grateful thanks to the Agricultural University in Wageningen for sponsoring the Workshop and to all who contributed to the success of the Workshop.

December 1983
Wageningen, The Netherlands   Antoon D. L. Akkermans,
Yellow Springs, Ohio, USA   Dwight Baker,
Umeå, Sweden   Kerstin Huss-Danell
Orono, Maine, USA   John D. Tjepkema

Plant and Soil 78, 1–6 (1984).
© 1984 *Martinus Nijhoff/Dr W. Junk Publishers, The Hague.*

Ms. Fr 30

The taxonomy of the genus *Frankia*

MARY P. LECHEVALIER
*Waksman Institute of Microbiology, Rutgers University, P.O. Box 759, Piscataway,
NJ 08854, USA*

Key word *Frankia* taxonomy

Summary A discussion covering the problems of *Frankia* taxonomy was held at the "*Frankia*
Workshop" in Wageningen, September 4–6, 1983. It was agreed that the genus *Frankia* can
be satisfactorily defined, but that solid criteria for species determination are not now available
and that use of specific names should be avoided for the present.

Introduction

The last (8th) edition of Bergey's Manual of Determinative Bacteriology (Bergey 8)[5] was published in 1974; the next edition will appear in about two years. As the new edition will probably not be superseded by another for at least ten years, the new description of the genus *Frankia* is of concern to all persons engaged in research on members of this taxon. As a consequence, a discussion on this subject took place during the "*Frankia* Workshop" held in Wageningen, September 4–6, 1983. The following summarizes some of the progress and some of the problems with the taxonomy of *Frankia* and the proposals which were made to deal with the latter.

Previous taxonomy

The description of the genus *Frankia* that appeared in the last Bergey's Manual was written by Becking[3]. Considering that no endophyte had yet been isolated in pure culture, he did a remarkable job of formulating the characters of the genus. He 1) had excellent sections of nodules showing the micromorphology of the endophyte, including vesicles and sporangia; 2) predicted that frankiae would be microaerophilic and 3) attempted a chemical analysis of the endophyte cells. His cell wall analysis of endophyte material from *Alnus glutinosa* nodules showed the presence of the *meso* isomer of diaminopimelic acid (DAP), arabinose, galactose and glycine as diagnostic constituents. Because the endophyte forms coccoid "bacteroids"

1

in planta (which we now know to be sporangiospores) and because of the content of *meso* DAP, galactose and arabinose, Becking hypothesized a relationship of frankiae to the *Nocardiaceae*. On the other hand, he noted that the presence of vesicles (which he thought might be degenerate sporangia) plus the content of glycine, might possibly indicate a relationship to the *Actinoplanaceae*. We now know[1,11] that the cell walls of frankiae are of type III (*meso* DAP, glutamic acid, alanine, muramic acid and glucosamine), the most common type among the actinomycetes[12]. Galactose is variably present, arabinose usually occurs only in traces and glycine is absent. Given the difficulties of isolating pure endophyte vesicle clusters from nodules, the glycine probably came from contaminating material of plant origin.

Becking proposed ten species of *Frankia* based on host plant relationships (Table 1) with *F. alni* as the type species. Unfortunately, his species were based on host cross-infectivity as determined using nodular inocula and these divisions do not always seem to be borne out by similar tests using pure cultures[7, 9, 16]. It should be noted however, that the possibility remains that techniques currently used to demonstrate infectivity of the isolates may require further refinement.

Present status of Frankia taxonomy

At present, actinomycetes are assigned to a genus on the basis of morphology (formation of organelles such as spore chains, endospores, motile spores, sporangia, sclerotia *etc.*) and cell chemistry (cell wall and whole cell composition, phospholipid and fatty acid (including mycolate) type, GC% and the like). Most species are characterized by morphology (spore surface, pigments, shape of spore chains *etc.*), physiology and menaquinone composition.

The criteria used in classifying an actinomycete as a member of the genus *Frankia* include the following: 1) morphology (sporangia and vesicle formation in submerged liquid culture); 2) chemistry (cell wall of type III, phospholipid of type PI)[15], 3) serology[2]; 4) DNA homology (An and Mullin, to be published) and 5) demonstration of infectivity and/or effectivity for a host plant. These criteria are useful and permit one, at least to date, to identify unequivocally an unknown as a frankia.

In contrast, it seems impossible to clearly define a species of the genus *Frankia* at the moment. The following are possible criteria for use in this endeavour along with comments made by a number of workers in the field.

Table 1. Species of the genus *Frankia*[a]

Plant family	Plant genus	*Frankia* species
1. Betulaceae	*Alnus*	*Frankia alni*
2. Casuarinaceae	*Casuarina*	*F. casuarinae*
3. Coriariaceae	*Coriaria*	*F. coriariae*
4. Elaeagnaceae	*Elaeagnus*	*F. elaeagni*
	Hippophaë	
	Shepherdia	
5. Myricaceae	*Myrica (Gale)*	*F. brunchorstii*
6. Rhamnaceae	*Ceanothus*	*F. ceanothi*
	Discaria	*F. discariae*
7. Rosaceae	*Cercocarpus*	*F. cercocarpi*
	Dryas	*F. dryadis*
	Purshia	*F. purshiae*

[a] = From[4]

1 Ecology

All isolates from the same host plant should belong to the same species. This seems logical, but is complicated by increasing evidence of great diversity in morphology, physiology, chemistry, serology, DNA homology and host range of isolates from the same host [4,6,13,14,16] (An and Mullin to be published).

2 Infectivity

All endophytes, regardless of source, capable of infecting a given plant host should belong to the same species. Unfortunately, boundaries between groups may vary with the strain. Although the *Myrica/Comptonia/Alnus* isolates appear to form the most solid host plant group, a strain from *Elaeagnus* that infects *Alnus* has been isolated (Lalonde, personal communication). Also, what happens when the strain is non-infective?

3 Morphology

Size and shape of sporangia or spores, vesicle formation on media with no or a fixed source of nitrogen, cellular or soluble pigments might be used. Such criteria have often been employed for actinomycete taxonomy in the past. In general, these characters are not sufficiently stable to be useful.

4 Cell chemistry

Protein patterns (*e.g.*[4]) may have potential, but the technique is highly specialized. Isoenzymes from frankiae have not yet been tested for purposes of taxonomy. Whole cell sugar patterns such as those found in physiological group B (see below)' seem to be consistent, but those in group A are very diverse and the meaning of this is not yet clear. All of these criteria may be of use, but will probably have to be employed in conjunction with others. Fatty acid and menaquinone patterns may have potential but have not yet been tried.

5 Physiology

Various tests such as production of acid during carbohydrate dissimilation, utilization or decarboxylation of organic acids, protease, urease or amylase production, effect of Tween 80 on growth *etc.* may all be useful. The groupings A and B (see Table 2) seen in *Alnus*[14] and *Comptonia*[13] isolates may also help, but other groups are certainly going to be found.

6 Serology

Serological analysis has good potential and needs to be further developed. It is felt by some workers that the technique is too sensitive to differentiate between stains whereas others find it too insensitive.

7 Phage grouping

No frankiaphage has been reported yet.

8 DNA homology

This may be very helpful in discerning relationships among strains but the technique is too difficult for routine taxonomy.

9 16s RNA nucleotide catalogs

Aside from its complexity, this technique is probably not useful in delineating species but rather for discerning phylogenetic relationships.

The following suggestions were made to solve the present dilemma concerning the use of species names:

1) Make all strains members of a single species and subdivide later.

2) Avoid the use of or creation of any species names until further work can be done.

3) Use numerical code to describe each strain and avoid species names altogether.

In general, the consensus of those present appeared to favor option (2); thus for the moment we will continue to refer to *Frankia* spp. rather than using specific names. It was pointed out that the plan to catalog various *Frankia* strains which was adopted at the previous *Frankia* meeting held in Madison, WI in 1982, has been described in a recent publication[10].   In this catalog, and in all future publications

Table 2. Physiological types A and B in frankiae

	A	B
Cellular pigments	+	—
Vesicles	V	+
	(on complex media)	(on nitrogen-free media + succinate)
Synergism		
Tween + carbohydrate	—	+
Depressed growth		
Tween + carbohydrate	+	—
Carbohydrate uptake		
at 0.5%	+	—
Maintenance on slants	+	—
Infective on original host	—	+ (usually)
Serological group	II or other	I
Whole cell sugar	Various: Fucose, Madurose, Xylose, Glucose, Galactose	Xylose

V = Variable; + = Positive; — = Negative.

on frankiae, a simple combination of three letters (standing for the culture collection involved) and numbers (up to 10) will replace the present systems of designating strains. Various data on each strain including previous synonyms will be included in the catalog. Persons wishing to have their strains cataloged should write the author of this article for a copy of the proper cataloging form.

Acknowledgements I wish to acknowledge the helpful suggestions, comments and information provided by A. Akkermans, D. Baker, J. H. Becking, D. Benson, A. Burggraaf, M. Lalonde, B. Mullin, J. G. Torrey and all other persons at the Workshop who participated in the discussion.

References

1 Baker D, Lechevalier M P and Dillon J T 1981 Strain analysis of actinorhizal microsymbionts (genus: *Frankia*) *In* Current Perspectives in Nitrogen Fixation. Eds. A H Gibson and W Newton, Australian Acad. Sci., Canberra, p 479.

2 Baker D, Pengelly W and Torrey J G 1981 Immunochemical analysis of relationships among isolated frankiae (Actinomycetales). Int. J. Syst. Bacteriol. 31, 148–151.

3 Becking J H 1974 Family III. *Frankiaceae* Becking, 1970, 201. *In* Bergey's Manual of Determinative Bacteriology. Eds. R E Buchanan and N E Gibons. Williams and Wilkins Co., Baltimore, pp 701–706.

4 Benson D R and Hanna D 1983 *Frankia* diversity in an alder stand as estimated by SDS-PAGE of whole cell protein. Can. J. Bot. 61, 2919–2923.

5 Buchanan R E and Gibbons N E 1974 Bergey's Manual of Determinative Bacteriology. Williams and Wilkins Co., Baltimore, pp 1268.

6 Burggraaf A and Valstar H 1984 Phenotypic heterogeneity in *Frankia* LD Agpl studied on clones and reisolates. Plant and Soil 78, 29–43.

7   Callaham D, Newcomb W, Torrey J G and Peterson R L 1979 Root hair infection in actinomycete-induced root nodule initiation in *Casuarina, Myrica* and *Comptonia* Bot. Gaz. 140 (Suppl.), S1–S9.

8   Dillon J T and Baker D 1983 Variations in nitrogen-fixing efficiency among pure-cultured *Frankia* strains tested on actinorhizal plants as an indication of symbiotic compatibility. New Phytol 92, 215–219.

9   Gautier D, Diem H G and Dommergues Y 1981 Infectivité et effectivité de souches de *Frankia* isolées de nodules de *Casuarina equisetifolia* et d'*Hippophaë rhamnoides*. C. R. Acad. Sci., Paris 293, 489–491.

10   Lechevalier M P 1983 Cataloging *Frankia* strains. Can. J. Bot. 61, 2964–2967.

11   Lechevalier M P and Lechevalier H 1979 The taxonomic position of the actinomycetic endophytes. *In* Symbiotic Nitrogen Fixation in the Management of temperate Forests. Eds. J C Gordon, C T Wheeler and D A Perry. Forest Research Laboratory, Oregon State Univ., Corvallis pp 111–122.

12   Lechevalier M P and Lechevalier H A 1980 The chemotaxonomy of actinomycetes. *In* Actinomycete Taxonomy. Eds. A Dietz and D W Thayer. Society for Industrial Microbiology. Special Publication #6. Arlington, VA. pp 227–294.

13   Lechevalier M P and Ruan J S 1984 Physiological and chemical diversity of *Frankia* spp. isolated from nodules of *Comptonia peregrina* (L.) Coult and *Ceanothus americanus* L. Plant and Soil 78, 15–22.

14   Lechevalier M P, Baker D and Horrière F 1983 Physiology, chemistry, serology and infectivity of two *Frankia* isolates from *Alnus incana* subsp. *rugosa*. Can J. Bot. 61, 2826–2833.

15   Lechevalier M P, Horrière F and Lechevalier H 1982 The biology of *Frankia* and related organisms. Develop. Indust. Microbiol. 23, 51–60.

16   Zhang Z, Lopez M and Torrey J G 1984 A comparison of cultural characteristics and infectivity of *Frankia* isolates from root nodules of *Casuarina* species. Plant and Soil 78, 79–90.

Plant and Soil 78, 7–13 (1984).   Ms. Fr 12
© 1984 *Martinus Nijhoff/Dr W. Junk Publishers, The Hague.*

In vitro physiological approach to classification of *Frankia* isolates of 'the Alnus group', based on urease, protease and β-glucosidase activities

F. HORRIÈRE
Laboratoire d'Ecologie Microbienne, ERA-CNRS 848, Bât. 405, Université Lyon I, 43 Bd. du 11 Novembre 1918, F-69622 Villeurbanne Cedex, France

Key words *Alnus* Esculin hydrolysis *Frankia* Gelatin hydrolysis β-Glucosidase *Myrica* Protease Urease

Summary   Most of the *Frankia* strains isolated from *Alnus* and *Myrica* species are morphologically almost indistinguishable, when grown under standard culture conditions. They form similar vegetative hyphae while sporangia are produced in variable amounts from strain to strain.
   Physiological reactions were assessed in order to compare 20 strains isolated from various species of *Alnus* and one species of *Myrica* in Europe and North America. Among invariant negative or positive characteristics, differences in urease, protease and β-glucosidase activities appeared to be of significant value.

Introduction

As numerous *Frankia* strains are now isolated and cultured *in vitro*, methods for classifying and identifying them have to be developed. Morphological criteria, chemotaxonomic and serologic studies, as well as host compatibility traits have been used as classification parameters of these strains. Such techniques allowed *Frankia* isolates to be placed into several large groups[2, 3, 7, 8].

Physiological and biochemical characters have been useful in identifying certain actinomycetes and are included in numerical taxomonic studies[13]. Using such characters, we have therefore attempted to compare 20 strains of *Frankia*, isolated in Europe or North America from *Alnus* and *Myrica* host plants. However, standard methods have to be modified to give reliable and repeatable results when applied to a slow growing actinomycete like *Frankia*, and to date, only a few tests were found useful to differentiate strains belonging to the same chemotype and serotype.

This study compares the urease, protease and β-glucosidase activities of these strains and discusses the validity of these activities as physiological classification parameters.

Material and methods

Test strains
   20 strains were studied; 19 were isolated from different species of *Alnus* and 1 from *Myrica pensylvanica* (Table 1). For most of them, strains previously tested for cell chemistry and

serology had shown a whole cell sugar pattern D and belong to serogroup I[1,8]. Two strains at least, one isolated from *Alnus incana* ssp. *rugosa* (AirI2) and the other from *Alnus glutinosa* (Aglla) differed from the typical *Alnus* isolates.

Cultural conditions

Strains were maintained in static liquid culture on a medium containing (g/1): glucose, 10; casamino acids, 4; K_2HPO_4, 0.5; $MgSO_4 \cdot 7H_2O$, 0.2; $CaCl_2 \cdot 2H_2O$, 0.1; trace elements of Q mod medium[6]; and 1 ml/l of the following vitamin stock solution (mg/l): thiamin HCl, 10; nicotinic acid, 50; pyridoxine HCl, 50; pH, 6.8. Tween 80 was added to this basal medium at 0.2% except for two strains (AirI2 and Aglla), since it had the effect of depressing their growth. Before testing, each strain was grown on one of these two media in a culture tube containing 5 ml of medium, for 5 weeks at 28°C.

Urease test

In 5 week old cultures, growth medium was replaced by 3 ml of a strongly buffered urea broth, where production of ammonia could be detected by pH indicator (Difco B280)[14].

Table 1. Strains used in this study

Strain	Source plant	Source location	WCS[a]	SG[b]	by[c]
Ac21a	*Alnus cordata* ·	France, Corsica			ULY
Ag10ai	*Alnus glutinosa*	France	D[d]		ULY
Ag10f	*Alnus glutinosa*	France			ULY
Ag10g	*Alnus glutinosa*	France	D		ULY
Ag11a	*Alnus glutinosa*	France	X[e]		ULY
Ag12a	*Alnus glutinosa*	France			ULY
Ag21e	*Alnus glutinosa*	France, Corsica			ULY
Ag24a	*Alnus glutinosa*	France			ULY
Ag24b	*Alnus glutinosa*	France			ULY
Ag24c	*Alnus glutinosa*	France			ULY
Ail5a	*Alnus incana*	France			ULY
AirI1	*Alnus incana* ssp. *rugosa*	U.S.A., VT	D	I[f]	MPL
AirI2	*Alnus incana* ssp. *rugosa*	U.S.A., VT	D	H[g]	MPL
ArI3	*Alnus rubra*	U.S.A., WA	D	I	AB
ArI4	*Alnus rubra*	U.S.A., WA		I	DDB
ArI5	*Alnus rubra*	U.S.A., OR		I	DDB
Av16	*Alnus viridis*	France	D		ULY
AvcI1	*Alnus viridis* ssp. *crispa*	Canada, ONT	D	I	DDB
AcN1[Ag]	*Alnus viridis* ssp. *crispa*	Canada, QUE	D	I	ML
MpI1	*Myrica pensylvanica*	U.S.A., MA	D	I	MPL

[a]Whole cell sugar pattern[9]
[b]Serogroup[3]
[c]Isolated by: AB = Alison Berry, Harvard Forest; DDB = Dwight Baker, Middlebury College; ML = Maurice Lalonde, Laval University; MPL = Mary Lechevalier, Rutgers University; ULY = University of Lyon
[d]Contains glucose
[e]Contains fucose
[f]'Alder' serotype
[g]Heterologous reaction with antisera to CpI1 (Serogroup I) and EuI1 (Serogroup II) antisera.

Colonies were then maintained at 28°C and the time needed for a positive response, indicated by a change in colour from yellow (pH 6.8) to red (pH 8.1 and more alkaline), was recorded daily.

Gelatin hydrolysis test

Proteolytic activity was tested with gelatin as the enzyme substrate. In 5 week old cultures, growth medium was replaced by 0.5 ml of physiological saline and a piece of photographic film (Institut Pasteur 53 861) was added into the culture tube[10]. Colonies were then maintained at 35°C and the time needed for a positive response, indicated by the hydrolysis of the gelatin layer of the film, was recorded daily.

To determine the reproducibility of the results, tests of urease and gelatin hydrolysis were replicated at least twice.

Table 2. β-Glucosidase and urease activities, gelatinolytic activity and sporulation of frankiae isolates

Strain	ESC[a]	URE[b]	GEL[c]	Frequency of sporangia[d]
Ac21a		3	0	0
Ag10ai		3	3	+ +
Ag10f	+	0	0	0
Ag10g	+	0	0	0
Ag11a	+	1	5	0
Ag12a		3	3	+ +
Ag21e	–	3	0	0
Ag24a		2	4	+ + +
Ag24b	–	2	4	+ + +
Ag24c	–	2	4	+ + +
Ail5a[e]		1[e]	1[e]	+ + +[e]
AirI1		2	3	+ +
AirI2	+	0	5	+ + +
ArI3	–	3	4	+ + +
ArI4	–	2	2	+
ArI5	–	1	3	+ +
Av16a		2	4	+ + +
AvcI1	–	3	2	+
AcN1[Ag]	–	3	4	+ + +
MpI1	-	5	1	0

[a, b, c]enzymatic activities were evaluated as follows,

[a]esculine hydrolysis:
+ β-glucosidase positive; – β-glucosidase negative

[b]urease activity:
0 negative reaction within 1 month; 1 positive reaction after more than 12 days; 2 positive reaction between 8 and 12 days; 3 positive reaction between 5 and 8 days; 4 positive reaction between 2 and 5 days; 5 positive reaction within 2 days

[c]gelatinolytic activity:
0 negative reaction within 1 month; 1 positive reaction after more than 10 days; 2 positive reaction between 7 and 10 days; 3 positive reaction between 4 and 7 days; 4 positive reaction between 2 and 4 days; 5 positive reaction within 2 days

[d]the frequency of sporangia was evaluated visually:
0 no sporangia; + a few small sporangia; + + intermediate frequency; + + + numerous and large-size sporangia

[e]Ail5a became available late in this study and grew poorly on all media tested.

Esculin hydrolysis test

A 5 week old colony was used to inoculate an agar medium supplemented with 0.1% esculin and 0.1% ferric citrate (Institut Pasteur 54 310). The esculin hydrolysis, due to a β-glucosidase activity, gives rise to coumarin which react with ferric citrate to form a brownish black complex.

Results and discussion

Urease activity

A large number of bacteria are capable of urease production. Using a strongly buffered urea broth, it was possible to differentiate strains producing large or small quantities of ammonia by urea hydrolysis and we suggest a graded rate response in their ability to hydrolyze urea under the conditions used (Table 2). Three strains (Ag10f, Ag10g and AirI2) were urease negative; in contrast, MpI1 exhibited a particularly strong positive reaction.

These reactions were consistent through repeated testing, and the variations never exceeded two days for the strain AcN1[Ag] tested five times over the year. Nevertheless, a good standardization has been

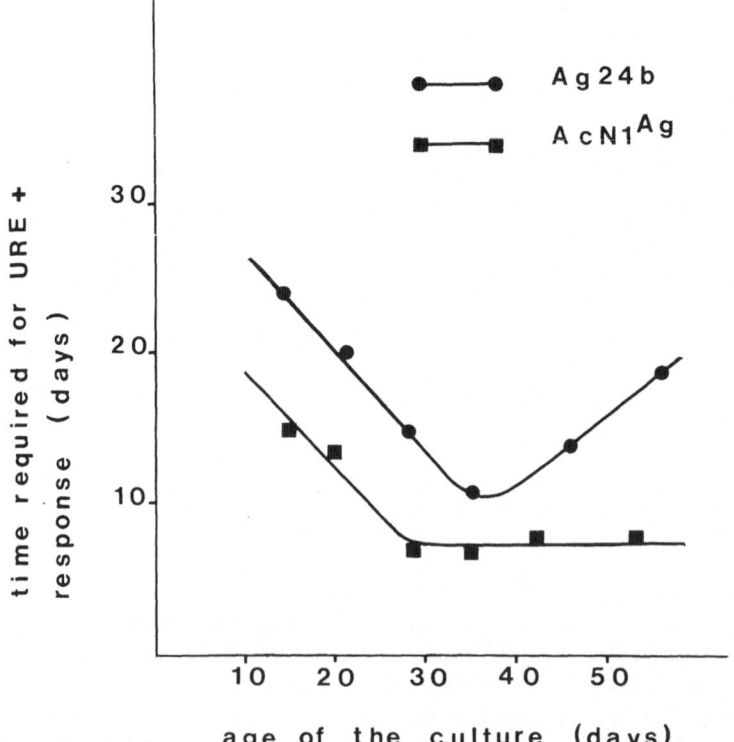

Fig. 1. Influence of the age of the culture on the urease activity of two strains of *Frankia* (Ag24b and AcN1[Ag]).

found necessary to give reliable results. Not only the cultural conditions, but also the age of the culture may markedly influence the urease production (Fig. 1). In our experiments, the sensitivity of urease activity to the age of the culture depended on the tested strains. After a phase of increasing reaction during the first month following the inoculation, this activity remained constant for AcNl[Ag], whereas it decreased drastically for Ag24b. Thus, two tests, performed on one resp. two months old cultures, would appear to be of better use to differentiate some strains from each other. Notwithstanding the difficulties in standardization, urease tests have been shown useful in identification of actinomycetes[11, 12].

Proteolytic activity

Gelatin was used as the enzyme substrate, because gelatin hydrolysis tests are very easy to perform, and because gelatin is very susceptible to pronase and is degraded by a lot of bacteria for which proteolytic activity has been revealed[15].

Pronounced differences in the ability of the tested strains to produce a protease capable of digesting gelatin under our conditions were observed (Table 2). AirI2, which has already been shown to produce caseinase[7], presented the strongest activity; in contrast, Ac21a, Ag10f, Ag10g and Ag21e gave negative to very weak results. The hydrolysis of gelatin was positively correlated with presence of numerous sporangia in the colonies, except for the strain Aglla (Table 2). A possible explanation is that extracellular proteases are mostly synthesized during sporogenesis. Similar differences between protease synthesis during growth and sporogenesis have been observed and studied for the genus *Bacillus*[4]. Both the ability of producing sporangia and the protease activity of each strain were not affected by numerous transfers over the year, under the conditions used.

β-Glucosidase activity

Most of the tested strains were not able to hydrolyze esculin and are, therefore, considered as β-glucosidase negative. However, Ag10f, Ag10g, Aglla and AirI2 were β-glucosidase positive (Table 2).

Validity of urease, protease and β-glucosidase activities as physiological classification parameters

On the basis of both chemotaxonomic and serologic results, all of the tested strains, except Aglla and AirI2, were so far indistinguishable. Physiologically, they are strictly microaerophilic, have rather similar carbohydrate and lipid-utilization pattern[7] and they could only be distinguished by nitrate-reductase activity[8]. With standard methods,

urease, gelatin and esculin hydrolysis tests gave reliable and reproducible results when applied to various *Frankia* isolates. These results
prove urease, protease and glucosidase activities as useful criterions
in the differentiation among strains isolated from *Alnus* or *Myrica*
host plants, since there were important variations in urease, protease
and β-glucosidase production from one strain to another.

The most significant results of this survey appear to be the very
strong urease activity of the strain MpI1, and the β-glucosidase activity
together with no urease activity of the two strains Ag10f and Ag10g.
The strains Ag11a and AirI2, which have already been separated from
the 'typical *Alnus* isolates', were shown once more physiologically
different.

The described tests can be a useful addition to the methods previously
used in differentiating frankiae isolates.

Acknowledgements We thank A. Berry, D. Baker, M. Lalonde and M. Lechevalier for supplying
strains.

References

1   Baker D 1982 A cumulative listing of isolated frankiae, the symbiotic nitrogen-fixing
    actinomycetes. The Actinomycetes 17, 35–42.
2   Baker D 1982 Serologic and host compatibility relationships among the isolated frankiae.
    Can. J. Bot. Unpublished manuscript.
3   Baker D, Pengelly W L and Torrey J G 1981 Immunochemical analysis of relationships
    among isolated frankiae (Actinomycetales). Int. J. Syst. Bacteriol. 31, 148–151.
4   Chaloupka J, Severin A I, Sastry K J, Kucerova H and Strnadova M 1982 Differences in
    the regulation of exocellular proteinase synthesis during growth and sporogenesis of
    Bacillus megaterium. Can. J. Microbiol. 28, 1214–1218.
5   Goodfellow M and Pirouz T 1982 Numerical classification of sporoactinomycetes con-
    taining meso-diaminopimelic acid in the cell wall. J. Gen. Microbiol. 128, 503–527.
6   Lalonde M and Calvert H E 1979 Production of *Frankia* hyphae and spores as infective
    inoculant for *Alnus* species. *In* Symbiotic Nitrogen Fixation in the Management of
    temperate Forests. Eds. J C Gordon *et al*. Oregon State university, Corvallis, pp 95–110.
7   Lechevalier M P, Baker D and Horrière F 1983 Physiology, chemistry, serology and
    infectivity of two Frankia isolates from *Alnus incana* subsp. *rugosa*. Can. J. Bot. 61,
    2826–2833.
8   Lechevalier M P, Horrière F and Lechevalier H 1982 The biology of Frankia and related
    organisms. Develop. Indust. Microbiol. 23, 51–60.
9   Lechevalier M P and Lechevalier H A 1979 The taxonomic position of the actinomycetic
    endophytes. *In* Symbiotic Nitrogen Fixation in the Management of temperate Forests.
    Ed. J C Gordon *et al*. Oregon State University, Corvallis, pp 111–122.
10 Le Minor L and Piéchaud M 1963 Une méthode rapide de recherche de la proteolyse de la
    gélatine. Ann. Inst. Pasteur 105, 792–794.
11 Scharfen J 1973 Urease as a useful criterion in the classification of microaerophilic
    actinomycetes. Zbl. Bakt. Hyg., 1. Abt. Orig. A 225, 89–94.
12 Schofield G M and Schaal K P 1980 Rapid micromethods for detecting deamination and
    decarboxylation of aminoacids, indole production, and reduction of nitrate and nitrite
    by facultatively anaerobic actinomycetes. Zbl. Bakt. Hyg., 1. Abt. Orig. A 247, 383–391.

13   Schofield G M and Schaal K P 1981 A numerical taxonomic study of members of the
      Actinomycetaceae and related taxa. J. Gen. Microbiol. 127, 237–259.
14   Stuart C A, van Stratum E and Rustignan R 1945 Further studies on urease production
      by Proteus and related organisms. J. Bacteriol. 49, 437–444.
15   Wikström M B 1983 Detection of microbial proteolytic activity by a cultivation plate
      assay in which different proteins absorbed to a hydrophobic surface are used as substrates.
      Appl. Environ. Microbiol. 45, 393–400.

Plant and Soil 78, 15–22 (1984).   Ms. Fr 11
© 1984 *Martinus Nijhoff/Dr W. Junk Publishers, The Hague.*

Physiology and chemical diversity of *Frankia* spp. isolated from nodules of *Comptonia peregrina* (L.) Coult. and *Ceanothus americanus* L.

MARY P. LECHEVALIER and JI-SHENG RUAN
Waksman Institute of Microbiology, Rutgers, University, P.O. Box 759, Piscataway,
NJ 08854, USA and Academia Sinica, Peking, Peoples Republic of China

Key words *Frankia* Physiology Taxonomy

Summary Two *Frankia* spp., isolated from the nodules of the plant host *Comptonia pere-grina*, were found to fall into two previously described physiological groups (A and B). Of five frankia isolates from *Ceanothus americanus* plants of the same provenance, three belonged to physiological group A and two to a novel group whose final disposition remains to be decided. The diversity in whole cell sugar chemistry, morphology and other growth characteristics of these strains is discussed.

Introduction

In the last few years, many strains of the nitrogen-fixing actino-mycete genus, *Frankia,* have been isolated in pure culture[1]. On the basis of our studies and those of others, one can make the three following observations: 1) putative members of the genus can, in most cases, be solidly identified as frankiae on the basis of micromorphology and some strains can be identified on the basis of their cell chemistry[11]; 2) physiological characteristics enable us to divide the genus into at least two subgroups[12]; 3) strains isolated from nodules of the same plant species can be exceedingly diverse in terms of morphology, cell chemistry, physiology, serology and infectivity for plants (*e.g.*[2,5,8,11,12,15]). We are thus confronted with reasonable homogeneity at the genus level and with considerable heterogeneity at all lower taxonomic ranks.

Recently[12] we compared two quite different frankiae, LLR 01321 (AirI1) and LLR 01322 (AirI2), which were isolated from field-collected nodules of the same host, *Alnus incana* ssp. *rugosa.* These strains differed in morphology, physiology, serology and infectivity for the host plant. Based on our physiological studies, we concluded that they represented two distinct types of frankiae which we have termed A and B.

Type A strains have pigmented cells, are relatively aerobic and can be maintained on slants, show rapid growth in the presence of various carbohydrates at a concentration of 0.5%, and are chemically and serologically diverse. Many produce lipases but growth in Tween-containing media is often inhibited because of an accumulation of released fatty acids. None of the group was known to be infective and effective on its original plant host, although, in one case, the formation of effective nodules on a host of the family Elaeagnaceae has been observed[7].

The second, or B type strains, have non-pigmented cells (although many can elaborate soluble pigments into the medium), are micro-aerophilic and thus cannot be maintained on slants, usually do not take up carbohydrates at a concentration of 0.5%, and show a characteristic growth response typically reflecting a synergism between certain carbohydrates and Tween 80. These strains are chemically homogeneous and have been shown to be serologically related[3]. All but one (Avs I2) produce effective nodules on their original plant host.

In this paper, we present the results of morphological, physiological and chemical studies on several recent isolates from the plant genera *Ceanothus* and *Comptonia*.

Materials and methods

Frankia sp. LLR 07011 was isolated from nodules of *Comptonia peregrina* collected in the Black Creek National Refuge, VT, USA. CpI1 was isolated by Callaham et al.[6] from the same plant host growing in Massachusetts. Strains LLR 03011, 03013, 03014, 03015 and DDB 030210 were isolated from nodules of *Ceanothus americanus* collected in Addison, VT. The last strain was kindly provided by Dr. D. Baker of Charles F. Kettering Research Laboratory, Yellow Springs, OH, USA.

Isolation of cultures was either by dilution[16] or by the sucrose density gradient technique[4] from nodules sterilized with NaOCl or with OsO_4[9]. LLR 07011, 03011 (Ca I1), 03014 and DDB 030210 are all maintained on medium "S"[12], LLR 03013 and 03015 are maintained on NZ-propionate[12] and CpI1 is maintained on L/2 medium[13] (old designations of strains are in parenthesis). Physiological testing was carried out as previously described[12,13]. The method of whole cell sugar analysis was that described by Lechevalier and Lechevalier[10].

Results and discussion

The two *Frankia* isolates from *Comptonia peregrina*, LLR 07011 and CpI1, were compared physiologically and chemically. Physiological tests show that LLR 07011 falls into our "A" group since it gave considerably reduced growth in media containing both glucose and Tween 80 over that in media with glucose alone (Fig. 1). Confirmation of this assignment was provided by its carbohydrate utilization pattern which shows uptake and utilization of arabinose, glucose and xylose at a concentration of 0.5% (Table 1). In contrast, CpI1, which is a "B" type *Frankia*, shows a marked synergistic effect on growth of the addition of 0.2% Tween 80 to glucose-containing media (Fig. 1) and no utilization of carbohydrates at 0.5% (Table 1). Also, the two strains

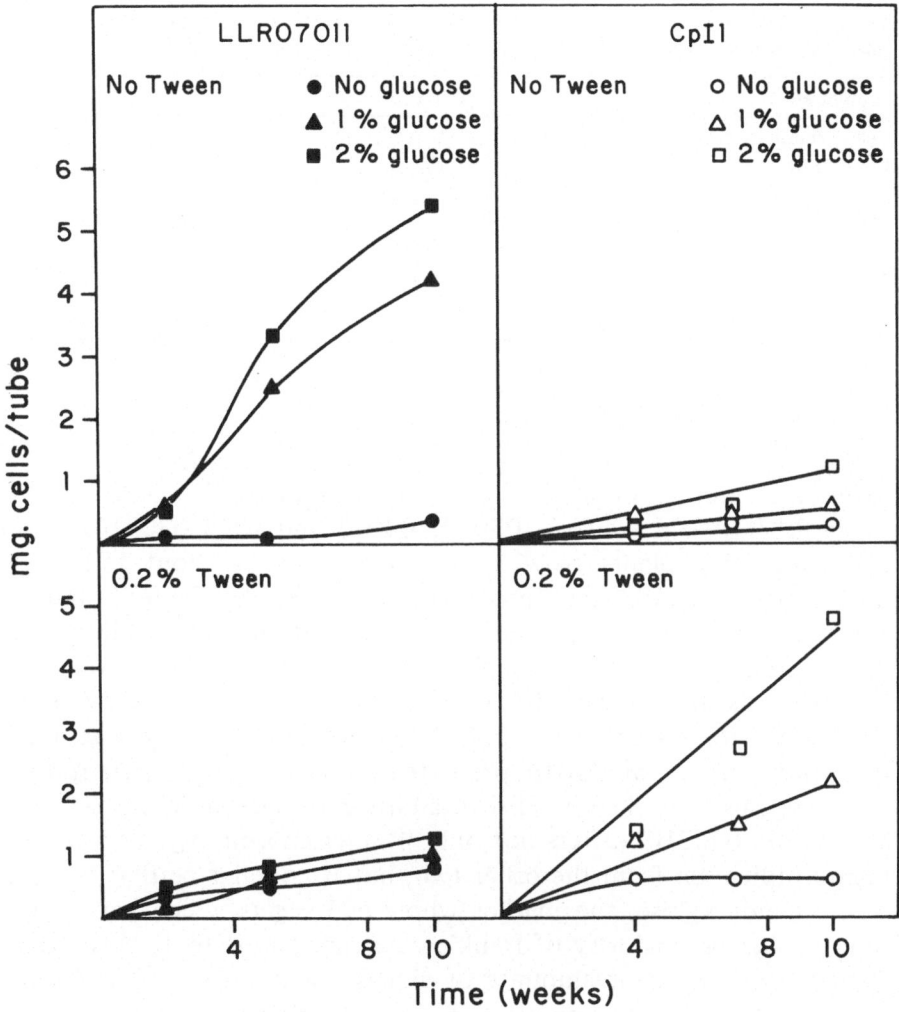

Fig. 1. Influence of Tween 80 and glucose on growth of *Frankia* spp. LLR 07011 and Cp I1 in "S" medium.

differ in whole cell sugar patterns. LLR 07011 has as characteristic whole cell sugar, fucose, but not xylose, whereas CpI1 contains xylose but no fucose, the latter pattern being typical of all B type strains examined to date. Fucose has previously been found in several other frankiae in our collection including several of the strains discussed below. Both *Comptonia* isolates contain rhamnose and the 2-O-methyl-D-mannose found by Mort *et al.*[14], to be a sugar common in cells of frankiae. In addition, typical of "A" types, LLR 07011 was found to be non-infective on *C. peregrina* (J.G. Torrey, personal communication) whereas CpI1 is known to be infective and effective[6], typical of "B"'s.

Table 1. Utilization of carbohydrates by *Frankia* spp. LLR 07011 and CpI1 (Isolates from *Comptonia peregrina*)

Carbohydrate	07011	CpI1
L (+) − Arabinose	U/A[1]	−[2]
D (+) − Glucose	U/A	−
Glycerol	−[2]	−
3-O-Methyl-D-Glucose	−	−
D (+) − Maltose	V[3]	−
Sucrose	U/NA[4]	−
D (+) − Trehalose	−	−
D (+) − Xylose	U/A	−

[1] U = Utilization A = Acid production.
[2] No utilization at 0.5%.
[3] Variable results; see note 1, Table 3.
[4] NA = No acid produced.

The five frankiae isolated from nodules of *Ceanothus americanus*, namely LLR 03011, 03013, 03014, 03015 and DDB 030210, were all isolated from plants growing within an area of about 15 m². A summary of data on these various strains is given in Tables 2 and 3. As can be seen, the strains vary in pigmentation, best medium for growth, commonness of sporangia and vesicle formation, whole cell sugar patterns and in carbohydrate utilization. The data show that LLR 03011, 03014 and DDB 030210 are clearcut "A" type strains. The growth curves of 03014 presented in Fig. 2 are representative of the responses of these strains to addition of Tween 80 to ":S" medium. Strain 030210, differs not only in micromorphology and carbohydrate utilization from the other two, but it contains neither fucose, madurose nor xylose, the marker whole cell sugars which have been found in the vast majority of frankiae examined to date. Rather, DDB 030210 contains major amounts of glucose (and galactose) and thus more closely resembles strain PtI1 from *Purshia*[1]. All three strains contain 2-O-methyl-D-mannose and rhamnose. Neither 03011 nor 03014 were infective on the host plant; the other strain has not been tested.

The remaining two strains from *Ceanothus*, LLR 03013 and 03015, share with the other three isolates an A-type pattern in Tween-amended broths. Their micromorphology is more like DDB 030210 than the other two. However, they do not utilize any of the carbohydrates tested at 0.5% concentration. Increasing the concentration of glucose gives some increase in growth over controls (data not shown); thus these two strains, like the "B" types, apparently lack a transport mechanism for carbohydrates, but respond to increased osmotic pressure. In these strains, however, addition of Tween does not serve to facilitate transport of the carbohydrates into the cell as it does for

Table 2. Comparative data on *Frankia* spp. from *Ceanothus americanus*

	LLR 03011	LLR 03013	LLR 03014	LLR 03015	DDB 030210
Nodule sterilization method	OSO$_4$	NaOCl	NaOCl	NaOCl	NaOCl
Isolation method	Dil.[1]	Dil.	Dil.	Dil.	SDG[2]
Nodule collection	May	May	May	May	September
Time of isolation following nodule collection	2 mos.	2 1/2 yr.	2 1/2 yr.	2 1/2 yr.	1 1/2 yr.
Sporangia formation	Rare, Late	Common	Rare, Late	Common	Common
Vesicle formation	NO[3]	Present	NO[3]	Present	Present
Color of cells in "S" medium	Brownish	Off-white	Brownish	Off-white	Brownish
Best medium for growth	S	NZ-Propionate	S	NZ-Propionate	S
Rate of growth in 20S medium[4]	Rapid	V. slow	V. rapid	Slow	Moderate
WCS[5]	Fucose	Xylose	Fucose	ND[6]	None[7]

[1] Dilution method. [2] Sucrose density gradient method. [3] Not observed. [4] See Fig. 3. [5] Diagnostic whole cell sugar. [6] Not determined. [7] See text.

Table 3. Carbohydrate utilization by *Frankia* spp. isolated from *Ceanothus americanus*

Carbohydrate	LLR 03011	LLR 03013	LLR 03014	LLR 03015	DDG 030210
L (+) – Arabinose	–	–	–	–	U/NA
D (+) – Glucose	U/A	–	U/A	–	U/NA
Glycerol	–	–	–	N.D.	–
3-O-Methyl-D-Glucose	–	–	–	–	+[1]
D (+) – Maltose	+[1]	–	+	–	–
Sucrose	+	–	U/A	–	–
D (+) – Trehalose	U/A	–	U/A	N.D.	U/NA
D (+) – Xylose	U/A	–	U/A[1]	–	U/NA

For meaning of symbols see Table 1. [1] Late; probably not constitutive; thus may be variable (*i.e.* positive only after 3–4 months).

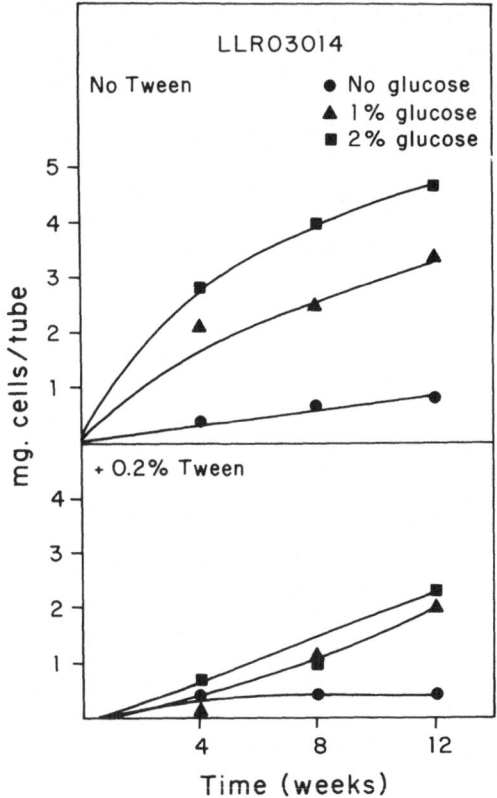

Fig. 2. Influence of Tween 80 and glucose on growth of *Frankia* sp. LLR 03014 in "S" medium.

the B strains; rather it depresses growth as in the A type strains. If these are then "B types", it may be that a different "facilitator" is needed in *Ceanothus* isolates. Thus, at present it is impossible to state with certainty whether 03013 and 03015 represent an intermediate group between the A and B strains or are a specialized B type which reflects the difference between the *Alnus/Myrica/Comptonia* cluster and the *Ceanothus* cluster.

In sum, we now have further evidence that two different physiological types of frankiae (A and B) may be isolated and are probably common from the same host species. In addition, we have observed considerable heterogeniety among isolates from the same provenance. This observation complements that of Benson and Hanna[5] who studied the gel-protein patterns of various endophyte isolates from *Alnus incana* ssp. *rugosa* nodules collected within a small area and found them to fall into several groups. Whether our strains will differ in their gel-electrophoresis protein patterns remains to be seen, but the differences in morphology, carbohydrate utilization and growth patterns

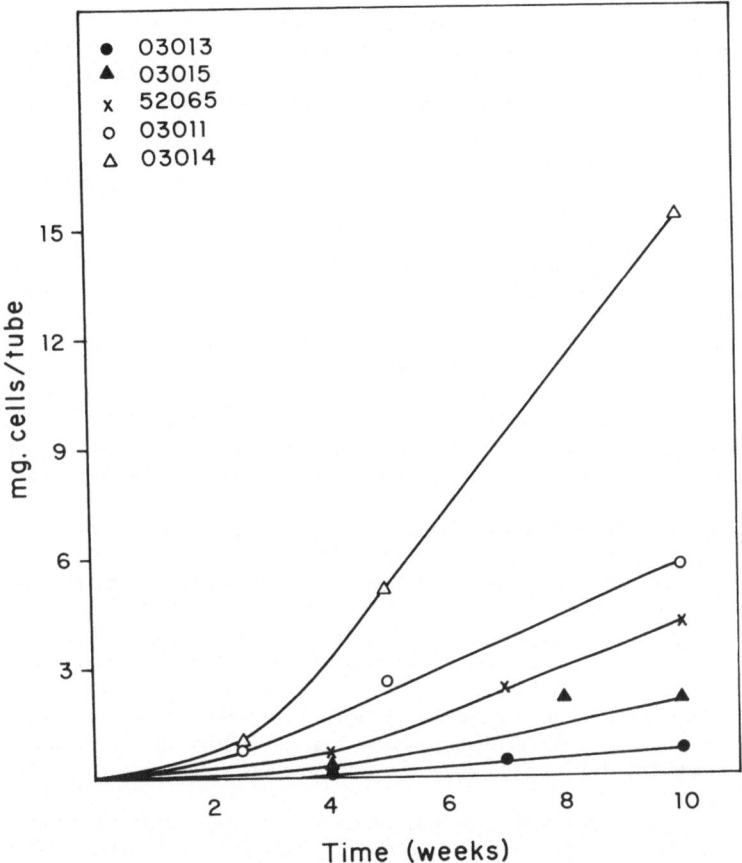

Fig. 3. Growth of *Frankia* spp. isolated from *Ceanothus americanus* in "20 S" medium (S medium made from 2.0% glucose).

in various media would lead us to expect differences in protein patterns also.

The problems for *Frankia* taxonomy pointed up by the differences (*i.e.* A *vs.* B) between strains isolated from the same plant host in different locales are made even greater when we consider the apparent diversity of isolates from the same provenance. It is already clear that the host-related species concepts that are used in the taxonomy of rhizobia will not work at all for the frankiae. Furthermore, the criteria classically used to delineate actinomycete species, notably morphology and physiology, also appear to have little value in this regard, at least as assessed the battery of tests we presently use. We are thus confronted with the need to modify the old techniques or develop new approaches to the problem of species within the genus *Frankia*.

Acknowledgements   We thank Drs. J. G. Torrey and D. Baker for providing cultures and testing infectivity and effectivity of strains, and Ms. Magda Gagliardi for expert assistance. The work reported in this paper was supported in part by the Charles and Johanna Busch Fund, Grant No. 80-CRCR-1-0439 from the U.S. Department of Agriculture and a Smith, Kline and French special Fellowship to J.-S. Ruan.

References

1   Baker D 1982 A cumulative listing of isolated frankiae, the symbiotic, nitrogen-fixing actinomycetes. The Actinomycetes 17, 35–42.
2   Baker D, Lechevalier M P and Dillon J T 1981 Strain analysis of actinorhizal microsymbionts (genus: *Frankia*) *In* Current Perspectives in Nitrogen Fixation. Eds. A Gibson and W Newton. Aust. Acad. Sci., Canberra, p 479.
3   Baker D, Pengelly W and Torrey J G 1981 Immunochemical analysis of relationships among isolated frankiae (Actinomycetales). Int. J. Syst. Bacteriol. 31, 148–151.
4   Baker D, Torrey J G and Kidd G H 1979 Isolation by sucrose-density fractionation and cultivation *in vitro* of actinomycetes from nitrogen-fixing root nodules. Nature London 281, 76–77.
5   Benson D R and Hanna D 1983 *Frankia* diversity in an alder stand as estimated by SDS-PAGE of whole cell-protein. Can. J. Bot. 61, 2919–2923.
6   Callaham D, Del Tredici P and Torrey J G 1978 Isolation and cultivation *in vitro* of the actinomycete causing root nodulation in *Comptonia*. Science 199, 899–902.
7   Gautier D, Diem H G and Dommergues Y 1981 Infectivité et effectivité de souches de *Frankia* isolées de nodules de *Casuarina equisetifolia* et d' *Hippophaë rhamnoides* C. R. Acad. Sci. Paris 293, 489–491.
8   Horrière F, Lechevalier M P and Lechevalier H A 1983 *In vitro* morphogenesis and ultrastructure of a *Frankia* sp. Ar I3 (Actinomycetales) from *Alnus rubra* and a morphologically similar isolate (Air I2) from *Alnus incana* subsp. *rugosa*. Can. J. Bot. 61, 2843–2854.
9   Lalonde M, Calvert H E and Pine S 1981 Isolation and use of *Frankia* strains in actinorhizae formation *In* Current Perspectives in Nitrogen Fixation. Eds. A Gibson and W Newton. Aust. Acad. Science. Canberra pp 296–297.
10 Lechevalier M P and Lechevalier H A 1980 The chemotaxonomy of actinomycetes. *In* Actinomycete Taxonomy. Eds. A Dietz and D W Thayer. Society for Industrial Microbiology. Special Publication #6. Arlington, VA pp 227–291.
11 Lechevalier M P and Lechevalier H A 1984 Taxonomy of *Frankia*. *In* Actinomycetes Biology. Eds. L F Bojalil and L Ortiz-Ortiz. Academic Press, New York. (*To be published*).
12 Lechevalier M P, Baker D and Horrière F 1983 Physiology, chemistry, serology and infectivity of two *Frankia* isolates from *Alnus incana* subsp. *rugosa*. Can. J. Bot. 61, 2826–2833.
13 Lechevalier M P, Horrière F and Lechevalier H 1982 The biology of *Frankia* and related organisms. Dev. Indust. Microbiol. 23, 51–60.
14 Mort A, Normand P and Lalonde M 1982 2-O-Methyl-D-mannose, a key sugar in the taxonomy of *Frankia*. Presented at the Annual Meeting of the Canadian Soc. Microbiologists, June 1982.
15 Normand P and Lalonde M 1982 Evaluation of *Frankia* strains isolated from provenances of two *Alnus* species. Canad. J. Microbiol. 28, 1133–1142.
16 Quispel A and Tak T 1978 Studies on the growth of the endophyte of *Alnus glutinosa* (L.) Vill. in nutrient solutions. New Phytol 81, 587–600.

Plant and Soil 78, 23–28 (1984).
© 1984 *Martinus Nijhoff/Dr W. Junk Publishers, The Hague.*

A modified sucrose fractionation procedure for the isolation of frankiae from actinorhizal root nodules and soil samples

DWIGHT BAKER
Charles F. Kettering Research Laboratory, 150 E. South College St., Yellow Springs, OH 45387, USA

and DAVID O'KEEFE
Plant Science Division, University of Wyoming, Laramie, WY 82071, USA

Key words Actinorhizal plants Bacterial isolation techniques *Frankia* Nitrogen fixation Root nodules Soil Sucrose density fractionation

Summary The isolation and pure culture of the symbiotic nitrogen-fixing frankiae has always been difficult. In the past the isolation of these actinomycetes directly from soil samples has proven impossible and isolations from root nodules of many genera has been only poorly successful. We report here a modified sucrose fractionation procedure which increased the success of isolations from root nodules and which permitted the isolation of *Frankia* directly from soil samples. Crushed nodule suspensions or soil suspensions were incubated briefly in 0.7% phenol (carbolic acid) just before application to a sucrose density gradient. This phenol incubation decreased the number of contaminating eubacteria and fungi but more importantly increased the number of *Frankia* developing on the isolation plates. If the phenol incubation was used solely without sucrose fractionation no *Frankia* were isolated, suggesting the death of the organisms due to phenol toxicity. The use of selective nitrogen-deficient media proved important for the isolation of frankiae from soils.

Introduction

The isolation and cultivation of *in vitro* of actinorhizal microsymbionts has always been problematic[1]. Reasons for difficulties in isolating the symbiotic frankiae are 1) the very slow growth of these bacteria in relation to other actinomycetes and eubacteria, 2) the preponderance of rapidly growing contaminating eubacteria and actinomycetes associated with a soil-borne structure, 3) the unknown media preferences of frankiae making the transition from a symbiotic to a heterotrophic or saprophytic metabolism, and 4) the inhibiting nature of phenolic compounds released from host plant cells during isolation procedures. Currently, several isolation procedures are employed which overcome one or more of these difficulties with variable success. The sucrose fractionation technique[2], the serial dilution or modified dilution techniques[5,7], the selective incubation or microdissection techniques[4,12], the osmium tetroxide technique[8] and the filter exclusion technique[3] are most commonly used to isolate frankiae from actinorhizal root nodules. These techniques have proven particularly useful with the host genus *Alnus* and to a lesser degree with *Myrica*,

Comptonia and *Elaeagnus*. No procedure has proven successful in isolating frankiae directly from soils to date.

Isolations of other soil-borne actinomycetes have been improved if phenol (carbolic acid) is incubated with the soil suspension[9,11]. In this study a modified sucrose fractionation technique was devised utilizing phenol which not only increased the relative success of isolations from root nodule material but also permitted the isolation of frankiae directly from soils. Recommendations are presented for culture media preferences of several *Frankia* types.

Methods

Collection procedures

Root nodules and root zone soil samples were collected from actinorhizal plants from natural populations in Table 1. Nodules and soil samples were chilled with ice at the time of collection and then frozen in the laboratory until isolations were undertaken.

Isolations from nodules

Small portions of root nodule tissue were surface sterilized in a 2% solution of glutaraldehyde containing Tween 20 as a surfactant. The nodule pieces were then rinsed twice with sterile distilled water (SDW) and crushed in a sterile mortar and pestle to form a crude nodule suspension. To this point no variation from the original sucrose fractionation technique[2] was made. The crude nodule suspension was then made to contain 0.7% phenol (v/v) using liquified phenol and incubated for 10 min. Control suspensions contained no phenol. After incubation a small sample of the nodule suspension was applied to a sterile discontinuous sucrose density gradient composed of three layers: bottom layer, 60% (2.5 M); middle layer, 45% (1.6 M), top layer 30% (1 M). Concentrations of sucrose solutions were determined by refractometer. The use of a refractometer was important in this step. Other workers[12] have not used a refractometer and failed to achieve proper separations. The gradients were centrifuged to equilibrium using either a low speed or an ultracentrifuge equipped with a swinging bucket rotor. At completion the gradient tubes were pierced and the lower interface (45/60) collected. This fraction was used to inoculate selective culture media (see below) by a pour plate procedure at a dilution of 1:100. The solidified plates were sealed with Parafilm (American Can Corp., Greenwich, CT USA) and incubated at 28°C. Observations were made at weekly intervals after inoculation and results recorded at 8 weeks as number of *Frankia* colonies per isolation plate. Morphological criteria were used to identify the organisms as *Frankia*[2].

Isolations from soils

Five grams of soil were suspended in 50 ml SDW and incubated for 30 min at room temperature. The suspension was then made to contain 0.7% phenol as described above and incubated for an additional 10 min. A small sample of the soil suspension was applied to a discontinuous sucrose gradient and the procedure continued as for the isolation from nodules.

Culture media

A defined propionate minimal medium (DPM) containing (g/l) KH_2PO_4, 1.0; $MgSO_4 \cdot 7H_2O$, 0.1; $CaCl_2 \cdot 2H_2O$, 0.01; sodium propionate, 1.2; and Hoagland's microelement stock, 1 ml/l and $FeSO_4$-EDTA stock 1.8 ml/l[6] adjusted to pH 6.8, was used for all soil and nodule samples. In addition a second complex medium was used for each soil and nodule sample as follows: *Frankia* agar (FA)[1] for samples from *Alnus, Myrica, Cercocarpus* and *Purshia*; Czapeks medium supplemented with 0.2% yeast extract (YCz)[10] for samples from *Casuarina*; or "S" agar[10] for samples from *Ceanothus*. Cycloheximide was added to media before sterilization at a concentration of 100 µg/ml, whenever necessary to inhibit growth of fungi.

Table 1. Source information for nodules and soils used in this study

Sample No.	Soil type	Plant taxon	Source location	Date of collection
Root nodules				
WA 26		*Alnus viridis* ssp. *sinuata*	North Bend, WA	7 July 1980
HA 2		*Casuarina equisetifolia*	Kahana, HA	8 Dec. 1980
WYG 73		*Purshia tridentata*	Woods Landing, WY	21 June 1982
BAH 3		*Casuarina* spp.	Freeport, Bahamas	7 Nov. 1982
FLA 1		*Myrica cerifera*	Venice, FL	10 Jan. 1983
CA 003		*Alnus rhombifolia*	Idyllwild, CA	27 Mar. 1983
CA 004		*Alnus rhombifolia*	Idyllwild, CA	27 Mar. 1983
AR 013		*Cowania mexicana* var. *stansburiana*	Flagstaff, AZ	2 Apr. 1983
VA 001		*Myrica cerifera*	Williamsburg, VA	17 Apr. 1983
Soils				
WYG 06	Sub	*Cercocarpus montanus*	Laramie, WY	9 June 1982
WYG 82	Sub	*Ceanothus velutinus*	Woods Landing, WY	21 June 1982
J 1	Sub	*Cercocarpus montanus*	Laramie, WY	23 June 1982
BAH 2	Sand	*Casuarina* spp.	Freeport, Bahamas	7 Nov. 1982
BAH 3	Sand	*Casuarina* spp.	Freeport, Bahamas	7 Nov. 1982
VA 001	Org. top	*Myrica cerifera*	Williamsburg, VA	17 Apr. 1983

Results

The results of the isolation trials are summarized in Table 2. Two out of six soil trials results in successful *Frankia* isolation as did all the nine nodule trials reported here.

The following visual observations were made during the experiments: 1) The phenol incubation reduced the numbers of eubacterial and fungal contaminants; 2) Minimal media formulations sometimes caused sporulation of contaminating actinomycetes and fungi; 3) Actinomycete contaminants were in general self-limiting but fungal contaminants were not; fungal contaminants were better controlled with cycloheximide; 4) Minimal media formulations were more useful than complex media formulations for isolations from soil; 5) Soils having an apparent high organic content had a higher population of contaminating microorganisms; surface soils (0–15 cm) for this reason were less useful in *Frankia* isolations than subsurface soils (20–30 cm); 6) Wet sand soils (beaches) were particularly high in contaminating microorganisms; 7) Colonies of *Frankia* from phenol-modified isolations appeared to demonstrate more robust growth than control-derived colonies although no quantitation of growth was made.

Discussion

The modified isolation procedure with a phenol incubation, in general, improved the success of a number of isolation trials from root

Table 2. *Frankia* isolation results*

Sample No.	Complex medium		Minimal medium (−N)	
	with phenol	without phenol	with phenol	without phenol
Root nodules				
WA 26	8.9	2.7	N.A.	N.A.
HA 2	5.2	4.6	7.0	9.8
WYG 73	6.8	0.2	13.6	0.4
BAH 3	25.1	33.3	0	6.6
FLA 1	4.5	5.9	16.2	8.4
CA 003	1.4	0.8	5.4	7.6
CA 004	0.4	0.2	20.8	18.0
AR 013	0.2	0.2	0.4	1.2
VA 001	Cont.	Cont.	2.8	4.4
Soils				
WYG 06	0	0	2.6	0
WYG 82	0	0	Cont.	Cont.
J 1	0	0	32.0	0
BAH 2	Cont.	Cont.	0	0
BAH 3	0	0	0	0
VA 001	Cont.	Cont.	0	0

*Figures presented are mean number of *Frankia* colonies per isolation plate
NA = not attempted
Cont. = plates overgrown with contaminating bacteria

nodules. For the host genera *Alnus* and *Purshia,* the phenol procedure was consistently better for isolations on complex media and equal or better on the minimal media. Isolation trials from *Casuarina, Cowania* and *Myrica* were sometimes improved using the phenol procedure but no consistent trend was observed. Such a result suggests that isolation success is greatly influenced by the ability of the bacterium to grow out of the nodule rather than by the choice of isolation technique. Under no circumstance was any nodule microsymbiont totally inhibited using the phenol-gradient fractionation procedure. Because of the observation that more robust growth of colonies resulted from the phenol procedure the routine use of this modified procedure is recommended.

The use of the phenol/sucrose-fractionation procedure for the isolation of frankiae directly from soil samples was shown to be essential. The phenol incubation kills bacterial and some fungal contaminants which would quickly overgrow the isolation plates. The use of cyclo-heximide was helpful in further eliminating fungal contaminants. Likewise, the use of a nitrogen-deficient minimal medium (DPM) was shown to be essential for soil isolations. Using several different more complex media we were unable to obtain any isolated *Frankia*. This is attributed to rapid overgrowth of the plates or inhibition of *Frankia* growth by the contaminating microorganisms which survived the isolation procedure. For the same reason surface soils (0–18 cm) were not useful for isolation purposes due to the higher populations of contaminating microorganisms.

From these results it is not clear what role phenol plays in the isolation procedure. Studies were undertaken to grow several pure-cultured *Frankia* strains in the presence of 0.05% phenol. It was found that growth was dramatically inhibited (C. Agbor, unpublished observations). In a related study we tested the original phenol isolation technique of Lawrence[9] for isolating frankiae. We inoculated a culture medium directly with the phenol-incubated nodule or soil suspensions. The resultant phenol concentration in the culture medium was only 0.017% (v/v), however no growth of any frankiae was observed. The same suspensions further treated by sucrose fractionation gave good growth of *Frankia*. Therefore the fractionation steps were necessary for removing phenol from the bacteria.

We feel that the phenol modified sucrose fractionation procedure is an improved method for the isolation of frankiae. As such it is the only method for isolating frankiae directly from soil samples. At present it is useful for subsurface soils but with minor modifications may be useful for surface soils. We feel that direct soil isolations are far superior for *Frankia* enumeration and ecology studies than are baiting techniques which use a host plant as intermediate. Direct soil isolations will give a more accurate reflection of soil frankiae populations because they do not involve symbiotic selection by a host plant. This new technique may prove useful in establishing the role of *Frankia* in the soil environment and whether it is capable of growth *ex planta*.

Acknowledgements The authors thank Dr. H.G. Diem for his stimulating discussions and ideas during the early part of this research. In addition the authors are indebted to the following persons for assistance with nodule collections: R. Abrahamson (Bahamas), L. Baker (Florida), M. Poth (California), T. Righetti (Arizona) and S.E. Williams (Wyoming). S. Marsh and C. Agbor provided excellent technical assistance during the experimental procedures. We thank Dr. T.V. Bhuvaneswari for critically reading the manuscript.

References

1   Baker D and Torrey J G 1979 The isolation and cultivation of actinomycetes root nodule endophytes. *In* Symbiotic Nitrogen Fixation in the Management of Temperate Forests. Eds. J C Gordon, C T Wheeler and D A Perry. Oregon State University Press, Corvallis, OR pp 38–56.

2   Baker D, Torrey J G and Kidd G H 1979 Isolation by sucrose-density fractionation and cultivation *in vitro* of actinomycetes from nitrogen-fixing root nodules. Nature London 281, 76–78.

3   Benson D 1982 Isolation of *Frankia* strains from alder actinorhizal root nodules. Appl. Environ. Microbiol. 44, 461–465.

4   Berry A and Torrey J G 1979 Isolation and characterization *in vivo* and *in vitro* of an actinomycetous endophyte from *Alnus rubra* Bong, *In* Symbiotic Nitrogen Fixation in the Management of Temperate Forests. Eds. J C Gordon, C T Wheeler and D A Perry. Oregon State University Press, Corvallis, OR pp 69–83.

5   Diem H G, Gauthier D and Dommergues Y R 1982 Isolation of *Frankia* from nodules of *Casuarina equisetifolia*. Can. J. Microbiol. 28, 526–530.

6   Hoagland D R and Arnon D I 1950 The water culture method for growing plants without soil. Circ. Calif. Agric. Exp. Sta. 347 (revised edition).

7   Lalonde M 1979 A simple and rapid method for the isolation, cultivation *in vitro* and characterization of *Frankia* strains from *Alnus* root nodules. *In* Symbiotic Nitrogen Fixation in the Management of Temperate Forests. Eds. J C Gordon, C T Wheeler and D A Perry. Oregon State University Press, Corvallis, OR p 480.

8   Lalonde M, Calvert H E and Pine S 1981 Isolation and use of *Frankia* strains in actinorhizae formation. *In* Current Perspectives in Nitrogen Fixation, Eds. A H Gibson and W E Newton, Aust. Acad. Sci., Canberra pp 296–299.

9   Lawrence C H 1956 A method of isolating actinomycetes from scabby potato tissue and soil with minimal contamination. Can. J. Bot. 34, 44–47.

10 Lechevalier M P, Baker D and Horriére F 1983 Physiology, chemistry, serology and infectivity of two *Frankia* isolates from *Alnus incana*. subsp. *rugosa*. Can. J. Microbiol. 61, 2826–2833.

11 Panthier J J, Diem H G and Dommergues Y 1979 Rapid method to enumerate and isolate soil actinomycetes antagonistic towards rhizobia. Soil Biol. Biochem. 11, 443–445.

12 Quispel A and Burggraaf A J P 1981 *Frankia,* the diazotrophic endophyte from actinohizas. *In* Current Perspectives in Nitrogen Fixation. Eds. A H Gibson and W E Newton, Aust. Acad. Sci., Canberra pp 229–236.

Plant and Soil 78, 29–43 (1984).
Ms. Fr 5
© 1984 *Martinus Nijhoff/Dr W. Junk Publishers, The Hague.*

Heterogeneity within *Frankia* sp. LDAgpl studied among clones and reisolates

A. J. P. BURGGRAAF and J. VALSTAR
Department of Plant Molecular Biology, Biological Nitrogen Fixation Research Group, Botanical Laboratory, State University Leiden, The Netherlands

Key words *Alnus glutinosa* Endophyte *Frankia* Infectivity Nitrogen fixation Sporulation Variability

Summary *Frankia* sp. LDAgpl, an isolate from spore positive nodules of *Alnus glutinosa*, only slowly infects its host plant. Reisolates obtained from occasional nodules caused by infection with LDAgpl, are capable of infecting the alder much more rapidly. A variability analysis of LDAgpl has been performed to obtain more insight into the question whether these reisolates constitute a different genotype within LDAgpl and if the plant is exerting an influence during plant passage. High dilutions of mildly sonicated *Frankia* suspensions were plated to obtain genetically homogeneous colonies. Clones thus generated showed differences in growth pattern, sporulation and C_2H_2-reduction on media containing propionic acid as sole C-source (P-medium). Differences in sporulation on P-medium indicate that LDAgpl was a highly heterogeneous strain. Comparisons of sporulation on several different media gave evidence that the differences in sporulation between LDAgpl clones are the result of differences in efficiency of propionic acid utilization.

The differences observed between the reisolates and LDAgpl clones indicate that the reisolates constitute a different genotype, which could be selected for by the plant during the infection process. Comparison with similar changes in phenotype occuring in a spore negative type strain from *A. glutinosa* is discussed.

Introduction

Since the isolation and cultivation of numerous diazotrophic free living *Frankia* cultures, an increasing part of the research has been dedicated to the physiological and biochemical evaluation of the isolated strains[1,2,3,4,5,14,18,20,22,23]. Yet relatively little work has been done on the internal strain variation[3,19] and change of the character of *Frankia* during the free living propagation. In preliminary studies[9,10] it was shown that *Frankia* sp. LDAgpl (a spore positive isolate from *Alnus glutinosa* root nodules) has lost most of its nodulation potential, giving only little nodulation after 5–6 weeks. In certain experiments it could be shown, that the nodulation potential could be enchanced by mild sonication of the inoculum, especially in inocula containing many sporangia. Reisolates from nodulated alders with LDApl had regained 'full' infectivity and formed numerous nodules after 2–3 weeks. The question arises whether these reisolates constitute a different genotype already present in the original strain or that a certain modification has been exerted by the plant passage.

The results presented here deal with an analysis of the phenotypic heterogeneity of LDAgpl and concern the following aspects:
— growth on different carbon sources,
— growth behaviour in stationary batch cultures,
— production of pigments in pure culture,
— sporulation in pure culture,
— N_2-fixation and growth on nitrogen limited media,
— infective potential of pure cultures.

Material and methods

Frankia isolates, inoculum, growth conditions and clone preparation
   Isolation LDAgpl was isolated from *A. glutinosa* spore positive nodules (1979, Hoogmade, Neth.). The strain designation has been described before[10].
   Reisolates were obtained after the first plant passage from young alder plants 3 months after inoculation with the free living LDAgpl culture and are designated r1, r2, r3 and r4. The axenic alder plants were grown and inoculated in either tubes or petri dishes[9], which minimized environmental contamination with other *Frankia* strains. Isolation was performed as follows: 1) extensive rinsing with soap (Adix) 5 min., 2) 5 min. in ethanol 70%, 3) rinsing in H_2O, disinfection of the outer cell layers with OsO_4[17] 1.5%, 2 min., 4) rinsing in H_2O, 5) homogenization to small nodule pieces in 5 ml H_2O, 6) rinsing with H_2O until solution is clear, 7) 6 nodule fragments put on a bottom agar and poured with a topagar[7]. Incubation was performed at 28°C in the dark for about 6 weeks. After sufficient outgrowth of *Frankia* from the nodules pieces, the material was homogenized in 5 ml of suitable growth medium and further incubated. Isolations were done on a mineral medium containing propionic acid medium (0.5 g/l), NH_4Cl (0.1 g/l) and biotin (0.002 g/l) (P-medium)[8] supplied with 5 g/l casamino acids. *Frankia* was routinely subcultivated with a 7 day transfer interval in 50 ml nutrient solutions (P-medium) in 100 ml Erlenmeyer flasks incubated without agitation at 28°C. Other conditions of inoculum subcultivation (= precultivation) concerning time and/or medium are indicated in the results.
   Experiments were carried out in glass tubes, 160 × 30 mm, containing 10 ml of nutrient solution. In experiments dealing with different C-sources, propionic acid was replaced by the desired carbon source (see results for concentrations). Growth was determined as the total protein amount in each culture using the Coomassie Brilliant Blue dye[6,8]. For analysis of the heterogeneity of LDAgpl, this isolate was mildly sonicated (90 W, 30 sec., sonifier B-12, Branson Power Co., Conn., USA) to generate small hyphal fragments and spores. Solutions were checked microscopically and 0.2 ml aliquots (10 ml sonicate) were plated on P-agar in a 10-fold dilution series (6 dilutions). Plates were sealed with Parafilm and stored at room temperature for a variable period of time (from one up to 15 months) before colonies were used for subcultivation in liquid medium. Colonies showing clear differences in colony type or development were picked off the plates, homogenized in P-medium (6 ml) and left to grow for about two to three weeks after which they were similarly transferred to 10 ml amounts of P-medium and further incubated for another two to three weeks. At this point routine subcultivation was started as mentioned above. Homogenization was performed as described previously[8].

Plant material and growth conditions
   Inoculation of *A. glutinosa* plants grown on agar both in petri dishes and in tubes was described previously[9]. Inoculation of water cultures was as described[21]. In each experiment 6 non-inoculated axenic plants were run as controls to test for contamination. Never was any nodulation on these plants observed.

Acetylene reduction and growth on N-limited media
Nitrogenase activity was estimated by the acetylene reduction assay as described[10]. Acetylene reduction was followed during the first 32 hours after addition of acetylene and determined between 10 and 24 hours. After 5 hours acetylene reduction was linear with time in the stationary batch cultures. Cultures were grown on nitrogen limited P-medium.

Results

From a stock of 100 clones, 9 clones were selected for further study together with the original strain LDAgpl and four of its reisolates. These clones were derived from roughly 3 different colony types, which developed in the higher dilutions of the sonicated samples: a) colonies with almost exclusively hyphal growth, b) colonies with concentrated hyphal growth and sporulation and often showing after two months a diffuse zone of hyphal growth around the initial colony, c) colonies with diffuse hyphal growth and sporulation.

Further selection of clones was based upon clear differences in growth during the first stages of subcultivation following transfer from the agar plates.

Growth behaviour on P-medium

After approximately 3 months of regular subcultivation on P-medium, clones were tested for their growth (Fig.1). Differences in the rate and the amount of protein production are visible. Relatively fast growing 'strains' (*e.g.* clone 3 and LDAgpl) have a logarithmic growth phase during the first 120 hours with a doubling time (dt) of 20 hours. Yet some slower growing 'strains' (*e.g.* clones 65 and 91), achieving lower levels of protein after 120 hours, do show rapid growth at the very start of the growth period (dt = 10 hours). This is correlated with the presence of numerous spores and sporangia in the inoculum of these two clones.

The log phase of the fast growing 'strains' is followed by a decaying phase without a clear stationary phase. The decomposition of protein leads to remarkably low levels of growth if determinations of protein are limited to samples of 18 days (see following section). Intermediary 'strains' such as clone 77 and 91 are showing a growth pattern similar to the 'fast' growing 'strains', although the decomposition of protein is less drastic after 120 hours of growth. Clones 64, 65, 95, 96 and the four reisolates exhibit relative low levels of growth.

Difficulties in reproducing the results can arise in these experiments especially if the precultivation conditions of the inoculum are not fully standardized, they may also indicate additional heterogeneity within the studied clones. Such heterogeneous behaviour has been found with clones 95 and 96. In this case the utilization of older

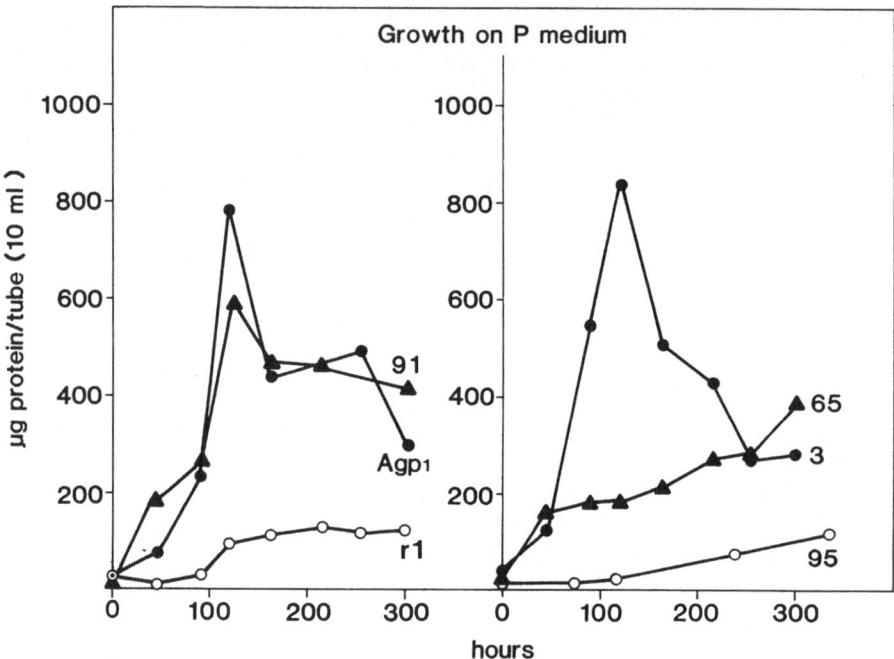

Fig. 1. Growth pattern of LDAgpl, clones and reisolates on P-medium. (Details see text).

inocula (20 days, containing more sporangia) was sufficient to enhance the growth of these clones to a level, where little difference with the faster growing clones was visible.

Growth on different carbon sources

Four carbon sources and one combination were tested for growth (Table 1). In previous studies it was shown, that *Frankia* was capable of utilizing those sources as sole carbon source depending on the tested strain[22].

None of the reisolates is able to grow on glucose, this is contrary to the growth of all tested clones on this C-source. None of the clones on the other hand is able to grow on Tween 80 as sole carbon source (including LDAgpl) in contrast to the reisolates. This result conflicts with earlier ones[22] in which LDAgpl was able to grow well on Tween 80, which is indicative of lipase activity. We suspect that the long and regular subcultivation on P-medium might be one of the causes for the loss of lipase activity. GP-medium is clearly enhancing the amount of biomass measured after 18 days of cultivation. There seems to be only a slight effect of this medium with the reisolates. Low values of protein with the clones on P-medium are likely to be due either to weak growth

Table 1. Growth of *Frankia* of different C-sources[1]

Strain	C source[2]				
	P	GP	G	Ac	Tw
Experiment 1					
LDAgp1	143[3]	316	78	208	<20
r1	171	219	<20	110	99
r2	113	224	<20	97	32
r3	109	217	<20	113	61
r4	330	295	<20	223	157
Experiment 2					
LDAgp1	164	484	298	156	<20
3	132	516	327	187	<20
4	133	484	231	ND	<20
20	143	368	108	143	<20
64	166	508	133	225	<20
65	234	416	107	65	<20
77	176	564	224	89	<20
91	187	436	54	106	<20
95	275	560	117	38	<20

[1] Growth determined after 18 days (μ g protein/10 ml medium)
[2] Basal medium P-medium without propionic acid
P – propionic acid (0.5 g/l); GP – propionic acid (0.5 g/l) + glucose (10 g/l); G – glucose (10 g/l); Ac – acetate (1.54 g/l); Tw – Tween 80 (2 g/l).
[3] mean, n = 3

or to decomposition of protein occuring in the batch culture (Fig. 1). Both propionic acid and acetate are universal as sole carbon source for the clones and reisolates tested here.

Sporulation in pure culture

Marked differences in sporulation on P-medium become visible between 14 and 18 days cultivation in 10 ml stationary batch cultures (Table 2). An example of these sporulation differences is given in Figs. 3 and 4. These differences are less pronounced when younger cultures of 11–14 days are investigated (Table 2C).

Two clones, 3 and 91, showing large differences in sporulation on P-medium were studied in respect to their response in sporulating capacity when the precultivation time was drastically changed. Precultivations were done on P-medium ranging from 4 to 24 days and each was followed by subcultivation intervals of 7 or 14 days. With clone 91 these different precultivation times caused the inoculants to consist of either hyphae (*e.g.* 4 days old inoculum) or inocula with increasing amounts of sporangia (*e.g.* 10 days or older inocula). These differences were not present with clone 3 since this clone did not sporulate on this medium. The differences in sporulation present in

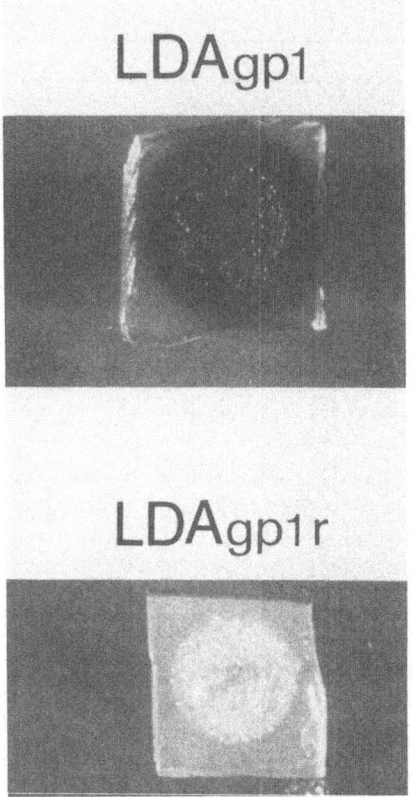

Fig. 2. Comparison of the original LDAgp1 strain producing a dark brown pigment and a non-pigmented reisolate, LDAgp1r, on P/tyrosin-agar.

clone 3 and 91 were not influenced by these changes of precultivation, thus the age and consistency of the inoculum did not influence the differences in sporulation observed in this study. This indicates that with these clones the differences in sporulation, which are stable over a long period of subcultivation on liquid P-medium, are likely to bear a genetic background.

Table 2A. Influence of C-source on sporulation[1] of LDAgp1 reisolates[2]

Strain	Medium[3]				
	P	GP	G	Tw	Ac
LDAgp1	−[4]	+(+)	−	ng[5]	−
r1	+(+)	+	ng	−	−
r2	++(+)	+/−	ng	+/−	+/−
r3	+(+)	+/−	ng	+(+)	+(+)
r4	−	−	ng	+/−	+/−

[1] determined after 18 days; [2] precultivation 7 days; [3] see details Table 1; [4] see details Table 2C; [5] ng = no growth

Table 2B. Influence of C-source on sporulation[1] of LDAgp1 clones[2]

Strain	Medium[3]				
	P	GP	G	Tw	Ac
LDAgp1	−[4]	++	−	ng[5]	−
3	−	++	+/−	ng	−
4	−	++	(+)	ng	ND[6]
20	−	++	+	ng	+
64	+	+++	++	ng	ND
65	+	+++	++	ng	+
77	−	++	(+)	ng	+/−
91	(+)	+++	+	ng	−
95	+/−	++	−	ng	−

[1] determined after 14 days; [2] precultivation 7 days; [3] see details Table 1; [4] see details Table 2C; [5] ng = no growth; [6] ND = Not determined

To obtain more evidence for this idea, the homogeneity of the LDAgp1 clones was tested, they were mildly sonicated again and plated on P-agar plates to generate subclones. Colonies thus obtained after 24 days were cultivated in liquid P-medium for another 14 days and then examined for sporulation. Subclones 3, 64, 77, 91, 95 and 96 thus examined, showed neither differences in sporulation between subclones, nor with the original clone. There were also no indications that these subclones differed in growth rate on P-medium (visual examination) when compared with the parent clones. Protein measurements of the growth of three subclones of 91 gave exactly the same

Table 2C. Sporulation of LDAgp1, its clones and reisolates on P- and P-N medium[1]

Strain	Medium		
	P-N[2,4]	P[3,4]	P[3,5]
LDAgp1	+[5]	−	−
3	+	−	−
4	+++	+/−	+/−
20	+++	+	+
64	+++	+(+)	++
65	+++	+(+)	++
77	+++	+/−	+(+)
91	+++	+/−	++
95	ND[6]	+	+++
96	+++	++	+++
r1	+	+/−	+/−
r2	+	+/−	+
r3	+/−	+/−	+/−
r4	+/−	−	−

[1] P-N medium is P-medium without NH_4Cl; [2] precultivation 5 days; [3] precultivation 7 days; [4] determined after 11 days; [5] determined after 18 days − None, +/− few, (+) low, + moderate, +(+) moderate to many, ++ many, ++(+) many to very many, +++ very abundant; [6] ND = Not determined

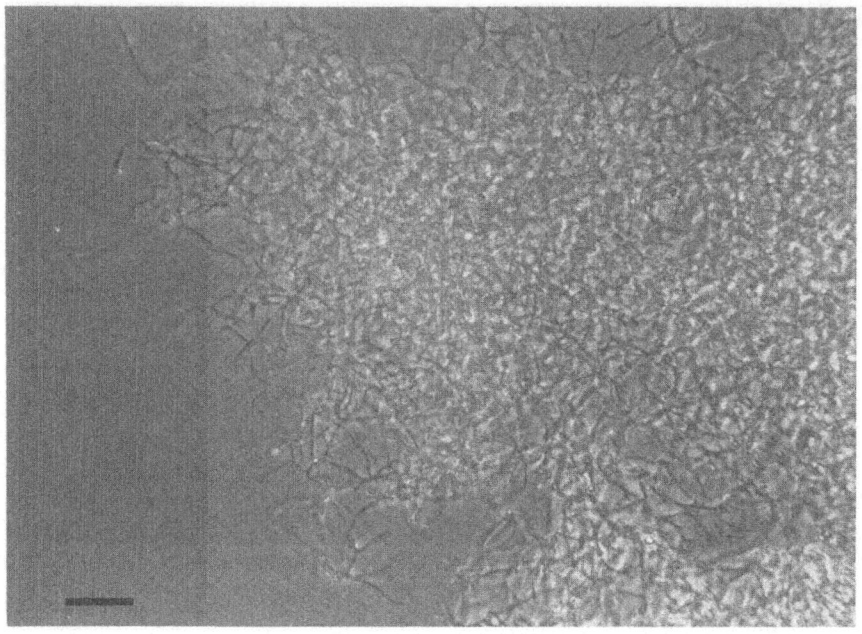

Fig. 3. Non-sporulating clone 3. Hyphal culture, 7 days old on P-medium. Bar = 10 μm.

Fig. 4. Sporulating clone 91. Culture with hyphae and sporangia, 18 days old on P-medium. Bar = 20 μm.

growth pattern (results not shown). Subclones of 95 and 96 were growing very slowly shortly after the first subcultivation from the agar plates, as did their parent clones (see section 'growth behaviour on P-medium').

Reisolates of the original LDAgp1 strain showed a low level of sporulation, except r2, which sporulated moderately (Table 2C), yet these results differ from those of Table 2A (P-medium), which indicates to heterogeneity comparable to that found in the original strain.

To investigate this, clones of the four LDAgp1 reisolates were similarly prepared as stated above and sporulation was examined again on liquid P-medium (Table 3).

It is interesting that clones of r1 generated 15 months before the present study all sporulated extensively after 3 weeks cultivation on liquid P-medium (compare Table 2C and 3). This indicates that a loss of genetic function coding for the expression of sporulation on liquid P-medium has taken place for r1 during repeated subcultivations on P-medium.

Some insight into the problem as to what causes the differences in the expression of sporulation on liquid P-medium has been obtained from the changes in sporulation on other media (Table 2B and C). A large stimulation was found when glucose was added to the P-medium (Table 2B). A similar effect was observed when NH_4Cl was omitted from the P-medium (P-N medium, Table 2C). Since in both GP- and P-media N-limitation, as judged from the vesicle production in the

Table 3. Sporulation in clones from reisolates of LDAgp1 (for details see text)

Strain	Colony type	Sporulation
r1[1]	I[3]	— to +/—
	II	—
r1[2]	I	++
	II	++
r2	I	++
	II	—
r3	I	—
	II	—
r4	I	—
	II	—

[1] colonies from 5 week old P-agar plates
[2] colonies from 15 month old P-agar plates
[3] colony type:
    I – round colony with concentrated hyphal growth, enlarging in vertical direction, orange coloured.
    II – colony with diffuse hyphal growth, enlarging in a horizontal plane, white coloured.

batch cultures, was preceding sporangia formation, it is likely that nitrogen limiting conditions in both P-, P-N- and GP-media are stimulating sporulation. Apparently clones such as 3 and 4 are normally not capable of synthesizing sporangia under nitrogen limitation on P-medium. This is probably due to both a limitation in propionic acid and NH_4Cl, since glucose enhances both growth (Table 1) and sporulation (Table 2B). Thus the differences observed in sporulation between clones are likely to be due to the expession of differences in the efficiency of propionic acid utilization. This is also reflected in the growth curves of Fig. 1. Stimulation of sporulation by GP-medium is not found with the reisolates. In this case an inhibition seems to be exerted by this medium.

Pigment production

Clear differences were observed in the production of a dark brown pigment in reaction to the addition of 2 mM tyrosin to the P-medium (Fig. 2). It is likely that this pigment is melanoid in nature and is therefore either reflecting laccase or tyrosinase activity[12,13], which are both involved in the production of melanin of which tyrosin is a precursor.

All reisolates are deficient in the production of this pigment, while LDAgp1 and all tested clones were capable of producing this pigment on P/tyrosin-agar plates after one week. In many instances the production of the pigment seems to start from single spots in the inoculated area. This indicates that the activities in the surrounding areas on the agar are being induced. Preliminary studies with methionine (2 mM) added as an inducer[13] for melanine production gave negative results.

N_2-fixation and growth on N-limited P-medium

An experiment monitoring N_2-fixation and growth in N-limited medium was carried out with nine LDAgp1 clones and its four reisolates. Fig. 5 shows the results of 3 representative clones and one reisolate. Clear differences were observed in growth on P-N-medium, N_2-fixation/tube (acetylene reduction) and the specific N_2-ase activities. Clone 3 was most successful in both growth and N_2-fixation. The reisolates were inferior and no specific N_2-ase activities above 300 nMol $C_2H_4 h^{-1}$ mg protein^{-1} were found. This was also correlated with the smaller amount of protein production observed. Increase in N_2-fixation was limited to the first stages of growth in these batch cultures, afterwards it decreased. There are indications that here growth as well as acetylene reduction are influenced by the precultivation conditions. A puzzling fact is the relatively high amount of protein reached by the reisolates despite a low level of fixation. Clone 3

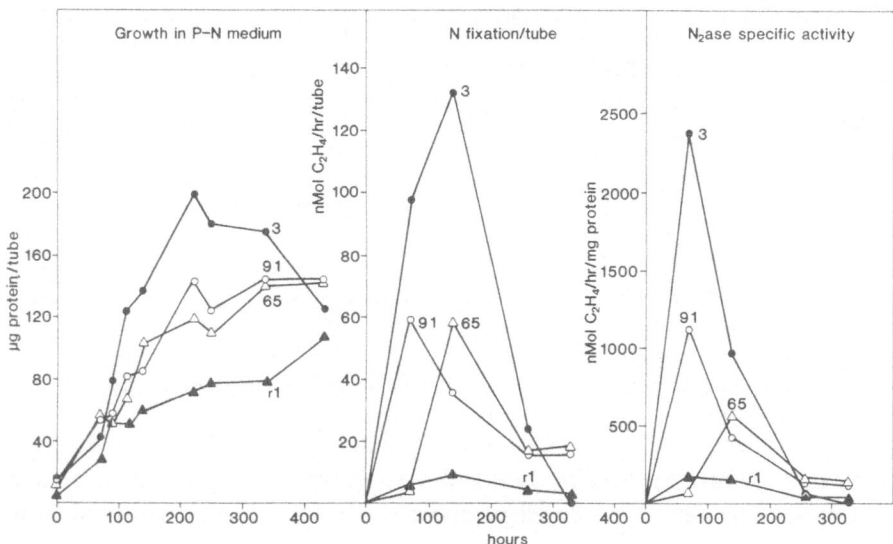

Fig. 5. Growth pattern and acetylene reduction on nitrogen limited P-medium (P-N medium, details see text).

compared with r1 shows a 3-fold difference in protein and 13-fold difference in acetylene reduction. This could indicate high endogenous N-reserves in r1 at the onset of the experiment.

Infectivity of pure cultures

Infectivity of 100 clones from strain LDAgp1 was tested after the first subcultivation in liquid medium (14 days) on alder seedlings grown on agar in test tubes[10]. None of these clones showed any nodulation in this system after two months. This could be due to difficulties in homogenization of some of the material, or to the absence of adequate numbers of infection loci on the agar grown plant root systems. Further tests were performed with 2 clones, 3 and 95 and the original strain LDAgp1. Ten day old cultures were subcultivated for 40 and 80 hours on P-medium and after this period homogenized and inoculated on water cultures of *A. glutinosa*.

The results (Table 4) show an enhancement of nodulation by clone 95 after 40 hours of subcultivation. This clone showed approximately 10 nodules after 4 weeks, while only few nodules were reported after 5 weeks for the 80 h subcultivation treatment of clone 95 and LDAgp1. Clone 3 is clearly less infective with only one nodule after 12 weeks.

Yet there is a significant increase in the number of nodules between 7 and 12 weeks with LDAgp1. Whether this phenomenon is due to any

Table 4. Infectivity of clone 3, 95 and LDAgp1

Strain	Growth period[1] in P-medium (h)	Weeks after inoculation			
		4	5	7	12[2]
LDAgp1	40	−	+	+[3]	36
	80	−	+	+	21
3	40	−	−	+	1
	80	−	−	−	0
95	40	+	+	+	15
	80	−	+	+	13

1) precultivation 10 days
2) no. of nodules on the same 3 plants counted after the weeks indicated
3) before 12 weeks no accurate determination of the number of nodules was possible without damaging the root system

growth of *Frankia* in the rhizosphere, or of liberation of infective particles from already formed nodules or of delayed nodule formation, is not known.

These data indicate that the infectivity changes with the clone used and that an enhancement in nodulation time can be reached if relative young inocula is used. Further research is needed to elucidate whether this is indeed correlated with any sporulation in the precultivation stage and is thus influenced by the precultivation medium.

All four reisolates were infective in water cultures of the alder, forming numerous nodules after 2 to 3 weeks.

Discussion

Frankia sp. LDAgp1 has a marginal nodulation capacity. Yet the reisolates gained from the occasional nodules formed on alder plants infected with LDAgp1, show normal nodulation within 2–3 weeks. This observation raises the question whether these reisolates constituted different genotypes within LDAgp1 and were selected by the plant during nodulation. To gain a better understanding of this problem several aspects of the variability of LDAgp1 were studied. Upon mild sonication of the pure LDAgp1 cultures several notebly different colony types developed on P-medium. The different colony forms were the starting point for isolation of different clones. Yet no differences in sporulation, growth or any other characters studied were found to be clearly correlated with these colony types of either LDAgp1 clones or reisolates. It therefore seems plausible to assume that the colony differentiation observed is not a consequence of genetic heterogeneity, but more likely due to phenotypic differentiation caused by the differences between *Frankia* cells in pure culture. These differences become more marked in the higher dilutions of the mildly sonicated pure *Frankia* cultures plated on P-agar. This could be caused by the diffusion barrier present in the agar system as compared to the liquid culture, allowing for separate development of several phenotypes.

The differences of the clones in sporulation were not influenced by precultivation conditions of the inoculum on P-medium. They also remained constant over a period of several months during regular subcultivations on P-medium. This indicates that these differences are genetically based. This idea is further corroborated by the results from the subclonation of the LDAgp1 clones. The subclones were homogeneous in respect to sporulation and did not deviate from the original clones. Thus if the differences in sporulation are genetically based, it proves that LDAgp1 is genetically heterogeneous in this respect. Full proof of this conclusion is of course only given when it is shown, that sporulation on P-medium is a character, which can be transferred from one strain to the other in conjugation or transduction events. The suspected heterogeneity forms the explanation for the observation that sporulation is nearly completely abolished in LDAgp1 after a years subcultivation on P-medium.

The previously employed mild sonication of LDAgpl and the storage of a dilution series of the sonicate on P-agar plates was thus essential to maintain a genetically heterogeneous stock and prevent selection occuring during subcultivation in liquid medium.

The loss of sporulation has also been observed in the strains CpI1, AvcI1 and the spore negative type strain LDAgn1[10,21] after prolonged cultivation on P-medium and might indicate a similar genetic selection for a non-sporulating type out of genetically heterogeneous material. With both CpI1 and LDAgn1 this was accompanied by a great loss of infectivity (unpublished results). Recently it was shown[11] that spores of CpI1 are the predominant infective stage on P-medium. It is still too soon to offer any speculations about the possible genetic background of the differences in sporulation. Possibilities are: 1) presence of plasmids, 2) presence of heteroclones and 3) more than one chromosome per cell.

Recently several workers have been able to isolate plasmids from both the free-living[20] and the endophytic *Frankia*[15]. All three possibilities offered above are being found with the related actinomycete *Streptomyces*[16].

The question posed at the start of this section, 'does the plant select a genotype from the apparent genetically heterogeneous LDAgp1 strain' remains. Fundamental differences between the reisolates and the LDAgp1 clones have been found in both growth pattern, nitrogen fixation, pigment production and infectivity. They tend to be stable and reproducible over a long period of time, which indicates that a different genotype is at work here. Since these reisolates were obtained in a period when LDAgp1 was still regularly sporulating on P-medium, it is possible that the plant selects by means of infectivity for certain segregation products within the LDAgp1 strain. The chance of occurrence of these segregation products in the infective inoculum is strongly enhanced by the mild sonification of the inoculum. This is just one explanation why the reisolates are capable of infecting more rapidly than the original LDAgp1 strain. There might be a specific plant influence exerted during plant passage or a reversion (mutation?) of the LDAgp1 genotype involving also the other aspects mentioned above, which could all be caused by a change in cell wall composition and thus its function. It is interesting that with the infective spore negative strain LDAgn1 occasionally revertants were found, which were very similar to the old LDAgp1 phenotype, that is production of pigments, growth on glucose and Tween 80, very slow and sparse nodulation and rapid growth on P- and P-N media. Yet the LDAgn1 strain shows more phenotypic analogies with the LDAgp1 reisolates, that is no rapid pigment production, rapid nodulation, and no growth on glucose (neither on Tween 80).

Thus the type of change observed in the LDAgp1 reisolates is probably also found, although in a reversed direction, with the spore negative type *Frankia* from *A. glutinosa*.

Acknowledgment We thank Prof. Dr. A. Quispel for his helpful discussions of this manuscript.

References

1 Baker D and Torrey J G 1980 Characterization of an effective actinorhizal microsymbiont *Frankia* sp. AvcI1 (Actinomycetales). Can. J. Microbiol. 26, 1066–1071.
2 Baker D, Pengelly W L and Torrey J G 1981 Immunochemical analysis of relationships among isolated frankiae (Actinomycetales). Int. J. Syst. Bacteriol. 148–151.
3 Benson D R and Hanna D 1983 *Frankia* diversity in an alder stand as estimated by sodium dodecyl sulfate polyacrylamide gel electrophoresis of whole cell proteins. Can. J. Bot. 61, 2919–2923.
4 Berry A and Torrey J G 1979 Isolation and characterization in vivo and in vitro of an actinomycetous endophyte from *Alnus rubra* Bong. *In* Symbiotic Nitrogen Fixation in the Management of Temperate Forests. Eds. J C Gordon *et al.* Oregon State University Corvallis, pp. 69–83.
5 Blom J 1981 Utilization of fatty acids and NH_4 by *Frankia* AvcI1. FEMS Microbiol. Lett. 10, 143–145.
6 Bradford M M 1976 A rapid and sensitive method for the quantitation of microgram quantities of protein utilizing the principle of protein binding. Anal. Biochem. 72, 248–254.
7 Burggraaf A J P, Quispel A, Tak T and Valstar J 1981 Methods of isolation and cultivation of Frankia species from actinorhizas. Plant and Soil 62, 157–168.

8   Burggraaf A J P and Shipton W A 1982 Estimation of *Frankia* growth under various pH and temperature regimes. Plant and Soil 69, 135–147.

9   Burggraaf A J P, van der Linden J and Tak T 1983 Studies on the localization of infectible cells on *Alnus glutinosa* roots. Plant and Soil 74, 175–188.

10 Burggraaf A J P and Shipton W A 1983 Studies on the growth of *Frankia* isolates in relation with infectivity and nitrogen fixation (acetylene reduction). Can. J. Bot. 61, 2744–2782.

11 Burggraaf A J P, van Vianen A, van der Linden and Tak T 1984 Infectivity of pure *Frankia* cultures from *Alnus glutinosa*. *In* Advances in Nitrogen Fixation Research. Proc. 5th Intern. Symp. on Nitrogen Fixation. Noordwykerhout, The Netherlands, August 28–September 3, 1984. Eds. C Veeger and W E Newton. Nijhoff/Junk Publishers, The Hague and Pudoc, Wageningen.

12 Chet I and Huttermann A 1982 De novo synthesis of polyphenol oxidase (laccase) during formation of sclerotia in *Sclerotium rolfsii*. FEMS Microbiol. Lett. 14, 211–215.

13 Crameri R et al. 1982 Secretion of Tyrosinase in *Streptomyces glaucescens*. J. Gen. Microbiol. 128, 371–379.

14 Dillon J T and Baker D 1982 Variation in nitrogenase activity among pure cultured *Frankia* strains tested in actinorhizal plants as an indication of symbiotic compatibility. New Phytol. 92, 215–219.

15 Dobritska S A 1982 Extrachromosomal circular DNA's in endosymbiont vesicles from *Alnus glutinosa* root nodules. FEMS Microbiol. Lett. 15, 87–91.

16 Hopwood D A and Merick M J 1977 Genetics of antibiotic production. Bacteriol. Rev. 41, 595–635.

17 Lalonde M and Calvert H E 1979 Production of *Frankia* hyphae and spores as an infective inoculant for *Alnus* species. *In* Symbiotic Nitrogen Fixation in the Management of Temperate Forests. Eds. J C Gordon *et al.* Oregon State University, Corvallis, pp 95–110.

18 Lechevalier M, Baker D and Horriere F 1983 Physiology, chemistry, serology and infectivity of two *Frankia* isolates from *Alnus incana* (L.) Moench. subsp. *rugosa* (Duroi) Clausen. Can. J. Bot. 61, 2826–2833.

19 Normand P and Lalonde M 1982 Evaluation of *Frankia* strains isolated from provenances of two *Alnus* species. Can. J. Microbiol. 28, 1133–1142.

20 Normand P, Simonet P, Butour J L, Rosenberg C, Moiroud A and Lalonde M 1983 Plasmids in *Frankia* sp. J. Bacteriol. 155, 32–35.

21 Quispel A van Tak T 1978 Studies on the growth of the endophyte of *Alnus glutinosa* (L.) Vill. in nutrient solutions. New Phytol. 81, 587–600.

22 Shipton W A and Burggraaf A J P 1982 A comparison of the requirements for various carbon and nitrogen sources and vitamins in some *Frankia* isolates. Plant and Soil 69, 149–161.

23 Tjepkema J D, Ormerod W and Torrey J G 1980 Vesicle formation and acetylene reduction activity in *Frankia* sp. CpI1 cultured in defined nutrient media. Nature London 287, 633–635.

Plant and Soil 78, 45–59 (1984).
© 1984 *Martinus Nijhoff/Dr W. Junk Publishers, The Hague.*

Morphology, physiology and infectivity of two *Frankia* isolates An 1 and An 2 from root nodules of *Alnus nitida*

FAUZIA HAFEEZ, ANTOON D. L. AKKERMANS* and ASHRAF H. CHAUDHARY
Department of Biological Sciences, Quaid-i-Azam University, Islamabad, Pakistan

Key words   Actinorhizae   *Alnus nitrida*   *Frankia*   Host-specificity   Nitrogen fixation   Root nodules

Summary   Two different strains, An 1 and An 2, were obtained from root nodules of *Alnus nitida* Endl., collected from one locality in the area of its natural habitat near Bahrin, District Swat, Pakistan. The light and electron microscopy of the isolates revealed the occurrence of septate and branched hyphae bearing sporangia and vesicles. The strains differed in their growth requirements, nitrogen-fixing ability and production of extracellular pigments, thus indicating the existence of more than one *Frankia* strain in the same locality. In the absence of combined nitrogen in the medium strain An 1 formed vesicles and fixed N_2 (up to 200 nmol C_2H_4.mg protein^{-1}.h^{-1}), while strain An 2 under the experimental conditions formed only few vesicles and fixed N_2 at a very low rate (*ca* 10 nmol C_2H_4.mg protein^{-1}.h^{-1}). The nitrogenase activity of strain An 1 was strongly affected by the O_2 concentration. *Frankia* An 1 and An 2 were infective and effective on *A. nitida* and *A. glutinosa* but not on *Datisca cannabina* and *Elaeagnus umbellata*. Both An 1 and An 2 strains were more infective and effective on *A. glutinosa* than *Frankia* strains AvcI1 and CpI1.

Introduction

Microorganisms that are able to induce N_2-fixing root nodules (actinorhizae) on actinorhizal plants have been classified in the genus *Frankia*. Since the first success in reproducible isolation of the endophyte from actinorhizae[10] in 1978, a number of isolates of *Frankia* spp. have been obtained from root nodules of different *Alnus* species, *e.g. A. glutinosa* from Europe[9,20,25], *A. incana* ssp. *rugosa* from USA[5,18], *A. incana* from Finland (A. Weber, in prep.) and *A. viridis* ssp. *crispa* from USA[3,25]. In addition *Frankia* isolates have been obtained from root nodules of other actinorhizal plants, *e.g. Casuarina* spp.[12,14] (Baas and Akkermans, in prep.), *Colletia* spp. (Baas and Akkermans in prep.), *Comptonia* sp.[10], *Elaeagnus* spp.[3] (Baas and Akkermans in prep.) and *Hippophaë* spp.[9,14]. Most isolates originated from native or introduced plants in northern America and Europe. Very little is known about the occurrence of *Frankia* spp. in Asia and its possible relationships with strains from other areas. In the present paper we report the isolation of *Frankia* strains from root nodules of *Alnus nitida*

* Department of Microbiology, Agricultural University, Wageningen, The Netherlands

Endl., an alder species with a restricted distribution in the Himalaya region primarily of Pakistan[11,23]. A detailed description of the morphological and physiological features of the isolates and their ability to infect plants is given.

Materials and methods

Isolation of Frankia

Actinorhizal nodules were collected from *Alnus nitida* plants growing in Bahrin, District Swat, Pakistan. A description of the sites has been given elsewhere[17]. The nodule lobes were thoroughly washed in water containing a drop of soap, broken apart into individual lobes and immersed in 3% aquous solution of OsO_4 for 2–10 min[20]. The OsO_4-treated nodule parts were washed 4–5 times with sterile water and incubated individually on agar slants containing Qmod medium[19] or Tween-NH_4^+ medium[7]. After two weeks of incubation the apparently axenic nodule parts were selected for isolation of *Frankia*. These nodule parts were crushed with a sterile forceps and transferred into tubes containing 10 ml of Qmod medium or Tween-NH_4^+ medium. The tubes were incubated at 29°C in the dark and regularly checked for growth of *Frankia*. Slow-growing *Frankia* colonies were visible within two months of incubation on both types of media. Best results were obtained after 2–5 min of sterilization with 3% OsO_4. The two isolates, *viz.* strain An 1, obtained after 2 min sterilization and cultivation on Tween-NH_4^+ medium and An 2 isolated after 5 min sterilzation and cultivation on Qmod medium, were subcultivated for confirmation of purity.

Both strains were cultivated on P + N agar (see below), after disintegration of the flocks by passing through a needle. Single colonies were picked from the plates and subcultivated in tubes containing P + N or P − N medium (see below). The codes P + N and P − N are used for media containing propionate (1.0 g.l^{-1}) with and without NH_4Cl (0.1 g.l^{-1}), respectively. P + cas medium contained (g.l^{-1}) basic medium with sodium propionate (1.0) and casamino acids (1.0). From each strain An 1 and An 2, 40 colonies were subcultivated as clones. The number of the clone is placed behind the code of the strain (*e.g.* An 1.1., An 1.2, *etc.*).

Light microscopy

Photomicrographs of water preparates or glycerine fixed cultures were made using either a phase-contrast microscope or Normarski differential interference-contrast photomicroscope.

Scanning electron microscopy

Isolates were fixed in glutaraldehyde (1%) and dehydrated. The dehydrated material was treated by critical point drying and covered with a gold layer as described elsewhere[16].

Growth experiments

Isolates were cultivated on Qmod medium or on other synthetic media with different carbon and nitrogen sources. The basic medium contained (g.l^{-1}): $CaCl_2 \cdot 2H_2O$ (0.1), $MgSO_4 \cdot 7 H_2O$ (0.2), K_2HPO_4 (1.0), $NaH_2PO_4 \cdot 2 H_2O$ (0.67), biotin (0.002), Fe-EDTA (0.025) and trace elements according to Allen and Arnon[2]. The carbon source was sodium propionate (1.0 g.l^{-1}), unless mentioned otherwise. The nitrogen source was (g.l^{-1}): N_2, NH_4Cl (0.1), KNO_3 (0.2), Casamino acids (1.0), glutamic acid (1.0) or aspartic acid (1.0). The final pH was 6.8. Cells were grown in the dark at 25°C in stationary Erlenmeyer flasks containing 50 ml medium. The media were inoculated with a one month old culture of the strains. The inoculum in each case was washed with a sterile phosphate buffer (50 mM, pH 7.0) by centrifugation and homogenized through a sterile needle (size 26G × 11 mm). Erlenmeyer flasks containing 50 ml of medium were inoculated with 0.5 ml of washed cells, containing 0.032 mg cell protein.l^{-1}.

Cells were harvested by centrifugation (5 min, 2700 rpm) and washed once for protein determination or four times for TOC (total organic carbon)[6] determination with phosphate buffer (50 mM, pH 7.0). The content of base soluble protein was determined according to Moss and Bond[22] and the TOC content was determined using a Beckman TOC analyzer (model 915A)[6]. Before TOC analyses, the cell suspensions were homogenized by sonication, the pH was adjusted to 4.0 and the suspensions were bubbled with argon for 5 min to remove CO_2[7].

Nodulation experiments

Infectivity of the *Frankia* strains An 1, An 2, Avcl1, Cpl1, Agsp$^+$ and Cc 1 was tested on seedlings of *Alnus glutinosa, A. nitida, Datisca cannabina* and *Elaeagnus umbellata*. Strains Avcl1 and Cpl1, the isolates of *A. viridis* ssp. *crispa* and *Comptonia peregrina* respectively, were obtained from D. Baker and J. G. Torrey, Petersham, Mass. USA. Strain Agsp$^+$ was isolated from *A. glutinosa* nodules with spores (sp$^+$) by Burggraaf[9], and strain Cc 1 originates from *Colletia cruciata* (Rob Baas, Wageningen). The inoculum of each strain was prepared by homogenizing one month old cultures by forcing through a syringe needle (26G × 11 mm) after washing twice by centrifugation (5 min, 2700 rp.) with 50 mM phosphate buffer (pH 7.0). Seedlings of *A. glutinosa, A. nitida, D. cannabina* and *E. umbellata* were inoculated by adding 0.1 ml of a homogeneous culture to axenic plants growing on agar slants (*Alnus* spp.), on perlite (*D. cannabina*) or on filter paper strips in tubes with Hoagland solution (*E. umbellata*), while semi-axenic *Alnus* seedlings growing in water culture were inoculated by immersing the roots for three days in a nitrogen-free solution containing the *Frankia* strain. Subsequently all the seedlings were grown in half strength nitrogen-free Hoagland solution.

Nitrogenase activity

Nitrogenase activity of pure cultures and of nodulated plants was determined as acetylene reduction. Ethylene was measured on a Becker 409 gas chromatograph fitted with a Porapak R (100–120 mesh) column, 60 cm, i.d. 2.8 mm, at 70°C. Cells of *Frankia* An 1 and An 2 grown previously in stationary culture with P − N or P + cas medium were harvested anaerobically. The effect of O_2 on the acetylene reduction of An 1 and An 2 was measured by incubating a 5 ml of the concentrated suspension in 16.6 ml Hungate tubes. After flushing with argon, CO_2 was added to a concentration of 0.03% in the tubes and oxygen was then injected at different concentrations. The concentration of O_2 was monitored gas chromatographically[26] and did not change during the incubation time. The incubation temperature was 25°C.

Results

Morphology

Two strains, An 1 and An 2, were isolated from root nodules of *A. nitida* from the same locality (Swat). On Qmod agar medium colonies of both of the isolates remained small and the size of the colonies did not exceed 1–2 mm in diameter within two months. The colonies were transparent, more dense in the centre than at the periphery and sporangia were more abundant at the centre of the colony, *i.e.* the older part.

The formation of vesicles and sporangia in both strains was affected by the composition of the medium (Table 1, Fig. 1a, b). *Frankia* An 1 formed vesicles and few but big sporangia on propionate medium (Fig. 1b). On Qmod medium no vesicles were present and sporangium formation was restricted: only very small mature sporangia or hyphae with few terminal swellings were observed. Sporangia were either intercalary (Fig. 2b) or terminal (Figs. 2a, c, 4a) and spores were easily

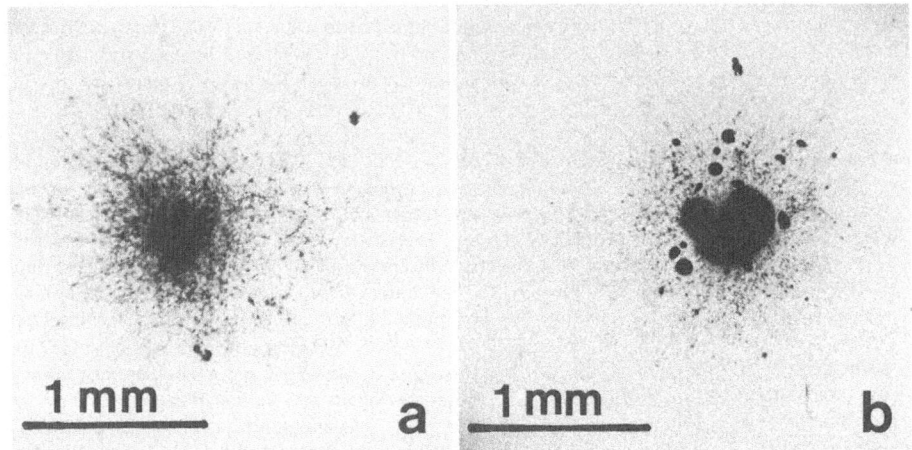

Fig. 1. Colonies of *Frankia* An 1 grown for 4 weeks on Qmod agar (a) and on P + N agar (b). Dark sites in the colonies are clusters of spores and sporangia.

Table 1. Influence of the medium on the formation of vesicles and sporangia in *Frankia* An 1 and An 2*

Medium	An 1		An 2	
	Vesicles	Sporangia	Vesicles	Sporangia
Qmod	−	±	−	+ +
P + N	+ +	+ +	−	+ +
P − N	+ +	+ +	±	+ +

* Relative amount of vesicles and sporangia: absent (−), occasionally present (±), many (+ +).

released from the sporangia (Fig. 2c). The hyphae on Qmod medium were smooth and smaller in diameter (Fig. 3a) than on P + N medium. In the latter case the hyphae contained lipid-like inclusions (Fig. 3b).

Vesicles were spherical (4–5 μm in diameter), formed terminally on short parental hyphae branching from hyphal strands (Fig. 2d, 4b). Each vesicle possessed a thick wall envelope around it which was not observed around the hyphae and sporangia. Hyphae of strain An 1 rapidly disintegrate by autolysis when grown for longer time without subcultivation. The vesicles often remained attached even when lysis had occurred. Autolysis of other *Frankia* strains, *e.g.* An 2 and Avc11, was much less pronounced, indicating that lysis is a feature of the strain.

Frankia An 2 formed sporangia both on Qmod and on propionate-containing media. The size of the sporangia was smaller but the number of sporangia was larger than in strain An 1 (Fig. 5). On Qmod medium the hyphal diameter of An 1 (0.8 μm) was smaller than that of An 2 (1.2 μm). The vesicles were formed only occasionally by strain An 2

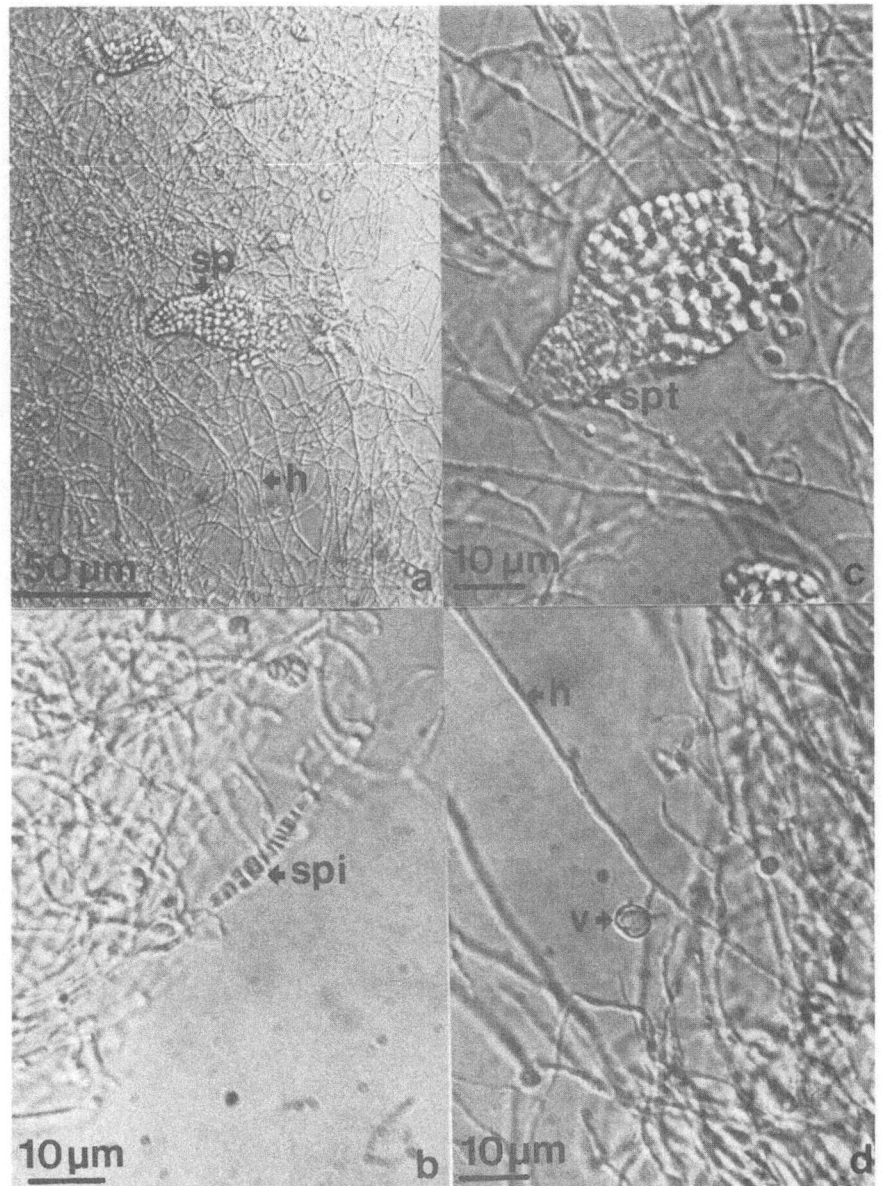

Fig. 2. Photomicrographs of *Frankia* An 1 grown for 2 weeks on P — N medium. **a.** Cluster of hyphae **(h)** with sporangium **(sp)**;**b.** Intercalary sporangium **(spi)**; **c.** Terminal sporangium **(spt)** with releasing spores and **d.** Vesicle **(v)** on short side branch of hyphen.

in P — N medium but not on other media. Strain An 2, in contrast to strain An 1, produced an extracellular brown pigment during the growth on all media and the amount of pigment was proportional to the cell yield.

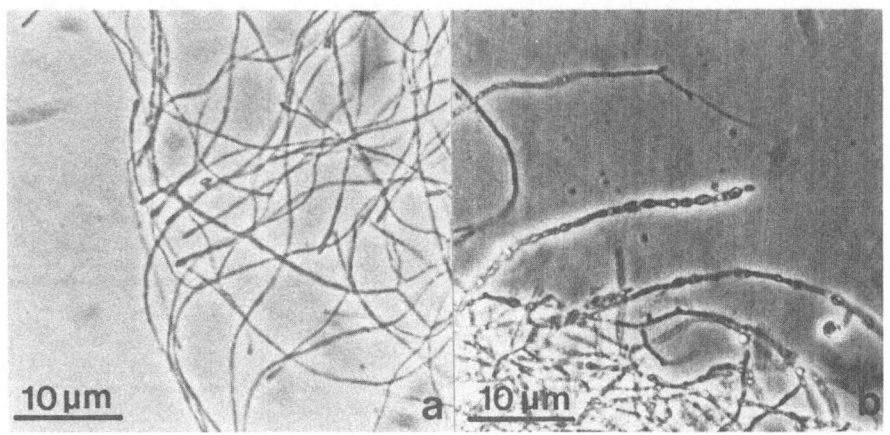

Fig. 3. *Frankia* An 1 grown for 4 weeks on Qmod medium **(a)** and on P + N medium **(b)**.

Table 2. Yield of *Frankia* An 1 and An 2 after growth for 21 days on various media

Medium	Yield (mg protein.l^{-1})	
	An 1	An 2
Qmod	7.5 (5.8–10.8)	13.0 (9.5–16.0)
P + N	5.2 (4.4–6.2)	1.1 (1.0–1.7)
P − N	3.3 (1.8–5.0)	3.3 (2.2–4.5)

Each value is the average of 3 replicates. Values between brackets denote extremes.

Table 3. Yields of *Frankia* An 1 and An 2 after growth for 24 days on media with different carbon sources

Carbon source*	Yield (mg.l^{-1})**			
	An 1		An 2	
	Protein	TOC	Protein	TOC
Acetate (1.6)	4.47	10.7	0.03	ND
Propionate (1.0)	3.37	8.42	0.13	ND
Butyrate (1.0)	0.05	ND	0	ND
Pyruvate (1.0)	0.07	ND	0	ND
Malate (1.0)	0	ND	0	ND
Succinate (2.2)	0	ND	0	ND
Glucose (1.0)	0	ND	0	ND
Tween-20 (1.0)	6.07	ND	0	ND
Tween-80 (1.0)	6.61	ND	0	ND

 * The basic medium was P + N in which propionate was replaced by other carbon sources.
   The concentration of the carbon compound (g.l^{-1}) is shown in brackets.
** Each value is the average of 3 replicates.
ND: Not determined.

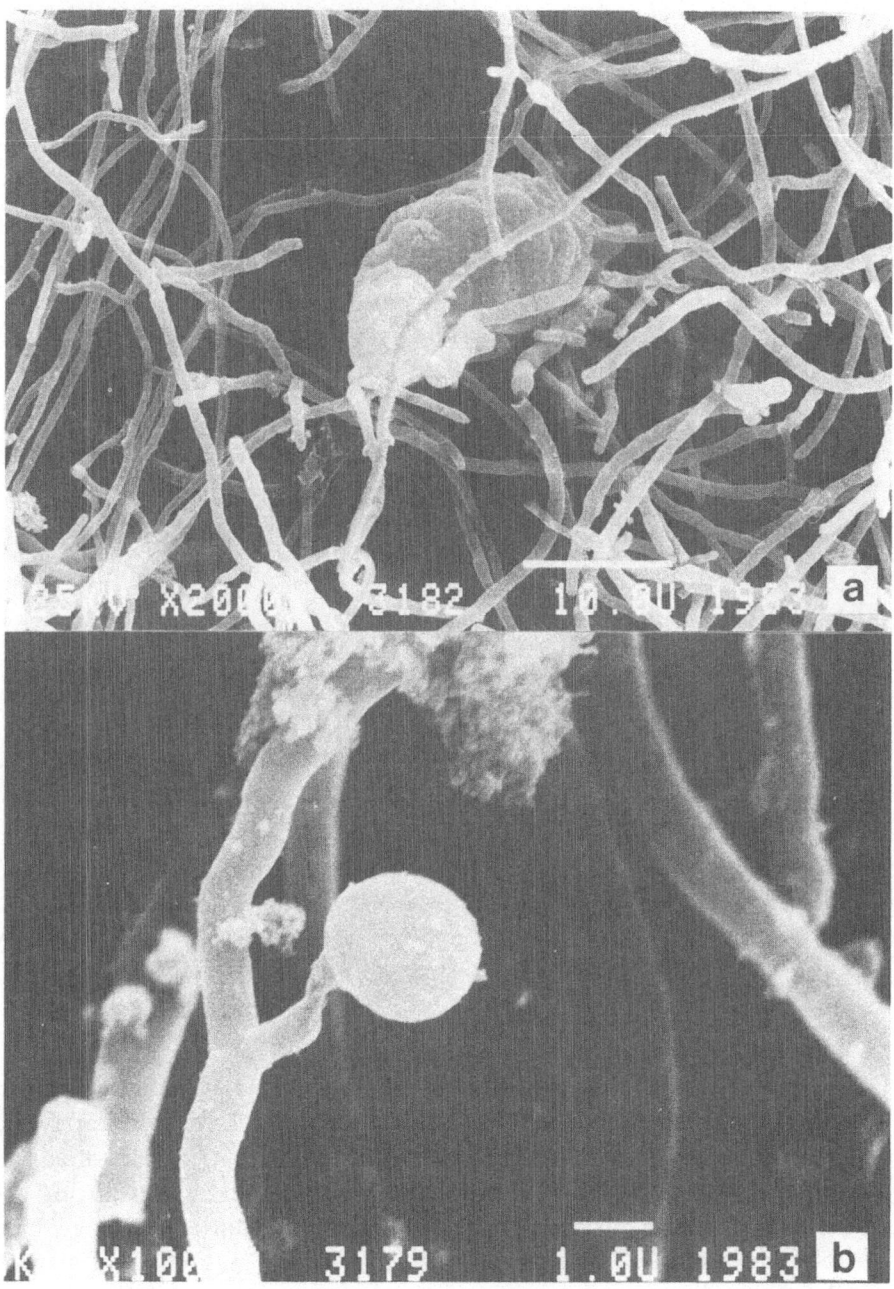

Fig. 4. Scanning electron micrograph of *Frankia* An 1 grown on P — N medium. **a.** Young sporangium within a network of hyphae. Spores visible inside sporangium. **b.** Vesicle on short side branch of hypha.

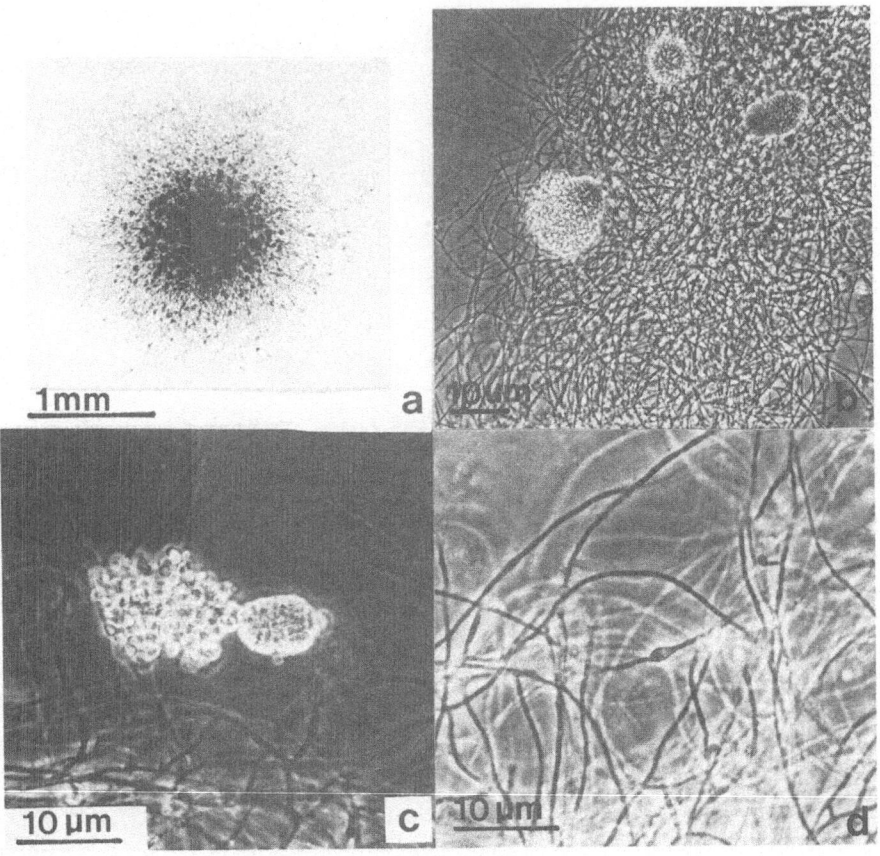

Fig. 5. Photomicrographs of *Frankia* An 2 grown for 3 weeks on Qmod agar (a) or on P + N medium (b, c, d). a. Colony on Qmod agar. Dark spots in colony are sporangia and clusters of released spores; b. Terminal sporangium; c. Terminal sporangium with releasing spores and d. Young intercalar sporangium.

Comparison of growth yield on various media

The growth yield was determined 21 days after inoculation and is based on the protein content of the cells (Table 2). Maximum yield was observed on Qmod medium. Growth of An 2 in P + N medium compared to growth in P − N medium was somewhat lower during the first three weeks of incubation (Table 2) and started after a lag phase of about one month. The phenomenon of initial reduction in growth on P + N medium in comparison to their growth on cultivation in P − N medium was also exhibited by 20 isolated clones of An 2.

Growth yield on various carbon sources

No growth of *Frankia* An 1 was observed in the media containing

Table 4. Yield of *Frankia* An 1 and An 2 after growth for 24 days on media with different nitrogen sources

Nitrogen source*	Yield (mg.l^{-1})**			
	An 1		An 2	
	Protein	TOC	Protein	TOC
N$_2$	1.53	3.02	0.29	ND
NH$_4$Cl (0.1)	3.37	8.42	0.13	ND
KNO$_3$ (0.2)	1.32	4.30	0.63	0.51
Casamino acids (1.0)	1.27	1.50	0.97	1.28
Glutamic acid (1.0)	0.25	0	0.72	0.60
Aspartic acid (1.0)	0.65	0.80	0.52	0.52

* The basic medium was P + N in which NH$_4$Cl was replaced by other nitrogen sources. The concentration of the nitrogen source (g.l^{-1}) is shown in brackets.
** Each value is the average of 3 replicates.

NH$_4$Cl (0.1 g.l^{-1}) with one of the following C-sources (g.l^{-1}): sodium malate (1.0), sodium succinate (2.2), glucose (1.0), pyruvate (1.0) or butyrate (1.0) (Table 3). The highest yield was observed in media containing tween 80 (1.0) or tween 20 (1.0) as C-source. Acetate (1.65) was a somewhat better C-source than propionate (1.0). Little growth yield of *Frankia* An 2 was observed in media containing propionate as C-source. The latter strain did not grow on any of the other tested C-sources.

Growth yield on various nitrogen sources

With propionate as the C-source, the highest yield of An 1 was found in media containing NH$_4$Cl while *Frankia* An 2 gave the highest yield on casamino acids (Table 4). Glutamate and, to less extent aspartate, gave a very low yield of An 1.

Nitrogenase activity

Both strains, An 1 and An 2, grew on nitrogen-free media with propionate as C-source (Tables 2 and 4). Addition of Fe-EDTA[27] was essential and no growth was found in P − N medium in which Fe-EDTA was replaced by Fe-citrate.

Strain An 1 fixed N$_2$ at a rate of up to 200 nmol C$_2$H$_4$ .mg protein^{-1} .h^{-1}. The nitrogenase activity was dependent on the O$_2$ tension in the gas phase. Cells precultivated in stationary culture in N-free medium showed temporary delay of the activity when shaken at O$_2$ tensions higher than 10% in the gas phase (Fig. 6). Strain An 2 fixed N$_2$ at a much lower rate (up to 10 nmol C$_2$H$_4$ .mg protein^{-1} .h^{-1}). Cells precultivated in P + cas medium showed activity *ca* 30 hours after transfer into P − N medium. The activity was dependent on the O$_2$ tension (Fig. 7).

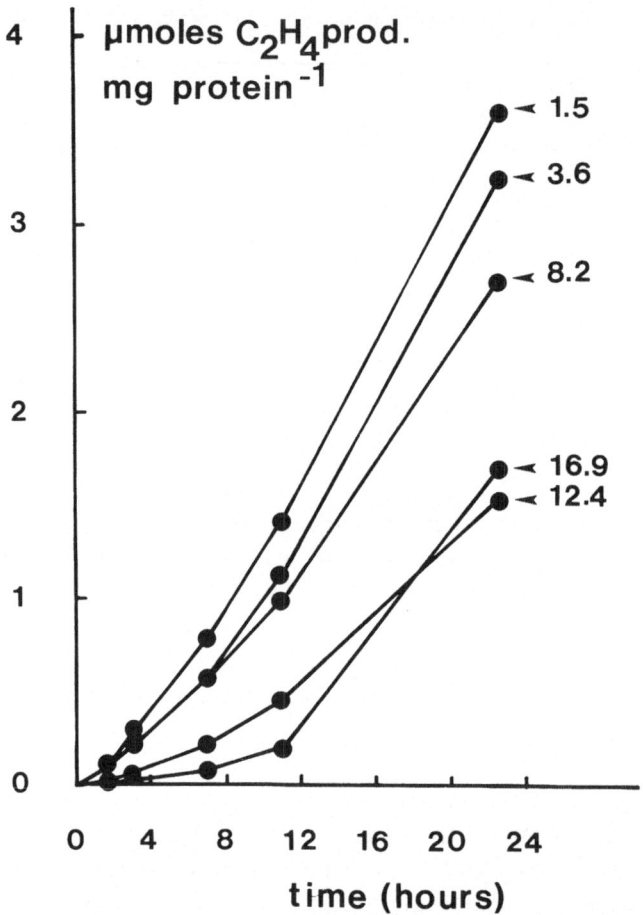

Fig. 6. Nitrogenase activity (C_2H_2 reduction) in *Frankia* An 1. Cells were pregrown in P – N medium and at t = 0 transferred into Hungate tubes (16.6 ml) with 5 ml fresh P – N medium and different concentrations of O_2 in the gas phase. The tubes were shaken at 25°C. The values in the graphs denote the % O_2 in the gas phase. The O_2 concentrations remained almost constant during the incubation time. Each point is the average of two replicates. Each tube contained 116 ± 5 μg cell protein.

Infectivity and effectivity of various Frankia strains on different actinorhizal plants

The infectivity of *Frankia* An 1 and An 2 and four other isolates (AvcI1, CpI1, Agsp$^+$, Cc 1) were tested on *A. glutinosa, A. nitida, D. cannabina* and *E. umbellata* (Tables 5 and 6). The nodule formation and the number of nodules were tested 8 weeks after inoculation. Both An 1 and An 2 were infective and effective on *A. glutinosa* and *A. nitida,* but not on the other plants. The infection of *A. nitida* was delayed on agar slants, possibly because this cultivation technique is less suitable for this species than for *A. glutinosa.* The infectivity

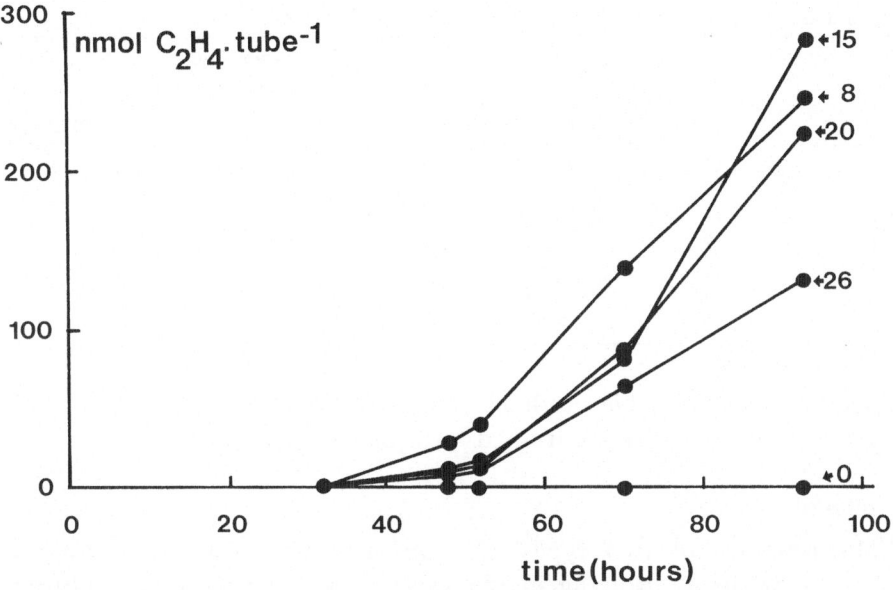

Fig. 7. Induction of nitrogenase (C_2H_2 reduction) in *Frankia* An 2.10, pregrown in P + cas medium for 2 weeks at $t = 0$ transferred in P – N medium. Cell suspensions (5 ml) were incubated in Hungate tubes (16.6 ml) in a gas phase of argon, CO_2 (1%), C_2H_2 (10%) and O_2 at a concentration of 0, 8, 15, 20 and 26%, respectively. The values in the graphs denote the % O_2 in the gas phase. Each tube contained $535 \pm 19 \mu g$ cell protein. The tubes were shaken at 25°C and the C_2H_4 production was followed over a period of 4 days. Each point is the average of 2 replicates.

Table 5. Infectivity and effectivity of various *Frankia* strains on *A. glutinosa* and *A. nitida*

Strain	*A. glutinosa*				*A. nitida*					
	Agar slants				Agar slants			Water cultures		
	Number of plants		N*	nmoles C_2H_4 . plant^{-1} .h^{-1}	Number of plants		N*	Number of plants		N*
	Total	Nodulated			Total	Nodulated				
								Total	Nodulated	
An 1	9	9	2	97	8	6	1	6	5	7
An 2	10	10	3	65	8	5	1	5	5	13
Avcl1	13	10	2	25	2	2	1	5	5	10
Cpl1	6	5	3	29	5	4	1	9	9	> 10
Agsp$^+$	5	0	0	0	5	0	0	9	0	0
Cc1	5	0	0	0	6	0	0	9	0	0
None	10	0	0	0	8	0	0	9	0	0

* N: Number of nodules per nodulated plant.

and effectivity of *Frankia* An 1 and An 2 was less in *A. nitida* when grown on agar slants than when grown in semi-sterile water cultures. With An 1 or An 2 as inoculum, no nodulation was observed on *D. cannabina* and *E. umbellata* seedlings. Eight other *Frankia* strains,

Table 6. Inoculation of *D. cannabina* and *E. umbellata* with various *Frankia* strains

Strain	D. cannabina Number of plants		E. umbellata Number of plants	
	Total	Nodulated	Total	Nodulated
An 1	15	0	5	0
An 2	14	0	5	0
AvcI1	15	0	5	0
CpI1	15	0	ND	0
Agsp+	15	0	ND	ND
Cc1	15	0	5	5
None	15	0	5	0

ND: Not determined.

isolated from *A. nitida* nodules showed a similar inefectivity and effectivity on the above mentioned host plants (data not shown).

Discussion

The isolation of two types of *Frankia* strains from root nodules of *A. nitida* plants growing in the same area, indicates that a natural population of *Frankia* may be a mixture of different strains. This confirms previous observations on the occurrence of spore positive and spore negative strains in root nodules of *A. glutinosa*[1,13] and the occurrence of strains with different growth requirements in *A. incana* ssp. *rugosa*[18], *A. crispa*[24], *A. rubra*[4] and *Casuarina equisetifolia*[15].

Both strains from *A. nitida* nodules produced vesicles and sporangia, though at different degrees. Moreover the differentiation was dependent on the composition of the medium. In strain An 1 but not in An 2, the formation of sporangia was suppressed in Qmod medium, which is possibly due to the presence of yeast extract in this medium. Reduction of spore formation has also been observed in strain CpI1 when yeast extract was added[21]. In general there was a negative correlation between spore formation by a strain and its growth rate. The cultures producing many spores formed compact flocs and grew slowly. With decreased spore formation the flocs became looser and the cells grew more rapidly. It is unknown whether the decreased growth is due to physiological changes during spore formation, *i.e.* increase of the number of growing top cells or due to a limited supply of O_2 and/or substrate(s) in the compact flocs.

Formation of sporangia not only seemed to be dependent on the composition of the medium but was also genetically determined. This is supported by the fact that faster growing clones (mutants?) with decreased spore production have now been obtained from the original An 2 strain. Similar clones have also been obtained from strain AvcI1 (Akkermans, unpublished results).

Both strains were able to form spherical vesicles and to grow in nitrogen-free medium, though to different degrees. The measurements of the nitrogenase activity confirm that these strains are real free-living N_2-fixers, like various other *Frankia* strains[5,9,12,14,15,27]. Nitrogenase activity of these strains was inhibited at high O_2 concentrations. This phenomenon has great similarity with data obtained in long-term experiments with a *Frankia* strain from *Casuarina equisetifolia* nodules[15]. The differences in the time period in the latter experiment and those described in the present paper can be explained on the basis of the differences in the growth rate of both strains. The initially low nitrogenase activity at high O_2 concentration can in part be explained by an effect on the nitrogenase, as has been demonstrated for *Azotobacter*[8].

The observation that vesicle formation was correlated with nitrogenase activity confirms previous observations on *Frankia* strain CpI1 and AvcI1[5,9,12,14,15,27]. Although these observations have been used as strong arguments for the claim that nitrogenase is localized within the vesicles, no proof can yet be given for this hypothesis!

Both strains were infective on *A. glutinosa* and *A. nitida,* but not on *D. cannabina* and *E. umbellata.* The differences in nodulation of both alder species may be related to difficulties in growing *A. nitida* seedlings on agar slants. The inability to nodulate *D. cannabina* confirms previous experiments with crushed *A. nitida* nodule homogenates as inoculum[17], indicating the occurrence of host specificity within the taxon *Frankia.*

Acknowledgements This work was supported by the United States Department of Agriculture under PL-480 programme grant No. FG-Pa-347 to Quaid-i-Azam University, Islamabad, and a fellowship to Fauzia Hafeez by the Agricultural University of Wageningen. Akkermans visit to Quaid-i-Azam University, Pakistan, was supported financially by United Nations Development Programme, UNDP (Project PAK 77/010). Wim Roelofsen is acknowledged for technical assistance. Electron micrographs were prepared at the Technical and Physical Engineering Research Service (TFDL), Wageningen. Thanks are due to the Plant Protection Service (Wageningen) for using the interference-contrast photomicroscope.

References

1 Akkermans A D L and Dijk C van 1981 Non-leguminous root nodules symbiosis with actinomycetes and *Rhizobium. In* Nitrogen Fixation vol 1, Ecology. Ed. W J Broughton, Oxford University Press, Oxford, pp 57–103.

2   Allen M B and Arnon D I 1955 Studies on nitrogen fixing blue green algae I. Growth and nitrogen fixation by *Anabaena cylindrica* Lemm. Plant Physiol. 30, 366–372.

3   Baker D, Torrey J G and Kidd G H 1979 Isolation by sucrose-density fractionation and cultivation *in vitro* of actinomycetes from nitrogen-fixing root nodules. Nature (London) 281, 76–78.

4   Berry A and Torrey J G 1979 Isolation and characterization *in vivo* and *in vitro* of an actinomycetous endophyte from *Alnus rubra* Bong. *In* Symbiotic Nitrogen Fixation in the Management of Temperate Forests. Eds. J C Gordon, C T Wheeler and D A Perry, Corvallis, Oregon, USA, pp 69–83.

5   Benson D R 1982 Isolation of *Frankia* strains from alder actinorhizal root nodules. Appl. Environ. Microbiol. 44, 461–465.

6   Blom J 1982 Carbon and nitrogen source requirements of *Frankia* strains. FEMS Microbiol Letters 13, 51–54.

7   Blom J, Roelofsen W and Akkermans A D L 1980 Growth of *Frankia* Avc I1 on media containing Tween-80 as C-source. FEMS Microbiol. Letters 9, 131–135.

8   Brotonegoro S 1974 Nitrogen fixation and nitrogenase activity of *Azotobacter chroococcum*. Ph.D. thesis. Agricultural University, Wageningen, The Netherlands.

9   Burggraaf A J P, Quispel A, Tak T and Valstar I 1981 Methods of isolation and cultivation of *Frankia* species from actinorrhizas. Plant and Soil 61, 157–168.

10 Callaham D, Torrey J G and Tredici P Del 1978 Isolation and cultivation *in vitro* of the actinomycete causing root nodulation in *Comptonia*. Science 199, 899–902.

11 Chaudhary A H, Khokhar S N, Zafar Y and Hafeez F 1981 Actinomycetous root nodules in angiosperms of Pakistan. Plant and Soil 60, 341–348.

12 Diem H G, Gauthier D and Dommergues Y R 1982 Isolation of *Frankia* from nodules of *Casuarina equisetifolia*. Can. J. Microbiol. 28, 526–530.

13 Dijk C van 1979 Endophyte distribution in the soil. *In* Symbiotic Nitrogen Fixation in the Management of Temperature Forests Eds. J C Gordon, C T Wheeler and D A Perry, Corvallis, Oregon, USA, pp 84–94.

14 Gauthier D, Diem H G and Dommergues Y R 1981 Infectivité et effectivité de souches de *Frankia* isolées de nodules de *Casuarina equisetifolia* et d' *Hippophaë rhamnoides* C. R. Hebd. Séances Acad. Sci. 293, 489–491.

15 Gauthier D, Diem H G and Dommergues Y R 1981 *In vitro* nitrogen fixation by two actinomycete strains isolated from *Casuarina* nodules. Appl. Environ. Microbiol. 41, 306–308.

16 Hafeez F, Akkermans A D L and Chaudhary A H Observations on the ultrastructure of *Frankia* sp. in root nodules of *Datisca cannabina* L. Plant and Soil *In press*.

17 Hafeez F, Chaudhary A H and Akkermans A D L 1984 Physiological studies on the N_2-fixing root nodules of *Datisca cannabina* L. and *Alnus nitida* Endl. from Himalaya region in Pakistan. Plant and Soil 78, 129–146.

18 Lechevalier M P, Baker D and Horriere F 1983 Physiology, chemistry, serology and infectivity of two *Frankia* isolates from *Alnus incana* sub sp. rugosa. Can. J. Bot. 61, 2826–2833.

19 Lalonde M, Calvert H E 1979 Production of *Frankia* hyphae and spores as an infective inoculant for *Alnus* species. *In* Symbiotic Nitrogen Fixation in the Management of Temperate Forests. Eds. J C Gordon, C T Wheeler and D A Perry, Corvallis, Oregon, USA, pp 95–110.

20 Lalonde M, Calvert H E and Pine S 1981 Isolation and use of *Frankia* strains in actinorhizae formation. *In* Current Perspectives in Nitrogen Fixation. Eds. A H Gibson and W E Newton, Australian Academy of Science, Canberra, pp 296–299.

21 McBride M J and Ensign J C 1983 Sporulation and germination of *Frankia* isolates of CpI1 and Acn 1. Unpublished manuscript.

22 Moss R D and Bond R C 1957 Protein determination. Manual of Microbiol. Methods. p 181.

23 Nasir Y J 1975 Betulaceae. Flora of West Pakistan. Eds E Nasir and S I Ali, 95, pp 1–5.

24   Normand P and Lalonde M 1982 Evaluation of *Frankia* strains isolated from provenances of two *Alnus* species. Can. J. Microbiol. 28, 1133–1142.
25   Quispel A and Tak T 1978 Studies on the growth of the endophyte of *Alnus glutinosa* (L.) Vill. in nutrient solutions. New Phytol. 81, 587–600.
26   Roelofsen W and Akkermans A D L 1979 Uptake and evolution of H_2 and reduction of C_2H_2 by root nodules and nodule homogenates of *Alnus glutinosa*. Plant and Soil 52, 571–578.
27   Tjepkema J D, Ormerod W and Torrey J G 1980 Vesicle formation and acetylene reduction activity in *Frankia* sp. Cp I1 cultured in defined nutrient media. Nature London 287, 633–635.

Plant and Soil 78, 61–78 (1984).
© 1984 *Martinus Nijhoff/Dr W. Junk Publishers, The Hague.*

Growth kinetics and nitrogenase induction in *Frankia* sp. HFPArI 3 grown in batch culture

MARCIA A. MURRY, MARK S. FONTAINE and JOHN G. TORREY
Cabot Foundation, Harvard University, Petersham, MA 01366, USA

Key words Batch culture *Frankia* HFPArl3 Nitrogenase Vesicles

Summary Kinetics of growth and nitrogenase induction in *Frankia* sp. ArI3 were studied in batch culture. Growth on defined medium with NH_4^+ as the N source displayed typical batch culture kinetics; however, a short stationary phase was followed by autolysis. Removal of NH_4^+ arrested growth and initiated vesicle differentiation. Vesicle numbers increased linearly and were paralleled by a rise in nitrogenase (acetylene reduction) activity. Nitrogenase activity ($10 \, nM \, C_2H_4 \cdot mg \, protein^{-1} \cdot min^{-1}$) was sufficient to support growth on N_2 and protein levels rose in parallel with nitrogenase induction. Optimal conditions for vesicle and nitrogenase induction were investigated. Maximum rates of acetylene reduction were obtained with 5 to $10 \, mM \, K_2HPO_4/KH_2PO_4$, $0.1 \, mM \, CaCl_2$ and $MgSO_4$. The optimum pH for acetylene reduction and respiration was around 6.7. The amount (5 to $10 \, \mu g$ protein/ml) and stage (exponential) of growth of the ammonium-grown inoculum strongly influenced the subsequent development of nitrogenase activity. Propionate was the most effective carbon source tested for nitrogenase induction. Respiration in propionate-grown cells was stimulated by CO_2 and biotin, suggesting that propionate is metabolized via the propionyl CoA pathway.

Introduction

Until recently, the physiology of nitrogen fixation in actinorhizal root nodules has been severely hampered by lack of understanding of the prokaryotic symbiont. Numerous early attempts to isolate the actinomycete from root nodules were largely unsuccessful (see ref. 6 for review), which has been attributed in part to the slow growth rates of the endophyte. In recent years, numerous isolations of the endophyte belonging to the genus *Frankia* (Frankiaceae, Actinomycetales)[9] have been reported from root nodules of various actinorhizal plants (see ref. [8] for review). Early success in cultivating *Frankia* isolates was achieved using complex media[5,12,13,21,30] often supplemented with lipids[13,28] or lipid extracts[34]. With the successful cultivation of various *Frankia* strains on defined media[11,14,18,19,35,37,39,40] it has become apparent that *Frankia* is more metabolically versatile than previously thought. These studies have shown that a wide variety of sugars, sugar alcohols, lipids and organic acids can serve as the sole carbon source for growth although marked differences in patterns of carbon utilization occur from strain to strain. Inorganic nitrogen sources such as NH_4^+ and nitrate support growth. Many strains are not dependent on exogenous vitamins or cofactors for growth[35].

The onset of nitrogenase activity in developing actinorhizal nodules has been correlated with the differentiation of specialized cells, the vesicles, from terminal swellings of the vegetative filaments of the endophyte[10,31]. The localization of reducing conditions within these specialized cells with tetrazolium dyes[2] and the isolation of vesicle clusters capable of acetylene reduction[46] lead to the proposal that the vesicles are the site of nitrogen fixation. This hypothesis has been substantiated by the correlation of vesicle differentiation with induction of nitrogenase activity in response to nitrogen limitation in aerobic *Frankia* cultures[19,25,40,41]. Since in all of the isolates tested, acetylene reduction occurs at ambient pO_2 levels, protection of the O_2-labile nitrogenase can be provided by the endophyte rather than by the nodules, as is the case in the legume/*Rhizobium* symbiosis[39], and has been attributed to localization of the enzyme in the structurally and presumably biochemically unique vesicle[41,42]. The present study is concerned with optimization of the physico-chemical and nutritional conditions which influence growth and nitrogen fixation in a *Frankia* isolate ArI3.

Materials and methods

Cultivation of organism

The *Frankia* strains used in this study are listed in Table 1. Axenic cultures were maintained at 28°C on a defined medium termed BAP which contained (in mM) KH_2PO_4, 7; K_2HPO_4, 3.4; NH_4Cl, 5; NaPropionate, 5; $MgSO_4$, 0.1; $CaCl_2$, 0.07; FeNaEDTA, 10 mg/1; biotin, 450 μg/1; and trace elements according to Tjepkema and coworkers[41]. The pH of the medium was adjusted to 6.7. Phosphate was added after autoclaving. Propionate or other carbon sources used were filter sterilized and added at a final concentration of 5 mM for organic acids (sodium salts) and 15 mM for sugars and sugar alcohols.

Cells for growth, nitrogenase induction and respiration experiments were grown at 28°C in 1- to 3-l glass cylindrical bottles that were magnetically stirred and sparged (200 to 600 ml/min) with filtered (0.45 micrometer Millipore filters) air dispersed at the bottom of the bottle through a scintered glass tube.

Nitrogenase induction

Nitrogenase induction was initiated by centrifuging an exponential phase culture (8 krpm, 10 min) and washing twice in N-free defined medium (termed induction or B medium). Filter sterilized MOPS (3-N-Morpholinopropanesulfonic acid, pH 6.7) was added to 20 mM to control pH. Concentrations of up to 50 mM had no inhibitory effect on short-term acetylene reduction or respiratory rates. The washed cells were resuspended in induction media at cell density of between 5 and 10 μg protein/ml and incubated on a rotary shaker (70 rpm) in 1- to 4-l cotton-plugged Erlenmeyer flasks.

Assays

Nitrogenase activity was measured under air with the acetylene reduction assay in standardized 6 to 10 ml serum vials. Two to four 2-ml replicate samples were removed aseptically from the culture and incubated with 10% acetylene (v/v) on a rotary shaker (70 rpm) at 28°C. Cells for the CO_2 experiments were buffered with 50 mM MOPS (pH 6.7) and sparged with a mixture of argon and 5% O_2 for 15 to 20 min to remove dissolved CO_2. Replicate 2-ml samples were withdrawn using an argon-flushed syringe fitted with a canula and injected into serum vials containing argon, 20% O_2 and CO_2 at the desired concentration. Ethylene production was measured using a Carle Model 9500 gas chromatograph equipped with a 1.2 m stainless steel column filled with a 25:75 (v/v) mixture of Porapak R (80–200 mesh) and N (50–80 mesh). Gas samples (100 µl) were injected directly onto the column. Acetylene was used as an internal standard to minimize injection error. The standard deviation of 2 to 4 replicate assays was generally less than 15% and is shown by the bars in each figure.

Respiration was measured concurrently with acetylene reduction by measuring CO_2 in the gas phase with a Carle Model 8700 thermal conductivity gas chromatograph. A 1.2 m column filled with the Porapak mixture described above was used at 50°C with helium as the carrier gas. The assay remained linear for at least 4 hours. Dissolved CO_2 was assumed to be in equilibrium with CO_2 in the gas phase and was included in calculations of CO_2 production. Overall precision of the assay was about 5%.

Respiration was also measured polarographically with a YSI-model 53 O_2 electrode (Yellow Springs, Ohio) thermostated at 29°C. Cells were concentrated to 40–80 µg protein/ml. For the CO_2 experiments, cells were sparged with N_2 to remove dissolved CO_2 and O_2 was added to air saturation levels. In ammonia-grown cells, rates were linear until very low O_2 tensions were reached. Variation between replicate samples was about 5%. CO_2 was added by bubbling the gas directly into the cell suspension. MOPS (50 mM, pH 6.7) was added to control the pH. For the pH experiments, cells were washed twice in BAP medium containing 0.5 mM phosphate and adjusted to the desired pH with 50 mM concentrations of organic buffer. MES (2-N-Morpholinoethanesulfonic acid) was used below pH 6.5; MOPS from pH 6.5 to 7.2 and TRIS (Tris (hydroxymethyl)aminomethane) above pH 7.2.

Cellular protein was estimated by the Bradford procedure[17] using cells that were sonicated (15 sec at 100 W) with a Braunsonic Model 1510 sonicator and boiled for 10 min in 0.3 N NaOH[24].

Vesicles were counted in a Petroff-Hausser chamber at 400X magnification with a phase-contrast microscope. Samples were briefly sonicated (15 s at 100 W) to detach vesicles from the vegetative hyphae. The criteria for mature vesicles were the presence of a refractile cell envelope and an attached 'stalk'.

Results

Growth kinetics

Fig. 1 shows a typical growth curve for batch culture of *Frankia* ArI3 grown in an air-sparged, stirred vessel in defined medium with propionate and NH_4^+ as the sole C and N sources, respectively. A lag period of variable length is normally seen following inoculation which is minimized by use of a large inoculum (10 to 20%, v/v) of log phase cells. An apparent exponential growth phase (as measured by total protein) ensues with a doubling time of less than 48 h. Under these conditions, exponential growth abruptly ceases and is followed by autolysis with soluble and cellular protein declining rapidly. Autolysis was not due to C-limitation since an additional 5 mM propionate added on day 5 to one culture did not prevent the decline in protein levels. Microscopic examination of filaments revealed cell wall 'ghosts' apparently devoid of cell contents. Autolysis is a common feature of microbial growth in batch culture and has been attributed to a variety of cultural factors[4, 23, 37, 44]. A similar pattern of apparent autolysis was also observed

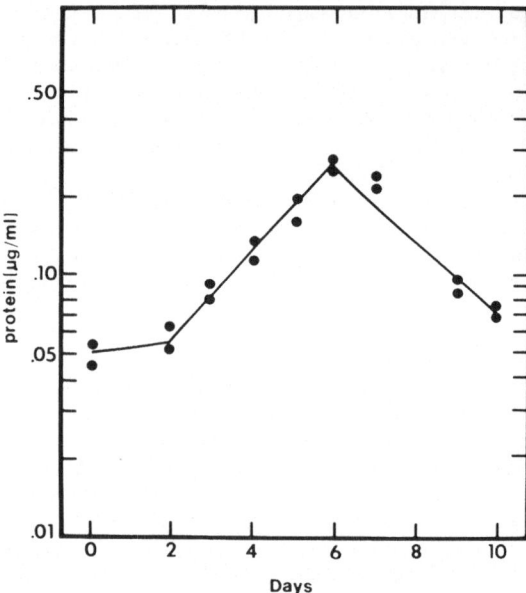

Fig. 1. Growth kinetics of ArI3 in batch culture. Cells were grown in 11 air-sparged cultures in defined medium with 5 mM NH_4^+ as the nitrogen source and 5 mM propionate as the carbon source. An additional 5 mM propionate was added to one culture at day 5.

by Burggraaf and Shipton[19] in several *Frankia* isolates grown under nitrogen-fixing conditions. We have observed autolysis also in the *Frankia* strain $EaNl_{pec}$ but only when grown in defined medium. Autolysis occurs in standing or shaking cultures and with a variety of carbon sources at concentrations up to 20 mM. However, in a rich organic medium a long stationary phase is observed and cells remain viable up to a year.

Vesicle and nitrogenase induction

A comparative time-course study of vesicle and nitrogenase induction was conducted on 8 *Frankia* isolates. Acetylene reduction and observations on sporangia formation were measured every 2–3 days after initiating induction by removing NH_4^+. There was considerable variation in maximum acetylene reduction rates among the 8 isolates (Table 1); however, rates correlated well with vesicle numbers in each case. Vesicles and nitrogenase activity were not detected in any of the isolates grown on NH_4^+ (5 mM). Sporangia and spores were produced in varying amounts by different isolates. EuIc, an ineffective symbiont in root nodules of its host, *Elaeagnus umbellata*[7], produced large numbers of active vesicles in culture. ArI3 isolated from its host *Alnus rubra*[12] was chosen for further studies since nitrogenase activity was greatest in this isolate. Furthermore, ArI3 grows homogeneously, thus

Table 1. Comparison of maximum acetylene reduction activity and sporangia formation in cultured *Frankia* strains

Strain	Source (ref. #)	Acetylene reduction[*] $nmol/C_2H_4 \cdot ml^{-1} \cdot h^{-1}$	Sporangial formation
ArI3	(12)	0.325	+
EuIIc	(7)	0.305	+++
EaNl$_{pec}$	(29)	0.304	0
CaI1	a	0.268	++
EuNf	b	0.116	++
AgII	c	0.106	++++
CpI1	(21)	0.103	++++
MpI1	a	0.091	+

[*] rates presented are the maximum rates observed during induction and occurred between days 8 and 10.
a. M. P. Lechevalier, Waksman Institute of Microbiology, Rutgers University, Piscataway, N.J.
b. M. Lalonde, Department of Ecology, Faculty of Forestry, Laval University, Quebec, Canada.
c. J. G. Torrey, Cabot Foundation, Harvard University, Petersham, MA

minimizing sampling error, and few spores are produced under these induction conditions.

Effects of nutrients. In a preliminary survey of a variety of sugars, sugar alcohols and organic acids used as the sole C source during induction, maximum acetylene reduction rates in ArI3 were obtained with 5 mM propionate. A comparison of specific acetylene reduction rates with vesicle numbers suggested that vesicle differentiation was enhanced by propionate. Sugars supported limited vesicle formation but high rates of acetylene reduction on a per vesicle basis (unpublished data).

Optimum concentration of several inorganic nutrients for nitrogenase induction was studied in time course experiments. Although there was considerable variation in maximum rates of acetylene reduction activities due to other suboptimal conditions in these initial studies, the optimum nutrient levels were consistent in several separate experiments. Maximum acetylene reduction activity occurred at day 10 with potassium phosphate levels between 5 and 10 mM. Since the media becomes alkaline when organic acids are utilized, 10 mM phosphate was routinely used to increase the buffering capacity of the media. Nitrogenase induction was limited by $CaCl_2$ concentrations less than about 0.05 mM while 0.35 mM was inhibitory (Fig. 2). A narrow concentration range for $MgSO_4$ was observed (Fig. 2).

Effects of biotin. Growth and induction of nitrogenase in ArI3 did not require exogenous vitamins. However, these processes were strongly

Table 2. Effect of biotin of respiration in ArI3*

| Substrate | Presence of biotin | | nmol $CO_2 \cdot mg^{-1}$ (protein) $\cdot min^{-1}$ |
	Growth	Assay	
Propionate	+	−	49.7 ± 0.2
	+	+	47.33 ± 0.7
	−	−	51.75 ± 2.07
	−	+	70.05 ± 1.25
Succinate	+	−	73 ± 9.4
	+	+	59.7
	−	−	78.7 ± 5.8
	−	+	57.47 ± 6.3

* Respiration was measured under air by CO_2 evolution in a 2 h assay. Biotin was added to the growth medium and the assay at a final concentration of 450 µg/1 where indicated.

stimulated by exogenous biotin when propionate was used as the sole C source (data not shown). Maximum rates of acetylene reduction were obtained with 450 µg biotin/1 in the induction media; higher levels proved inhibitory. Similar results were reported by Shipton and Burggraaf[36] for other *Frankia* strains, however, optimal levels were much higher. Table 2 shows that the short-term respiratory rate of ArI3 grown on propionate without exogenous biotin is stimulated by nearly 40% when biotin is added to the assay system. Biotin had a somewhat inhibitory effect on respiration on succinate-grown cells.

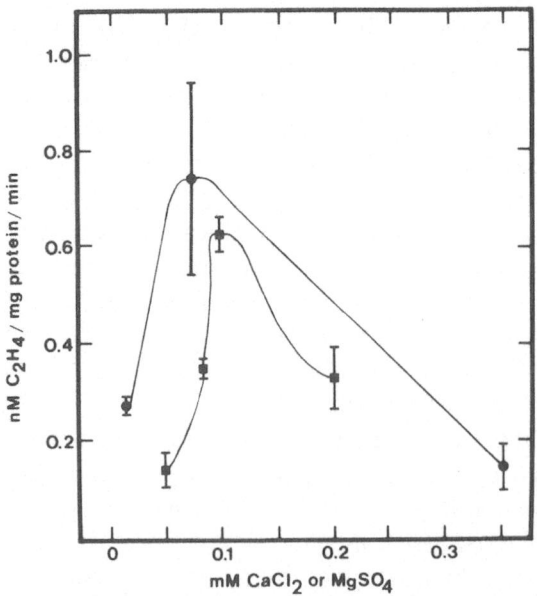

Fig. 2. Maximum rates of acetylene reduction in ArI3 induced with varying levels of $CaCl_2$ (●) and $MgCl_2$ (■). Acetylene reduction was measured on day 8.

Effects of pH and temperature. The pH of the medium was an important variable affecting induction. The response of acetylene reduction
and respiration of ArI3 to various pH values is shown in Fig. 3a and 3b,
respectively. The optimum pH for both processes was around 6.7. The
range for respiration was broader than that for acetylene reduction. The
type of organic buffers used to control pH did not influence the rates
of either respiration or acetylene reduction. The rates of both processes
varied by less than 10% when assayed at pH 6.5 with MES and MOPS
and at pH 7.2 with MOPS and Tris. The pH optimum for these two
processes in ArI3 is similar to that reported for growth of *A. glutinosa*
isolate LDAgpl[18], and nitrogenase induction of the *Comptonia
peregrina* isolate CpI1[41].

The optimum temperature for nitrogenase induction in ArI3 was
about 25 to 28°C but substantial activity was observed at 20 and at
36°C.

Effect of inoculum, size and growth phase. The size and growth
phase of the NH_4^+-grown inoculum strongly influenced the subsequent
development of vesicles and nitrogenase activity. Maximum rates of
acetylene reduction and the longest duration of nitrogenase activity
resulted when exponential phase cells were used as the inoculum

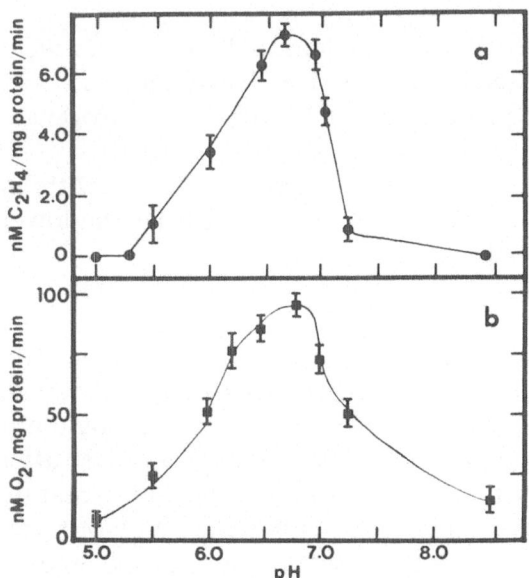

Fig. 3. Response of acetylene reduction activity (a) and O_2 uptake (b) to pH of the assay
media. Cells were grown at pH 6.5, and equilibrated at the indicated pH level for at least 1 h.
before assaying. Acetylene reduction was measured under atmosphere of Argon:O_2:CO_2
(79:20:1).

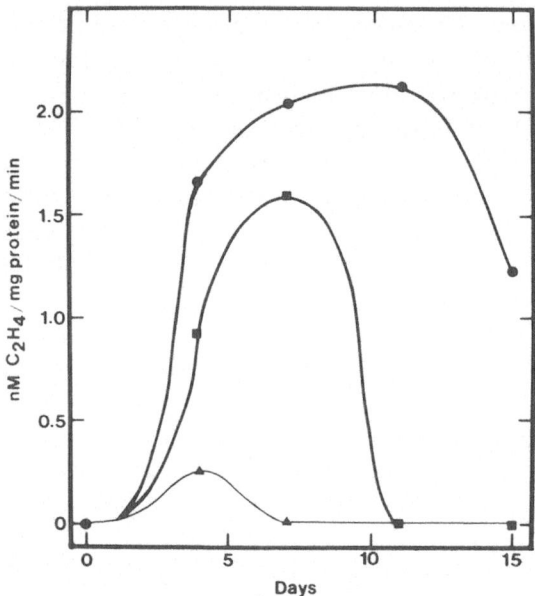

Fig. 4. Effect of growth phase of inoculum on subsequent nitrogenase induction in ArI3. Ammonium-grown cells were harvested in late lag-phase (■), mid-exponential phase (●) and declining phase (▲) of a batch culture, washed in N-free medium and resuspended at an initial cell density of 7 μg protein/ml. Each point is the mean value of 4 replicate assays from 2 separate experiments.

(Fig. 4). Lower, but still substantial acetylene reduction rates were obtained when late, lag-phase cells were used as inoculum. However, when cells from the declining phase were used, little or no nitrogenase activity or vesicles developed.

Maximum specific acetylene reduction rates were obtained with an initial cell density of between 5 and 10 μg protein/ml. Higher and lower cell densities inhibited the induction process.

Kinetics of nitrogenase induction

The time course of vesicle differentiation and nitrogenase induction in ArI3 cultured in the medium developed in this study is shown in Fig. 5. Vesicles were extremely rare and acetylene-reduction could not be detected in ammonium-grown filaments. Within 4 days after initiating induction by removing NH_4^+ from the medium, mature vesicles developed and increased rapidly in number. Induction of nitrogenase activity closely paralleled vesicle differentiation. In this experiment, propionate (initially 5 mM) became limiting by day 8 and nitrogenase activity and vesicle numbers decreased. Addition of 5 mM propionate at this time resulted in a linear increase in both vesicle numbers and specific nitrogenase activity until day 13. Propionate may

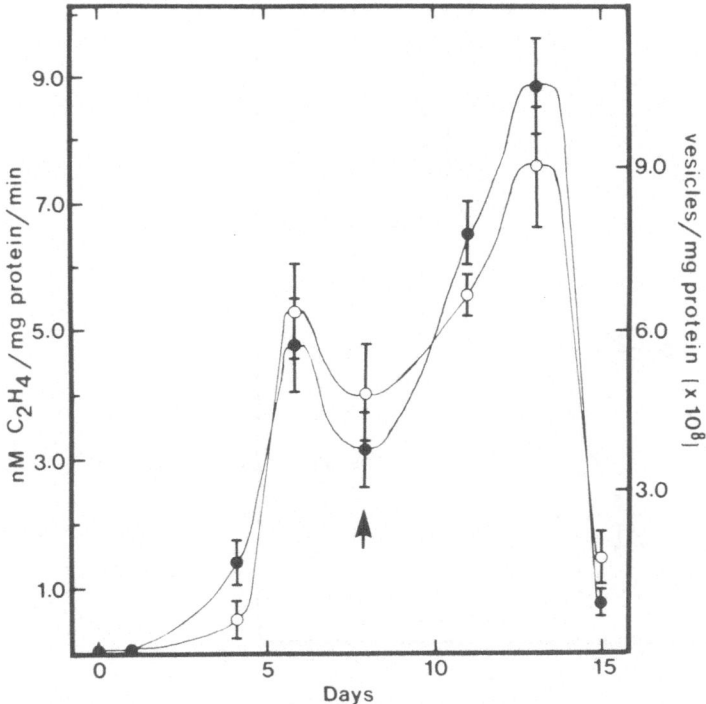

Fig. 5. Kinetics of vesicle differentiation (○) and nitrogenase activity (●) in ArI3. An additional 5 m*M* propionate was added at day 8.

have become limiting at this point and nitrogenase activity was nearly abolished. Autolysis apparently occurs since protein and vesicles decrease in parallel with nitrogenase activity.

Growth on N_2

Induction of nitrogenase resulted in resumption of growth at the expense of atmospheric N_2 (Fig 6). A substantial (10 to 40%) increase in protein was consistently seen within 24 to 48 h. after removing NH_4^+. This protein synthesis was supported by residual NH_4^+ or nitrogenous reserves, rather than by nitrogen fixation since acetylene reduction could not be detected at this point and since growth occurred under a N_2-free (argon) atmosphere (data not shown). However, as vesicles matured, acetylene reduction increased exponentially and cellular protein followed in parallel. Assuming a 3 to 4.5:1 ratio of acetylene:N_2, and that 6.25% of protein in N^{20}, then the rates of acetylene reduction between days 8 and 13 are more than sufficient to support the observed rise in protein.

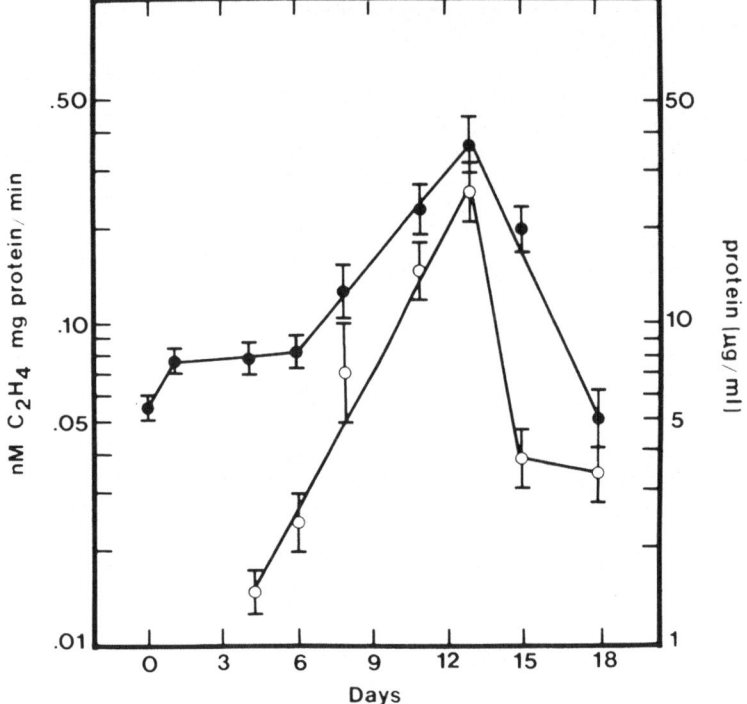

Fig. 6. Kinetics of nitrogenase induction (○) and growth (protein) (●) in ArI3.

Propionate metabolism

Our initial attempts to utilize propionic acid as a C source for growth and incubation of ArI3 were unsuccessful, which, as became apparent, was due to the restricted concentration range for oxidation of this organic acid. Fig. 7a shows the response to propionate concentration of respiratory rates in cells of ArI3 which were induced to form vesicles in 5 mM propionate, washed in C-free media and incubated for 18 h until endogenous respiration was lost. Maximum respiratory rates were supported by 0.5 to 5.5 mM propionate. Respiratory activity was inhibited by higher levels and abolished at 12.5 mM propionate. A similar trend was observed in ammonia grown cells (Fig. 7b) that were washed 3 times in C-free medium to remove exogenous propionate and assayed immediately. Under these conditions a substantial rate (nearly 60% of maximum) of O_2-uptake occurred without exogenous propionate indicating a high endogenous respiratory capacity. High propionate levels did not inhibit the endogenous rate which indicates that the inhibitory effect of propionate involves an effect of propionate uptake or subsequent metabolism rather than by uncoupling respiration.

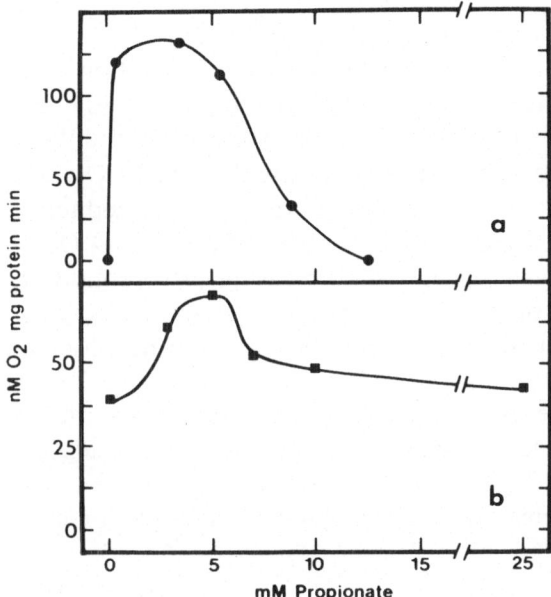

Fig. 7. Response of respiration (O_2 uptake) to exogenous propionate concentration. 7a. Actively-fixing cells were washed and starved for C 18 hours prior to assay. 7b. Ammonia-grown cells were washed 3 times to remove exogenous propionate and assayed within 1 h.

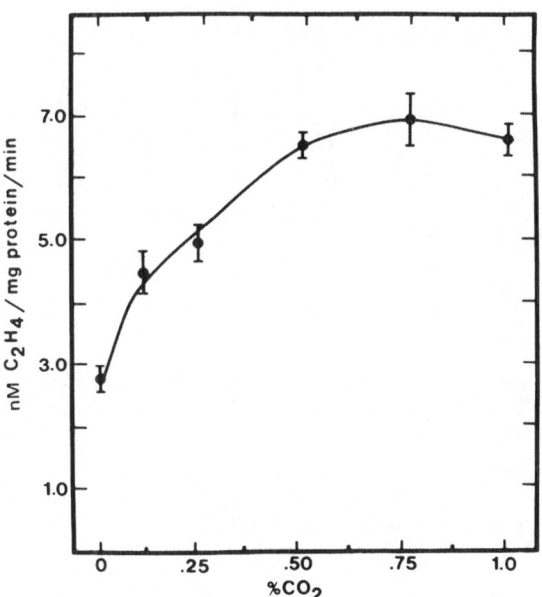

Fig. 8. Effect of exogenous CO_2 on acetylene reduction rates (2-h assay) of *Frankia* sp. ArI3 induced on propionate as the sole carbon source.

CO₂ stimulation of respiration and acetylene reduction

In our early studies, non-linear rates of acetylene reduction were often observed in propionate-induced cells particularly at low cell densities (less than 5 μg protein/ml) and in cells assayed under defined gas mixtures. This may have been due to limiting dissolved CO_2. The stimulatory effect of exogenous CO_2 on short term acetylene reduction activity in ArI3 that was sparged with a mixture of argon and O_2 to remove dissolved CO_2, is seen in Fig. 8. Increasing exogenous CO_2 stimulated acetylene reduction until the effect was saturated between 0.75 to 1% CO_2. Significant acetylene reduction rates were observed without exogenous CO_2 in the assay vessel. Although less than 1 nmol CO_2/ml was detected in the gas phase when the assay was initiated, CO_2 increased rapidly during the course of the assay and acetylene reduction rate accelerated (Fig. 9). In contrast, cells incubated with saturating CO_2 showed a linear rate of acetylene reduction for more than 2 h.

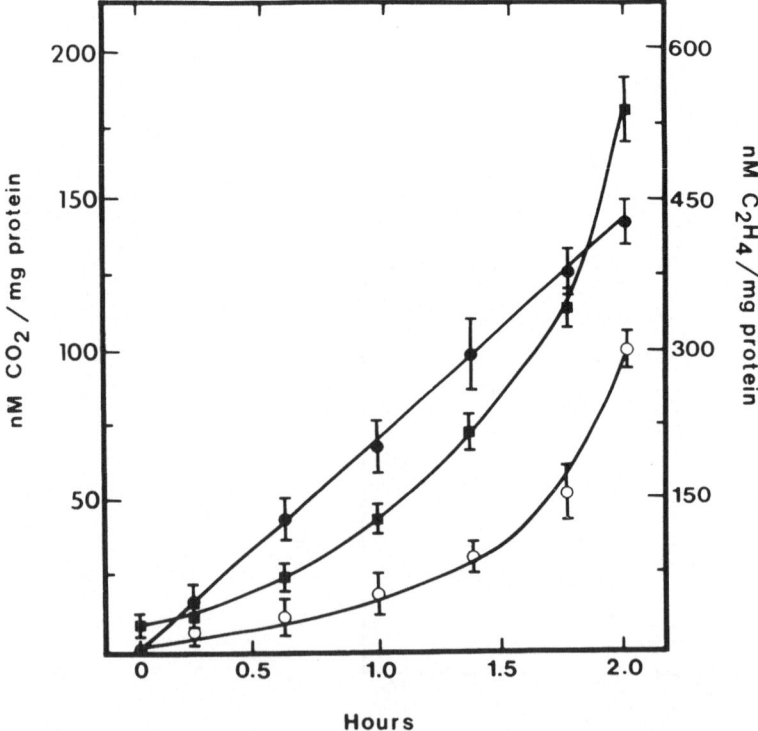

Fig. 9. Time course of acetylene reduction and CO_2 evoulution by *Frankia* sp. ArI3. Acetylene reduction under Argon + 20% O_2 in the absence of exogenous CO_2 (○) and with 1% CO_2 (●) in the gas phase. CO_2 evolution (■).

Table 3. Effect of exogenous CO_2 on respiration of ArI3 grown on various carbon sources

Carbon source	nmol $O_2 \cdot mg^{-1}$ (protein) $\cdot min^{-1}$	% CO_2 stimulation
Propionate	49.75 ± 2.1	64
Pyruvate	79.8 ± 2.85	22.5
Succinate	54.65	0
Acetate	51.77 ± 0.46	0
Malate	90.35 ± 3.5	0
Trehalose	63.55	0
Cellobiose	51.5 ± 0.7	0
Maltose	58.35 ± 4.6	0

A variable, but substantial (20 to 40%) CO_2 stimulation of respiration was also demonstrated in ammonia-grown cells that were sparged with N_2 to remove endogenous CO_2. This effect was not due to a change in pH of the medium since with the high buffer levels used (50 mM), pH values were maintained to within 0.1 pH unit after addition of exogenous CO_2. However, CO_2 stimulation of both acetylene reduction and respiration was seen only within the optimum pH range for these processes (between 6.3 and 6.9). This result suggests that at sub-optimal pH, factors other than CO_2 limit respiration and acetylene reduction. A specific requirement for one of the hydration products of CO_2, either H_2CO_3 or HCO_3^-, which predominate in this pH range, cannot, however, be discounted.

Respiratory rates of ArI3 grown with a variety of carbon sources are summarized in Table 3. A CO_2 stimulation of respiration was seen only when propionate or pyruvate was used as the sole carbon source for growth. The magnitude of CO_2 stimulation was variable. This was apparently due to variation in endogenous CO_2 levels in the assay. The high cell densities used for polarographic measurement of O_2 result in rapid CO_2 buildup. The CO_2-assimilation reaction may be quickly saturated and the stimulatory effect of CO_2 on respiration masked. However, the CO_2 response was substantial when either propionate or pyruvate was used as substrate. We could not detect a CO_2 stimulation of respiration even after repeated trials, when other organic acids or disaccharides were used as the sole C source.

Discussion

The growth kinetics of ArI3 in batch culture resemble those of other bacteria and filamentous fungi. An apparent exponential phase is observed with growth on NH_4^+ and N_2 which in mycelial organisms is determined by the balance between apical growth of the individual hyphae and the branching rate as described earlier[27]. A striking feature of growth in batch culture of ArI3 and other *Frankia* strains[19] in defined (but not organic) media, is the absence of a stationary phase; after exponential growth, hyphae progressively lyse. Addition of propionate reverses the autolytic process only at low cell densities (Fig. 5). Autolysis in response to carbon limitation has been described in other bacteria[23,34,37]. However, other factors, perhaps induction of an activating factor upon reaching a critical cell mass[4], must be involved, since autolysis in inorganic media was also observed using higher carbon levels (Fig. 1) and ArI3 can be washed free of carbon and starved for long periods without lysis (see details of cell preparation in Fig. 7a). Autolysis may be an important feature in symbiosis since hyphal disintegration is observed in aging nodules[32].

Nitrogenase induction was influenced by several physical and nutritional factors. The most important was the removal of NH_4^+ which at 5 mM completely repressed vesicle formation and acetylene reduction in the 8 strains tested. Removal of NH_4^+ resulted in the parallel induction of vesicles and nitrogenase activity which was followed by growth apparently at the expense of the N_2. Apparent growth on N_2 of several *Frankia* strains was shown by the data of Burggraaf and Shipton[19]. Thus, unlike some *Rhizobium* strains in culture[33,45] which excrete a majority of the N_2 fixed into the medium as NH_4^+, at least some strains of *Frankia* in culture are capable of assimilating the NH_4^+ fixed and using it for growth. The occurrence of glutamine synthetase and glutamate dehydrogenase in the plant but not endophyte fractions of *Alnus* nodule homogenates[15], indicated that, like the *Rhizobium*/ legume symbiosis[16], NH_4^+ is assimilated by the host rather than the symbiont in the nodule.

The concentration of several salts in the medium was critical for nitrogenase induction. Low levels of $CaCl_2$, $MgSO_4$ and K_2HPO_4/ KH_2PO_4 limited development of nitrogenase activity; higher levels of the Mg and Ca salts were inhibitory. Inhibition was probably due to precipitation with other essential nutrients rather than a direct inhibitory effect on cellular metabolism and may be related to the earlier observation[41] that the chelating agent EDTA was essential for vesicle formation under similar nutritional conditions.

Propionate supports rapid growth when used as the sole carbon source in all but one (see 38) *Frankia* isolate reported to date[10,11,14,36]. The enzymes required for its utilization appear to be constitutive in some isolates[38]. A substantial CO_2 stimulation of propionate-supported growth in *Frankia* sp. AvcII was reported by Akkermans and coworkers[3]. Although CO_2 is an essential metabolic intermediate required for bacterial growth, endogenous production of respiratory CO_2 normally masks this requirement. In ArI3 the CO_2 stimulation of acetylene induction and respiration apparently exceeds the basal CO_2 requirement for growth since relatively high levels of exogenous CO_2 (about 0.75% for acetylene reduction) were required to saturate the reaction(s).

A CO_2 requirement for expression of nitrogenase activity was also demonstrated on free-living *Rhizobium* which was distinct from its role in growth[1]. It is unclear whether the effect of CO_2 on acetylene reduction in *Frankia* ArI3 was specific or a consequence of its positive effect on respiration. The latter view seems more plausible since a differential effect of CO_2 on the two processes was not observed. Furthermore, CO_2 stimulation of respiration was dependent on the carbon source; of the substrates tested, only propionate and pyruvate oxidation was stimulated by CO_2 suggesting that oxidation of these substrates requires a carboxylation reaction.

Propionate utilization in *Rhizobium* occurs by conversion to succinate via propionyl CoA carboxylase and subsequent oxidation by the Krebs cycle[22] rather than by the β-oxidation pathway described in plant tissue[26]. Biotin stimulation of propionate oxidation *in vivo* suggests operation of the propionyl CoA pathway in *Frankia* sp. ArI3. Both propionyl CoA carboxylase and pyruvate carboxylase contain biotin as the prosthetic group. Although biotin is apparently synthesized by this organism, since growth of ArI3 does not require an exogenous source[36] respiration was stimulated by an exogenous supply of this vitamin.

The large scale culturing technique developed and described here for nitrogenase derepression in rapidly agitated liquid culture provides more precise gaseous and nutritional control and allows for less variability in sampling than has been obtained previously. The rates of acetylene reduction obtained (more than 10 nmol/mg/min) are sufficient to support growth on N_2 and are comparable to rates in other free-living nitrogen fixing organism. Expression of nitrogenase in free-living *Frankia* will allow investigation of the regulatory mechanisms controlling vesicle differentiation and nitrogenase synthesis and activity by key factors like O_2 and combined N and C sources which will be critical in understanding the symbiotic associations.

Acknowledgements This research was supported in part by a training grant to MM # 1042103580B1 from the U. S. National Institutes of Health, a research grant DEB-8106952 from the U. S. National Science Foundation, research grant DE-AC02-82ER12036 from the Department of Energy and by the Maria Moors Cabot Foundation for Botanical Research of Harvard University. The authors wish to thank Pat Young for technical assistance, Mary Lopez for helpful discussions and Frances O'Brien for typing the manuscript.

References

1   Aguilar O M and Favelukes G 1982 Requirement for carbon dioxide for nonsymbiotic expression of *Rhizobium japonicum* nitrogenase activity. J. Bacteriol. 152, 510–513.

2   Akkermans A D L 1971 Nitrogen fixation and nodulation of *Alnus* and *Hippophaë* under natural conditions. Ph.D. Thesis, University of Leiden, The Netherlands.

3   Akkermans A D L, Roelofsen W, Blom J, Huss-Danell K and R Harkink 1983 Utilization of carbon and nitrogen compounds by *Frankia* in synthetic media and in root nodules of *Alnus glutinosa, Hippophaë rhamnoides* and *Datisca cannabina*. Can. J. Bot., 2793–2800.

4   Arias J M, Fernandez-Vivas A, Montoya E 1983 Evidence for an activating substance related to autolysis in *Myxococcus coralloides* D. Arch. Microbiol. 134, 164–166.

5   Baker D, Torrey J G and Kidd G H 1979 Isolation by sucrose-density fractionation and cultivation *in vitro* of actinomycetes from nitrogen-fixing root nodules. Nature London 281, 76–78.

6   Baker D and Torrey J G 1979 The isolation and cultivation of actinomycetous root nodule endophytes. *In* Symbiotic Nitrogen Fixation in the Management of Temperate Forests. pp 38–56. Eds. J C Gordon, C T Wheeler and D A Perry. Forest Research Laboratory, Oregon State University, Corvallis, OR.

7   Baker D, Newcomb W and Torrey J G 1980 Characterization of an ineffective actinorhizal microsymbiont, *Frankia* sp. EuI1 (Actinomycetales). Can. J. Microbiol. 26, 1072–1089.

8   Baker D 1982–1983 A cumulative listing of isolated *Frankia,* the symbiotic nitrogen fixing actinomycetes. The Actinomycetes 17, 35–42.

9   Becking J H 1970 Frankiaceae fam. nov. (Actinomycetales) with new combination and six new species of the genus *Frankia* Brunchorst 1886. Int. J. Syst. Bacteriol. 20, 201–220.

10 Becking J H 1977 Endophyte and association establishment in non-leguminous nitrogen-fixing plants. In Recent Developments in Nitrogen Fixation, pp 551–567. Eds. W Newton, J R Postgate and C Rodriguez-Barrueco Academic Press, London.

11 Benson D R and Hanna D 1983 *Frankia* diversity in an alder stand as estimated by SDS-PAGE of whole cell proteins. Can J. Bot. 61, 2919–2923.

12 Berry A and Torrey J G 1979 Isolation and characterization *in vitro* of an actinomycetous endophyte from *Alnus rubra* Bong. *In* Symbiotic Nitrogen Fixation in the Management of Temperate Forests. pp 69–83. Eds J C Gordon, C T Wheeler and D A Perry. Oregon State Univ., Corvallis, OR.

13 Blom J, Roelofsen W and Akkermans A D L 1980 Growth of *Frankia* AvcI1 on media containing Tween 80 as C-source. FEMS Microbiol. Letters 9, 131–135.

14 Blom J 1981 Utilization of fatty acids and NH_4^+ by *Frankia* AvcI1. FEMS Microbiol. Letters 10, 143–145.

15 Blom J, Roelofsen W and Akkermans A D L 1981 Assimilation of nitrogen in root nodules of alder (*Alnus glutinosa*). New Phytol. 89, 321–326.

16 Boland M J, Farnden K J F and Robertson, J G 1980. Ammonia assimilation in nitrogen-fixing legume nodules. *In* Nitrogen Fixation, Vol II. Eds. W E Newton and W H Orme-Johnson. Univ. Park Press, Baltimore.

17 Bradford M M 1976 A rapid and sensitive method for quantification of microgram quantities of protein utilizing the principle of protein dye binding. Anal. Biochem. 72, 248–254.

18 Burggraaf A J P and Shipton W A 1982 Estimates of *Frankia* growth under various pH and temperature regimes. Plant and Soil 69, 135–147.

19 Burggraaf A J P and Shipton W A 1983 Studies on the growth of *Frankia* isolates in relation with infectivity and nitrogen fixation (acetylene reduction). Can. J. Bot. 61, 2774–2782.

20 Calam C T 1969 The evaluation of mycelial growth. *In* Methods in Microbiology, pp 567–591. Vol I. Eds. J R Norris and D W Ribbons. Academic Press, New York.

21 Callaham D, DelTredici P and Torrey J G 1978 Isolation and cultivation *in vitro* of the Actinomycete causing root nodulation in *Comptonia*. Science 199, 899–902.

22 DeHertogh A A, Mayeux P A and Evans H J 1964 The relationship of cobalt requirement to propionate metabolism in *Rhizobium*. J. Biol. Chem. 239, 2446–2453.
23 Dosŏil J, Sikyta B, Kašparova J, Doskočilova D and Zajiček J 1958 Development of the culture of *Streptomyces rimosus* in submerged fermentation. J. Gen. Microbiol. 18, 302–314.
24 Drews G 1965 Untersuchungen zur regulation der bacteriochlorophyll-synthese bei *Rhodospirillum rubrum*. Arch. Mikrobiol. 51, 186–198.
25 Gauthier D, Diem H G and Dommergues Y 1981 *In vitro* nitrogen fixation by two actinomycete strains isolated from *Casuarina* nodules. Appl. Environ. Microbiol. 41, 306–308.
26 Giovanelli, J and Stumpf P F 1958 Fat metabolism in higher plants. X. Modified β-oxidation of propionate by peanut mitochondria. J. Biol. Chem. 231, 411–426.
27 Kretschmer S, Riesenberg D and Bergten F 1981 Comparative analysis of mycelial growth. *In* Actinomycetes, Zbl. Bakt. Suppl. 11, 131–135.
28 Lalonde M and Calvert H E 1979 Production of *Frankia* hyphae and spores as an infective inoculation for *Alnus* species. *In* Symbiotic Nitrogen Fixation in the Management of Temperate Forests. pp 95–110. Eds. J C Gordon, C T Wheeler and D A Perry. Forest Research Laboratory, Oregon State Univ, Corvallis OR.
29 Lalonde M, Calvert H E, and Pine S 1981 Isolation and use of *Frankia* strains in actinorhizae formation. *In* Current perspectives in Nitrogen Fixation pp 296–299. Eds. A Gibson and W Newton. Australian Academy of Science Canberra.
30 Lechevalier M and Lechevalier H A 1979 The taxonomic position of the Actinomycetic Endophytes. *In* Symbiotic Nitrogen Fixation in the Management of Temperate Forests. pp 111–122. Eds. J C Gordon, C T Wheeler and D A Perry Oregon State Univ, Corvallis OR.
31 Mian S and Bond G 1978 The onset of nitrogen fixation in young alder plants and its relation to differentiation in the nodular endophyte. New Phytol. 80, 187–192.
32 Newcomb W, Peterson R L, Callaham D and Torrey J G 1978 Structure and host-actinomycete interactions in developing root nodules of *Comptonia peregrina*. Can J. Bot. 56, 502–531.
33 O'Gara F and Shanmugan K T 1976 Regulation of nitrogen fixation by *Rhizobium* export of fixed N_2 as NH_4^+. Biochim Biophys. Acta 437, 313–321.
34 Postgate J R and Hunter J R 1964 Accelerated death of *Aerobacter aerogenes* starved in the presence of growth-limiting substrates. J. Gen. Microbiol. 34, 459–473.
35 Quispel A and Tak T 1978 Studies on the growth of the endophyte of *Alnus glutinosa* (L.) Vill. New Phytol. 81, 587–600.
36 Shipton W A and Burggraaf A J P 1982 A comparison of the requirements for various carbon and nitrogen sources and vitamins in some *Frankia* isolates. Plant and Soil 69, 149–161.
37 Strange R E, Dark F A, and Ness A G 1961 The survival of stationary phase *Aerobacter aerogenes* spores in aqueous suspension. J. Gen. Microbiol. 25, 61–76.
38 Tisa L, McBride M and Ensign J C 1983 Studies of growth and morphology of *Frankia* strains $EANI_{pec}$, $EuII_c$, CpII and $ACNI^{AG}$. Can. J. Bot. 61, 2768–2773.
39 Tjepkema J D and Yocum C S 1973 Respiration and oxygen transport in soybean nodules. Planta 115, 59–72.
40 Tjepkema J D, Ormerod W and Torrey J G 1980 Vesicle formation and acetylene reduction activity in *Frankia* sp. CpII cultured in defined nutrient media. Nature London 287, 633–635.
41 Tjepkema J D, Ormerod W and Torrey J G 1981 Factors affecting vesicle formation and acetylene reduction (nitrogenase activity) in *Frankia* sp. CpII. Can. J. Microbiol. 27, 815–823.
42 Torrey J G and Callaham D 1982 Structural features of the vesicle of *Frankia* sp. CpII in culture. Can. J. Microbiol. 28, 749–757.
43 Torrey J G, Tjepkema J D, Turner G L, Bergersen F J and Gibson A H 1981 Dinitrogen fixation by cultures of *Frankia* sp. CpII by $^{15}N_2$ incorporation. Plant Physiol. 68, 983–984.

44 Trinci A P J and Righelato R C 1970 Changes in constituents and ultrastructure of hyphal
    compartments during autolysis of glucose-starved *Penicillicem chrysogenum.* J. Gen.
    Microbiol. 60, 239–249.
45 Tubb R S 1976 Regulation of nitrogen fixation in *Rhizobium* sp. Applied Environ.
    Microbiol. 32, 483–488.
46 van Straten J, Akkermans A D L and Roelofsen W 1977 Nitrogenase activity of endophyte
    suspensions derived from root nodules of *Alnus, Hippophaë, Shepherdia* and *Myrica* spp.
    Nature London 266, 257–258.

Plant and Soil 78, 79–90 (1984).
© 1984 *Martinus Nijhoff/Dr W. Junk Publishers, The Hague.*

A comparison of cultural characteristics and infectivity of *Frankia* isolates from root nodules of *Casuarina* species

ZHONGZE ZHANG, MARY F. LOPEZ and JOHN G. TORREY
Cabot Foundation, Harvard University, Petersham, MA 01366, USA

Key words   Actinorhizal plants   *Casuarina   Frankia*   Root nodules   Symbiosis

Summary   The isolations of three new strains of *Frankia* were made from root nodules of *Casuarina cunninghamiana* growing aeroponically. Two strains, HFPCcI1 and HFPCcI2 isolated by Lopez are typical *Frankia* strains, producing sporangia among filamentous mats in culture and, in the absence of combined nitrogen, forming vesicles and showing acetylene reduction. They are red-pigmented and, although failing to nodulate *Casuarina* hosts, effectively nodulated *Elaeagnus* and *Hippophaë*. A third strain HFPCcI3 isolated by Zhang from the same source, also a typical *Frankia,* can form sporangia and vesicles in culture and reduce acetylene, is unpigmented, fails to nodulate *Elaeagnus* but effectively nodulates *C. cunninghamiana* and *C. equisetifolia*. Comparisons are made among all of the *Casuarina* isolates in our collection from around the world (twelve in all) with regard to their cultural characteristics and capacity to infect host plant species. Questions are raised about the specificity of the various isolates and their possible affinities. Opportunities are suggested for inoculation of seedlings for forestry and field application using the infective, effective strains now available.

Introduction

An appreciation of the real and potential significance of *Casuarina* species as multipurpose tree crops has increased world wide, especially in the developing countries of the tropics and subtropics[18,19,21]. *Casuarina* is economically the most important genus among the actinorhizal plants, that is, among the diverse array of woody dicotyledonous plants nodulated by the filamentous soil bacterium, *Frankia* of the Actinomycetales. This symbiotic association leads to fixation of atmospheric nitrogen and accounts in large part for the capacity of Casuarinas successfully to occupy poor sites such as disturbed areas, desert or coastal dunes and for its use in land reclamation, dune stabilization, and shelter belt plantations.

Since the infective microorganism is not carried by the seeds, successful establishment of nodulated plants in the past has required transfer of soil or leafy litter containing viable bacteria or nodule material, fresh or dried, along with the seeds or seedlings. That successful transmission of *Frankia* and the seed has occurred is clear from the present world-wide distribution of nodulated *Casuarina* species throughout the tropics from their origins in Australia and the South Pacific Islands[22].

Considerable effort has been expended in a number of laboratories to isolate and culture *Frankia* from root nodules of *Casuarina* species so that controlled distribution and planned inoculation of seeds or seedlings could be practised easily and systematically. Recently published reports of such isolations have been made[8, 9, 10, 12, 13] and a number of isolations have been made without formal public notice. In general, these isolations have proved unsuccessful in that although the isolate is a typical *Frankia* in growth behavior and morphological structure, the organisms have proved non-infective on *Casuarina* seedlings. In several cases[14] the isolates have nodulated seedlings of members of the Elaeagnaceae, an affinity which still requires explanation. An isolate reported by Diem *et al.*[8, 9, 10, 11, 12] from *C. junghuhniana* designated by them ORS1106 has been reported to infect seedlings of several *Casuarina* species but has proved difficult for several laboratories to maintain in culture since it shows poor or slow growth in complex media.

Over the past two years we have achieved isolation of *Frankia* strains from *Casuarina* species on several occasions in our laboratory and have cultured them successfully in complex or defined synthetic media. We have also obtained cultures of *Frankia* strains reputed to be *Casuarina* isolates from a number of laboratories. These strains we have cultured, characterized morphologically and tested on host plants for infectivity and effectivity. This report presents a summary of our experiences together with a comparison of these different isolates.

Materials and methods

Sources of Frankia *strains isolated from root nodules of Casuarina species*

We have been sent cultures of several strains isolated by Diem, Gauthier and Dommergues from laboratories at ORSTOM Dakar, West Senegal. These include strains originally designated G2, D11[13] and one from the Phillipines designated P1[14]. In addition we were provided strains ORS1106 (formerly referred to as CJ1-82). These strains were cultured on media recommended by Dommergues.

Dr. J. Ruan working in the laboratiry of the Lechevaliers at the Waksman Institute at Rutgers University provided us with two isolates of *Frankia* from root nodules of *C. cunninghamiana* designated by him R3 and R43.

Dr Dwight Baker of the Kettering Research Laboratory, Yellow Springs, Ohio sent us two different strains of *Frankia* isolated by sucrose gradients from nodules of *Casuarina* spp. These were designated 53008 and 53024. Dr. Baker also provided us with *Frankia* strain 53001 which was a fast growing red-pigmented sub-strain derived in Baker's laboratory from ORS1106.

Mary Lopez of our laboratory made two different isolations of *Frankia* strains from plants of *C. cunninghamiana* grown aeroponically and originally inoculated with nodule suspensions from *C. equisetifolia* collected in Florida. These were designated HFPCcI1 and HFPCcI2.

Zhang Zhongze, visiting scholar from the Institute of Forestry and Pedology, Academia Sinica, Shenyang, working in our laboratory isolated a *Frankia* strain from *C. cunninghamiana* plants grown in aeroponics. His isolate, which is described in some detail in this paper, was designated HFPCcI3.

Culture of strains in the laboratory

Our practise on receipt of a new strain has been to culture the organism in the medium recommended by the donor for the particular strain. If successful in maintaining the culture on its original medium, we usually used that for maintenance and then made comparative tests on other standard complex or defined synthetic media used in our laboratory. The *Frankia* strains from *Casuarina* that we have grown and tested are listed in the Results below, together with information on each strain, the culture conditions, and some of their characteristics in culture.

A variety of nutrient media have been developed for the culture of isolated *Frankia* since the first complex medium referred to as *Frankia* broth was used in our laboratory[2]. A medium of special value in early stages of isolation is the complex QMod medium developed by Lalonde and Calvert[16]. Other media used successfully in culturing *Frankia* isolates from *Casuarina* root nodules include L/2 medium[17], YCz medium[3], S and TW broth (Ruan and Lechevalier, personal communication) and BAP medium (Murry *et al.* this volume). B medium is used to designate a modified BAP medium lacking combined nitrogen (see also Murry *et al.* this volume).

Trials for infectivity and effectivity

The characteristics of the *Frankia* strains, their growth rates, pigmentation, morphological expression, capacity to sporulate or form vesicles in culture and chemical properties are all important criteria for understanding and identifying them as distinct entities. For practical purposes, the capacity of the strain to infect a host plant[6] and to establish effective N_2-fixing nodules is one of the most interesting and crucial characteristics.

Tests for infection of hosts may fail for a number of reasons relating to the trial itself rather than to the host-microbial combination. Nodulation of a compatible host plant by an appropriate *Frankia* strain may be inhibited by many factors: low or high pH of the nutrient solution, excess available fixed nitrogen in the root medium, unfavorable root and/or shoot temperatures, unfavorable root substrates, age of host plants, and other factors[21].

Trials of *Frankia* isolates for infectivity of *Casuarina* have been carried out with young seedlings, started by germination in sand-vermiculite in plastic trays in a growth chamber on 16 hr light and 8 hr dark cycles at 25°C and 19°C respectively for 3–6 weeks, watered with 1/4-strength Hoagland's solution. When ready for a trial, seedlings were washed in tap water and the plants transferred to the test containers. Usual trials included water culture jars with 1/4-strength Hoagland's solution lacking nitrogen, or sand in "Rootrainers" (manufactured by Spencer-Lemaire Industries, Ltd., Edmonton, Alberta, Canada), plastic containers with 4 plants each in a separate root cavity which can be opened like a book for periodic examination of the root system, or in larger scale in aeroponics boxes[23]. Trials were carried out in growth chambers under artificial light or in the greenhouse under natural daylight or with supplemental lighting from high-intensity halide lamps.

Examinations of the root system for early nodulation was usually at weekly intervals[20]. Successful infection and effective nodule development was reflected by the green appearance of the shoot and shoot extension growth; failure was manifested by lack of growth of the shoot and a purple-red discoloration of the shoot tips followed by yellowing and then browning of the shoots.

Determination of effectivity was usually made as well by acetylene reduction assays of the root system or of excised root nodules using gas chromatography[5].

Results

Cultural characteristics of Frankia *isolates from* Casuarina cunninghamiana *made in Petersham*

The three separate isolates from root nodules of *C. cunninghamiana* made at our laboratories in Petersham have been the basis for the comparative study reported here. These strains and their isolations have not been described previously and therefore deserve a careful examination as preliminary to the comparison with strains isolated by others and grown in our laboratory collection.

Isolations of strains HFPCcI1 and HFPCcI2 were made by Mary Lopez from nodules taken from plants of *C. cunninghamiana* grown in an aeroponic tank in the greenhouse with 1/4-strength Hoagland's solution lacking nitrogen. The original inoculation of seedlings grown aeroponically was from nodule suspension of *C. equisetifolia* collected in Florida. Nodule lobes were surface-sterilized with 30% hydrogen peroxide, cut into pieces and incubated in QMod medium on agar plates[16]. Within 4 weeks filamentous colonies could be seen growing out from nodule lobes. Two strains which were isolated from separate nodule lobes grew slowly and were finally established as cultures with regular transfer into several different nutrient media.

CcI1 grew best on BAP medium but also on YCz and L/2 media (Fig. 1). The cultures were filamentous and formed vesicles in all media tested as well as sporangia. This isolate routinely showed large amounts of extracellular debris associated with the filaments. Usually the cultures showed red pigmentation but in different media were invariable in color. The organism showed acetylene reduction in culture B medium. CcI1 did not nodulate seedlings of *C. cunninghamiana* or *C. equisetifolia* but did nodulate *Elaeagnus umbellata* and *Hippophaë rhamnoides* effectively. Tests to determine the nature of the culture debris led to streaking and plating of cultures on synthetic media, in order to determine their purity. Red-colored colonies of filamentous bacteria which proved difficult to culture alone were presumed to be *Frankia;* smooth yellow filamentous bacterial colonies not identified grew on the plates and could be subcultured readily on yeast extract-casamino acid-dextrose agar. Thus, CcI1 appeared to be a difficult-to-culture mixed bacterial population, one strain of which was probably *Frankia.*

CcI2 is a filamentous bacterium which grew well on BAP medium but also on YCz, L/2 and QMod agar media (Fig. 2). On the last medium it formed small rough rounded red colonies. Vesicles were formed in all media and also large numbers of sporangia many of which released spores into the medium spontaneously. In YCz the sporangia were very reduced. In L/2 medium the red pigment was soluble. In other media an orange insoluble crystal was produced. From microscopic and cultural evidence, CcI2 was believed to be a pure culture of *Frankia.* When tested on seedlings of *C. equisetifolia* or *C. cunninghamiana,* no nodules were produced. CcI2 effectively nodulated

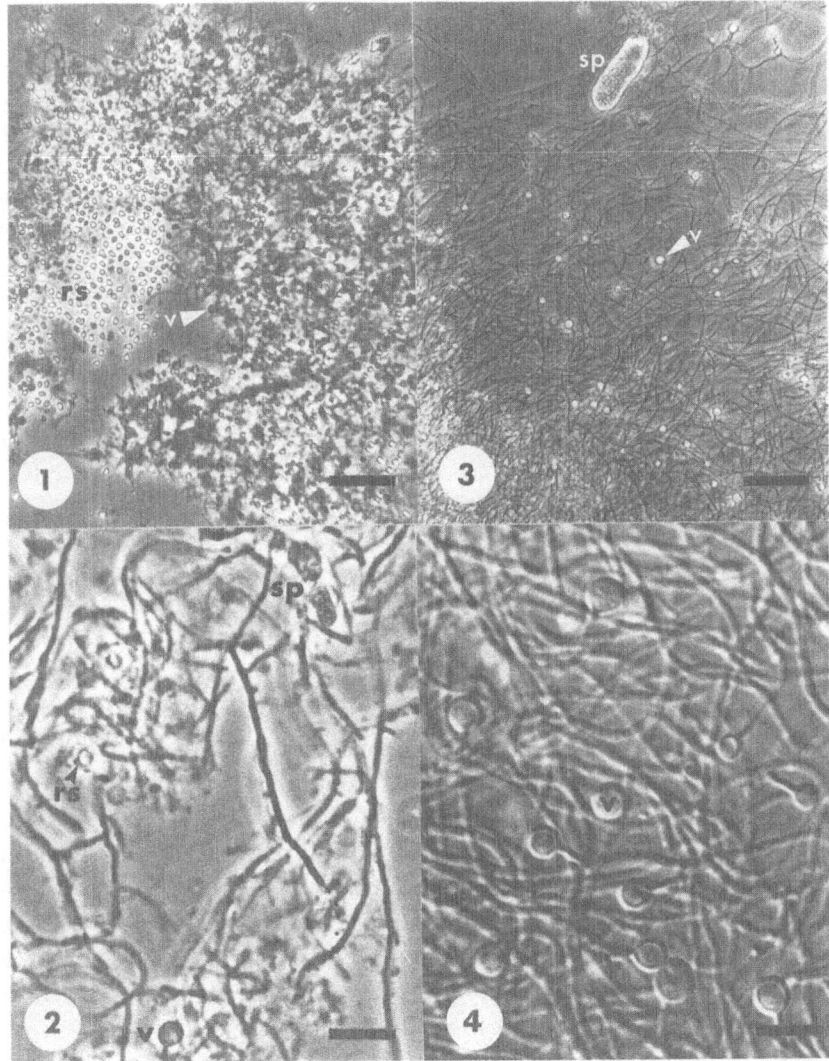

Figs. 1–4. Photomicrographs of cultures of *Frankia* strains isolated from nodules of *Casuarina cunninghamiana* growing in aeroponics. Abbreviations: released spores (rs), sporangia (sp), vesicle (v).

Fig. 1. *Frankia* isolated HFPCcI1 showing diffuse filamentous mat, large numbers of phase-bright released spores and scattered phase-dark vesicles. Phase optics Bar = 25 μ.

Fig. 2. *Frankia* isolate HFPCcI2 showing filaments, young sporangia and occasional vesicles, cultured on BAP medium. Phase optics Bar = 5 μ.

Fig. 3. Low-power view of *Frankia* isolate HFPCcI3 grown in BAP medium. The filamentous colony shows numerous phase-bright spherical vesicles. A large sporangium packed with spores appears at the top of the picture. Phase optics Bar = 25 μ.

Fig. 4. High-magnification view of *Frankia* isolate HFPCcI3 showing numerous vesicles attached to the filaments. Nomarski optics Bar = 5 μ.

seedlings of *Elaeagnus umbellata* and *Hippophaë rhamnoides* grown in aeroponics.

Strain HFPCcI3 was isolated by Zhang Zhongze in our laboratory following procedures outlined above with some modifications. After surface-sterilization with hypochlorite solution with detergent, individual nodule lobes were transferred separately to tubes containing 5 ml sterile yeast-extract-dextrose broth and incubated for 7–10 days at 28°C to assure they were free of contaminants. Then each lobe was homogenized in 2 ml QMod medium and the homogenate transferred into 10 ml of QMod medium in 25 ml Erlenmeyer flasks and incubated either standing or on a shaker at 80 rpm at 28°C in the dark. When visible filamentous growth had occurred, then cultures were again homogenized and transferred into fresh QMod medium for continued culture.

HFPCcI3 grew well on a number of media including M6B, QMod, YCz, BAP and B medium (Fig. 3, 4). Best growth was on BAP medium in which an unpigmented filamentous growth occurred. Optimum pH was 7 and culture was carried out routinely at 28°C. In this medium lacking NH_4^+, abundant vesicles were formed as well as sporangia. In BAP medium vesicle formation was a function of the concentration of NH_4^+ in the medium or of the nature of the combined N. Neither vesicles nor sporangia were seen on the other media listed. A detailed description of the cultural characteristics of CcI3 will be published elsewhere.

Seedlings inoculated with CcI3 in nodulation trials using water cultures and sand cultures included *C. cunninghamiana, C. equisetifolia, Allocasuarina lehmaniana, Elaeagnus umbellata, Hippophaë rhamnoides* and *Ceanothus americanus*. Nodules formed in 3–4 weeks on both *Casuarina* species but were not formed on other seedling species tried. Infection proceeded more quickly on plants grown at higher temperatures (16 h day–8 h night at 35°C and 28°C respectively versus 16 h day–8 h night at 24°C and 19°C respectively). In the higher temperature regime 100% of the plants were nodulated at 4 weeks; in the lower temperature regime 100% infection occurred by 6 weeks. Nodules were shown to be effective by acetylene reduction assays and seedlings remained green and elongated in normal growth. Uninoculated and unnodulated plants stopped growing, shoot tips turned purple and then plants became etiolated, browned and then died.

Comparative studies among Frankia *strains isolated from* Casuarina *spp. and available for culture and testing*

In Table 1 are listed the twelve strains of *Frankia* available to us

Table 1. *Frankia* isolates from *Casuarina* root nodules, showing designation, host source, cultural characteristics maintenance medium and publication citation

Designation	Host Source	Cultural Characteristics			Maintenance medium	Reference
		Sporangia	Vesicles	Pigment		
G2	C. equisetifolia (Guadelope)	+	+	+ (red)	YCz	13
D11	C. equisetifolia (W. Senegal)	+	+	+ (red)	YCz	13
P1 ("Phil")	C. equisetifolia (Phillipines)	+	+	+ (red)	YCz	14
ORS1106 (formerly CJ1-82)	C. junghuniana (Thailand)	+	−	−	YCz	12
R3	C. cunninghamiana (Greenhouse, Petersham)	+	+	−	S and TW broth	24
R43	C. cunninghamiana (Greenhouse, Petersham)	+	+	−	L/2	24
KRLC53008	Casuarina spp. (Bahamas)	+	+	−	YCz	D. Baker (unpublished)
KRLC53024	Casuarina spp. (Florida)	+	+	−	YCz	D. Baker (unpublished)
KRLC53001	Casuarina (derived from ORS1106)	+	+	+ (red)	YCz	D. Baker (CJ1-82)
HFPCcI1	C. cunninghamiana (Greenhouse, Petersham)	+	+	+ (red)	BAP and 5 mM propionate	M. Lopez (this paper)
HFPCcI2	C. cunninghamiana (Greenhouse, Petersham)	+	+	+ (red)	BAP and 5 mM propionate	M. Lopez (this paper)
HFPCcI3	C. cunninghamiana (Greenhouse, Petersham)	+	+	−	BAP	Zhang Zhongze (this paper)

for comparison which had their origins from root nodules of *Casuarina* species. The table lists the published host source together with a reference to the more detailed description of the isolate if available. Also listed in Table 1 are the cultural characteristics used for comparisons in this study. Important information for comparative studies which would include cell wall analyses, serological type, *etc.* (*c.f.* Baker[1]) for the most part were not available.

It is of interest to note that all strains in the appropriate medium formed vesicles and sporangia. Most isolates showed an orange-to-red coloration in culture, subject to the medium in which they were grown. They grew on a variety of nutrient media with YCz, L/2 and BAP representing the most generally usable maintenance media. The first is supplemented with yeast extract and the last is a simplified, synthetic medium completely defined chemically.

The fact that all *Frankia* isolates formed vesicles in culture is particularly interesting since vesicles are seldom, if ever, found in actively fixing root nodules of *Casuarina* species[21]. Although the evidence is less compelling and also less complete, essentially the same statement can be made for sporulation. The reasons for this lack of expression in the nodule remain to be fathomed.

In Table 2 are presented summaries of nodulation trials of the *Frankia* strains described in Table 1. Details of the several different procedures used in the trials are listed together with the host plants tested and special conditions of the trials.

The first striking fact is that of the twelve *Frankia* isolates of *Casuarina* studied to date only two strains are reported to be effective on the host of origin, *viz.*, strains ORS1106 from *C. junghuhniana* reported by Diem, Gauthier and Dommergues[12] and strain HFPCcI3 described here. In each case reported successful on *Casuarina* host plants, the strains nodulate species of *Casuarina sensu stricto*[15], but fail to nodulate members of closely related genera such as *Allocasuarina* and *Gymnostoma*. Our trials in this connection are still quite limited but fit a pattern established in other studies in which nodulation of *Allocasuarina* species failed almost completely using nodule suspension of *C. equisetifolia* or *C. cunninghamiana*. These results suggest strongly that different *Frankia* strains are involved specifically in root nodulation of different genera in the family Casuarinaceae. This possibility has been suggested before[7] based on specificity of nodulation among members of the family using nodule suspension.

The second striking and puzzling fact is that several of the *Frankia* strains isolated from root nodules of *Casuarina* nodulate members of the Elaeagnaceae. In their report on the isolation of G2 and D11,

Table 2. Nodulation trials of *Frankia* isolates collected from different research labs (See Table 1)

Designation	Host Source	Trials: Host seedlings tested	Method of seedling culture	Special Conditions	Nodulation Test Result	Source Information
G2	*C. equisetifolia*	*C. equisetifolia*	Water culture	Petri dish	Neg.	14
		Hippophaë rhamnoides	Sand		Pos.	14
D11	*C. equisetifolia*	*C. equisetifolia*	Water culture	Petri dish	Neg.	14
		Hippophaë rhamnoides	Plastic Pouches		Pos.	
P1 ("Phil")	*C. equisetifolia*	*C. equisetifolia*	Leonard Jars		Neg.	9
ORS1106	*C. junghuniana*	*C. cunninghamiana*				
		C. glauca			Pos.	12
R3	*C. cunninghamiana*	*C. cunninghamiana*	Water culture	Greenhouse	Neg.	This paper
		C. cunninghamiana	Sand in pots	Greenhouse	Neg.	Tests by Zhang
		Elaeagnus umbellata	Water culture	Greenhouse	Neg.	
R43	*C. cunninghamiana*	*C. cunninghamiana*	Water culture	Greenhouse	Neg.	This paper
		C. cunninghamiana	Sand in pots	Greenhouse	Neg.	Tests by Zhang
		Elaeagnus umbellata	Water culture	Greenhouse	Pos.	
KRLC53008	*Casuarina* spp.	Trials in progress				
KRLC53024	*Casuarina* spp.	Trials in progress				
KRLC53001	*Casuarina junghuniana* (derived from ORS1106 by D. Baker)	Trials in progress				
HFPCcI1	*C. cunninghamiana*	*C. cunninghamiana*	Aeroponics	Greenhouse	Neg.	This paper
		C. equisetifolia	Aeroponics	Greenhouse	Neg.	Tests by Zhang
		Elaeagnus umbellata	Aeroponics	Greenhouse	Pos.	Tests by Zhang
		Hippophaë rhamnoides	Aeroponics	Greenhouse	Pos.	Tests by Zhang
HFPCcI2	*C. cunninghamiana*	Trials in progress				
HFPCcI3	*C. cunninghamiana*	*C. equisetifolia*	Water culture, Sand	Greenhouse (Growth Chamber)	Pos.	This paper
		C. cunninghamiana	Water culture, Sand	Greenhouse (Growth Chamber)	Pos.	Tests by Zhang
		Allocas, Ichmanniana	Sand	Greenhouse	Neg.	Tests by Zhang
		Elaeagnus umbellata	Water culture	Greenhouse	Neg.	Tests by Zhang

Gauthier *et al.*[14] first described the infectivity of these strains of
Hippophaë. Since that report, we have tested our isolated strains from
Casuarina on several host plants. *Elaeagnus umbellata* is an excellent
host for HFPCcI1 from our laboratory as is also *Hippophaë*. Ruan's
isolate R43 nodulates *Elaeagnus umbellata* abundantly and effectively within three weeks of inoculation of seedlings grown in water
culture.

Several possible explanations can be given for nodulation of the
Elaeagnaceae by *Frankia* isolates derived from root nodules of a totally
unrelated family. First the most obvious is that the strains isolated
from *Casuarina* nodules were closely associated contaminants on the
root nodule surfaces. With our isolates and those of Ruan derived
from nodules obtained from the same source in our laboratory this
explanation could be possible since nodulated *Elaeagnus* and *Hippophaë* plants are common in our greenhouse and could provide stray
contamination. Dommergues has stated he had no *Hippophaë* within
great distances of their laboratory when G2 and D11 were first isolated.
Furthermore, we are convinced that our isolations were of organisms
from within the nodules and not surface contaminants.

A second possible explanation is that *Casuarina* nodules contain
more than one strain of *Frankia*. A report of such an occurrence
has been made by Benson and Hanna[4] for root nodules of *Alnus
incana* ssp. *rugosa*. In this case, however, both strains were identified
as *Alnus* isolates differing in minor proteins. Perhaps *Casuarina* nodules
are subject to secondary infections by non-specific *Frankia* that are
not capable of the initial infection process. Why they infect members
of the Elaeagnaceae is not clear.

A third explanation, suggested by Baker at the Madison meeting in
1982, is that members of the Elaeagnaceae are promiscuous hosts and
susceptible to *Frankia* strains not normally present as infecting agent.
This interesting idea needs testing. In our efforts to nodulate *Elaeagnus umbellata* seedlings we were unable to cause infections with
Frankia strains ArI3 and CpI1 tested under favorable conditions in
water culture. A suspension of *C. cunninghamiana* root nodules applied
without surface sterilization led to infection in half of the plants
tested – a result we attributed to contamination. The fact that a
bone fide isolate from *Casuarina* nodules that effectively nodulates
100% of the seedlings of *Casuarina* tested fails completely to nodulate *Elaeagnus* seems not to fit the idea that *Elaeagnus* is merely a
very susceptible host. The matter of the true identity of the *Frankia*
strains which nodulate members of the Elaeagnaceae but fail to nodulate *Casuarina* species remains a mystery.

At the present time there is reason for great optimism that the practical inoculation of *Casuarina* seedlings for forest plantations can be implemented. Two cultured strains of *Casuarina* are available ORS1106 and HFPCcI3 for further experimentation. Diem *et al.*[11] have reported successful inoculation of seedlings using their isolate. Having pure cultures available for dissemination by mail and for propagation in the laboratory should facilitate fundamental research on the special features of root nodule development in *Casuarina* and the development of optimum symbiotic associations for field application.

Acknowledgements   This research was supported in part by research grant DEB-8106952 from the U.S. National Science Foundation, research grant DE-AC02-82ER12036 from the Department of Energy and by the Maria Moors Cabot Foundation for Botanical Research of Harvard University. The authors wish to thank Shirley LaPointe, Ralph Lundquist Pat Young and Susan Lancelle for technical assistance, and Frances O'Brien for secretarial services.

References

1   Baker D 1982 A cumulative listing of isolated frankiae, the symbiotic nitrogen-fixing actinomycetes. The Actinomycetes 17, 35–42.
2   Baker D and Torrey J G 1979 The isolation and cultivation of actinomycetous root nodule endophytes. pp 38–56. *In* Symbiotic Nitrogen Fixation in the Management of Temperate Forests. Eds. J C Gordon, C T Wheeler and D A Perry. Forest Research Laboratory, Oregon State Univ., Corvallis.
3   Baker D and Torrey J G 1980 Characterization of an effective actinorrhizal microsymbiont, *Frankia* sp. Avc11 (Actinomycetales). Can. J. Microbiol 26, 1066–1071.
4   Benson D R and Hanna D 1983 *Frankia* diversity in an alder stand as estimated by sodium dodecyl sulfate-polyacrylamide gel electrophoresis of whole cell proteins. Can. J. Bot. 61, 2919–2923.
5   Burris R H 1974 Methodology. pp 9–33. *In* The Biology of Nitrogen Fixation. Ed. A. Quispel North-Holland Publ. Co., Amsterdam. The Netherlands.
6   Callaham D, Newcomb W, Torrey J G and Peterson R L 1979 Root hair infection in actinomycete-induced root nodule initiation in *Casuarina, Myrica* and *Comptonia.* Bot. Gaz. 140 (Suppl.), S1–S9.
7   Coyne P D 1973 Some aspects of the autecology of *Casuarina,* with particular reference to nitrogen fixation. Ph. D. Thesis. Dept. of Forestry, Australian National University Canberra, Australia.
8   Diem H G and Dommergues Y 1983 The isolation of *Frankia* from nodules of *Casuarina* Can. J. Bot. 61, 2822–2825.
9   Diem H G, Gauthier D and Dommergues Y R 1982 Isolement et culture in vitro d'une souche infective and effective de *Frankia* isolee de nodules de *Casuarina* sp. C. R. Acad. Sci. (Paris). (*In press*).
10 Diem H G, Gauthier D and Dommergues Y R 1982 Extranodular growth of *Frankia* on *Casuarina equisetifolia.* FEMS Microbial. Lett. 15, 181–184.
11 Diem H G, Gauthier D and Dommergues Y R 1983 Inoculation of *Casuarina* using a pure culture of *Frankia.* Nitrogen Fixing Tree Research Reports 1, 18–19.
12 Diem H G, Gauthier D and Dommergues Y R 1983 An effective strain of *Frankia* from *Casuarina* sp. Can. J. Bot. 61, 2815–2821.
13 Gauthier D, Diem H G and Dommergues Y R 1981 *In vitro* nitrogen fixation by two actinomycetous strains isolated from *Casuarina* nodules. Appl. Environ. Microb. 41, 306–308.

14   Gauthier D, Diem H G and Dommergues Y R 1981 *Casuarina equisetifolia* in Western Africa. III. Investigations carried out at ORSTOM Laboratory, Dakar. pp 10–19. Workshop, CSIRO, Forest Research Laboratory, Canberra, Australia.

15   Johnson L A S 1982 Notes on casuarinaceae II. J. Adelaide Bot. Gard. 61, 73–87.

16   Lalonde M and Calvert H E 1979 Production of *Frankia* hyphae and spores as an infective inoculant for *Alnus* species. pp 95–110. *In* Symbiotic Nitrogen Fixation in the Management of Temperate Forests. Eds. J C Gordon, C T Wheeler and D A Perry. Oregon State Univ., Corvallis, OR.

17   Lechevalier M P, Horriere F and Lechevalier H A 1982 The biology of *Frankia* and related organisms. *In* Developments in Industrial Microbiology. 23, 51–60.

18   National Academy of Sciences 1980 Firewood Crops. 237 p. National Academy Press, Washington, D.C.

19   National Academy of Sciences 1983 Casuarinas: Tree Resources for the Future. National Academy Press, Washington, D.C. (*In press*).

20   Torrey J G 1975 Initiation and development of root nodules of *Casuarina* (Casuarinaceae). Am. J. Bot. 63, 335–344.

21   Torrey J G 1982 Casuarina: actinorhizal nitrogen-fixing tree of the tropics. pp 427–439. *In* Biological Nitrogen Fixation Technology for Tropical Agriculture. Eds. P H Graham and S C Harris. CIAT, Cali, Colombia.

22   Turnbull J and Midgley S J (eds.) 1983 Proceedings of the *Casuarina* Workshop, CSIRO Division of Forest Research, Canberra, Australia (*In press*).

23   Zobel R W, DelTredici P and Torrey J G 1976 Method for growing plants aeroponically. Plant Physiol. 57, 344–346.

24   Lechevalier M P and Ruan J S 1984 Physiology and chemical diversity of *Frankia* spp. isolated from nodules of *Comptonia peregrina* (L.) Coult. and *Ceanothus americanus* L. Plant and Soil 78, 15–22.

Plant and Soil 78, 91–97 (1984).

Influence de basses températures sur la croissance et la survie de souches pures de *Frankia* isolées de nodules d'Aulnes

Influence of low temperatures on the growth and viability of pure strains of *Frankia* isolated from Alder nodules

A. MOIROUD, M. FAURE-RAYNAUD et P. SIMONET

ERA 848, Ecologie microbienne et Laboratoire de Microbiologie Physiologique et Appliquée, Département de Biologie Végétale, Université Claude-Bernard Lyon I, 43 Bd du 11 Novembre 1918, F-69622 Villeurbanne Cedex, France

Mots cles Croissance *Frankia* Température Viabilité

Resumé L'influence des basses températures sur la croissance et la survie de cultures pures de *Frankia* a été étudiée. Les souches, même isolées de stations à climat froid, sont incapables de se développer à des températures constantes inférieures à 15°C. Cependant elles résistent à ces températures défavorables et sont capables de reformer des colonies dès que la température se maintient à 28°C pendant quelques heures.

Summary Influence of low temperatures on the growth and survival of *Frankia* strains were studied. Even when isolated from cold areas, the strains were unable to grow below 15°C. However, the strains remained viable at the low temperatures and were able to give colonies as soon as the growth temperature raised up to 28°C during a few hours.

Introduction

Les espèces du genre *Alnus* sont largement répandues dans les zones tempérées et froides de l'hémisphère Nord[3]. Certaines d'entre elles sont même capables de coloniser des milieux aussi défavorables que les moraines récemment abandonnées par le retrait des glaciers[7,8,17]. Dans ces stations, qui restent soumises une grande partie de l'année à des températures très basses, les aulnes sont constamment nodulés. Pourtant certains auteurs ont montré que la croissance de *Frankia*, actinomycète responsable de la formation de nodules fixateurs d'azote chez l'aulne[2], était à peu près nulle pour des températures inférieures à 10°C[6]. Ainsi on peut se demander si certaines souches de *Frankia*, et particulièrement celles provenant de sols de montagne, ne présenteraient pas une adaptation aux basses températures, ce qui leur permettrait de coloniser et de mener une vie saprophytique active dans les sols froids. Une telle adaptation aux sols froids a déjà été reconnue depuis longtemps chez d'autres types de bactéries[12,15,16]. L'influence des basses températures sur la croissance et la survie de *Frankia* est aussi importante à connaître dans le cas d'inoculation artificielle. En effect, les souches introduites seront soumises à des températures souvent très inférieures à celles utilisées pour leur culture *in vitro* et à la concurrence des souches autochtones, peut-être mieux adaptées à ce facteur de milieu.

Materiel et méthodes

Souches de Frankia

Toutes les souches utilisées ont été isolées au laboratoire 11 selon la méthode décrite par Lalonde et col[14], à partir de nodules d'*Alnus glutinosa* (Ag) ou d'*A. incana* (Ai) récoltés dans des stations soumises à des conditions de températures très différentes. Les 4 souches étudiées provenaient respectivement de stations soumises à un climat de type:

méditerranéen : AgN21 (Corse)

continental peu marqué : AgN10ai (Lyon; 200 m d'altitude, température moyenne annuelle 11°C)

montagnard : AgN12a (Massif Central; 900 m d'altitude, température moyenne annuelle 8°C)

montagnard froid : AiN15a (Fond de France, Alpes du Nord; 1082 m d'altitude, température moyenne annuelle 6°C)

Cultures

Toutes les souches ont été cultivées sur milieu liquide de composition suivante: (g/l) K_2HPO_4 0.5; $MgSO_4 \cdot 7H_2O$ 0.2; $CaCl_2$ 0.1; Yeast extract (Difco) 0.05; hydrolysat de caséine (Sigma) 4; glucose 10; (ml/l) Tween 80 (Sigma) 2; citrate ferrique (solution à 1g de citrate ferrique dans 100 ml d'acide citrique à 1%) 1; solutions vitamine B12 (100 mg/100 ml) 1; pyridoxine monohydrochloride (50 mg/100 ml) 1; acide nicotinique (50 mg/100 ml) 1; solution d'oligo-éléments* 1. Après inoculation par une suspension homogène de *Frankia,* les tubes de culture ont été soumis soit à des températures constantes (28°C; 15°C; 12°C; 10°C; 4°C) soit à des températures alternées (4°C–10°C; 10°C–15°C; 4°C–28°C; 10°C–28°C). Dans les cas d'alternance de température, les temps de séjour à chaque température ont été les suivants: (Tableau 1).

Mesure de la croissance

La croissance des cultures a été appréciée par mesure de la D.O. à 570 nm (colorimètre digital Chemtrix) après homogénéisation pendant 30 secondes par ultrasons (Labsonic 1510) à 100 watts. A 4°C et 10°C les mesures de croissance on été faites après 6 et 10 semaines d'incubation. Pour toutes les autres températures la croissance a été évaluée après 3 et 6 semaines d'incubation. A chaque fois, les mesures ont été réalisées en triple.

Viabilité des inoculums

Après inoculation, des tubes de culture ont été placés sous une température constante de 4°C et de 10°C. Après 1 semaine, 3 semaines et 6 semaines à ces températures, 3 tubes de culture pour chaque souche ont été transférés sous une température constante de 28°C. La survie de l'inoculum a été appréciée par la reprise de la croissance et la mesure de la D.O. après 4 semaines d'incubation à 28°C.

Viabilité des cultures

Des cultures, préalablement développées à 28°C pendant 16 jours, ont été transférées à une température constante de 4°C et 10°C. Après 1 semaine, 3 semaines et 8 semaines

---

* Oligo-éléments (g/l): H_3BO_3 1.5; $MnSO_4 \cdot 7H_2O$ 0.8; $ZnSO_4 \cdot 7H_2O$ 0.6; $CuSO_4 \cdot 7H_2O$ 0.1; $(NH_4)_6Mo_7O_{24} \cdot 4H_2O$ 0.2; $CoSO_4 \cdot 7H_2O$ 0.01.

Table 1. Temps de séjour des cultures de *Frankia* à chaque thermopériode

Alernance de températures		Temps de séjour à:
4°C	10°C	10°C–12 h 4°C– 5 h
15°C	10°C	15°C–12 h 10°C– 5 h
4°C	28°C	28°C–12 h 4°C– 5 h
10°C	28°C	28°C–12 h 10°C– 6 h

Table 1. Incubation periods of Frankia *cultures at each temperature*

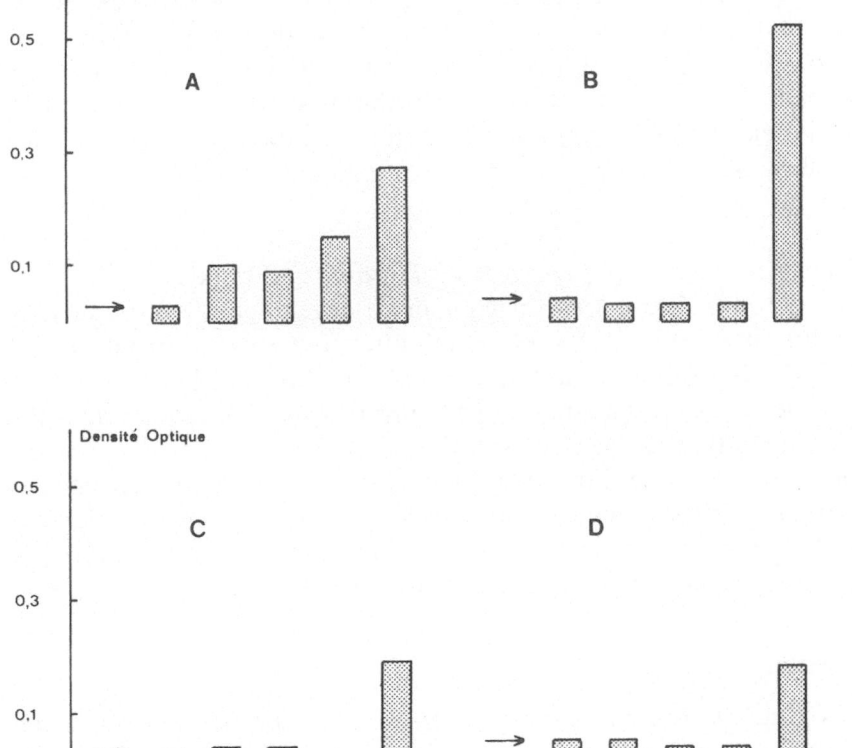

Fig. 1. Croissance de diverses souches de *Frankia* (A: AgN10ai–B: AgN21 C: AgN12a–D: AiN15a) cultivées à température constante; DO prises en fin de culture.
(→ : niveau de l'inoculum)
Fig. 1. Growth at constant temperatures of the various Frankia *strains studied; OD at the end of the culture (→ OD of the inoculum).*

de séjour à ces températures, la viabilité des cultures a été contrôlée par leur aptitude à réduire un sel de tétrazolium, le p-iodonitrotétrazolium violet (Sigma), selon la méthode de Bitton et Koopman[5].

Resultats

Influence des basses températures sur la croissance de Frankia

Les résultats obtenus montrent (Fig. 1) que sur 4 souches de *Frankia* utilisées, 3 sont incapables de croître à des températures égales ou inférieures à 15°C, même après 10 semaines d'incubation. Par contre, la croissance des cultures témoins à 28°C est rapide; la biomasse maximum étant généralement obtenue après 3 semaines d'incubation à cette température.

Une alternance de températures basses (4°C–10°C ou 10°C–15°C) ne permet pas non plus le développement de *Frankia* (Tableau 2). Par contre toutes les souches étudiées présentent un développement important lorsque, au cours d'une thermopériode, la température se maintient à 28°C pendant au moins 12 heures (Tableau 2). Une alternance de températures 4°C–28°C conduit même à la plus forte production de biomasse (Tableau 2).

Influence des basses températures sur la survie de Frankia

Des cultures de *Frankia* préalablement développées à 28°C pendant 16 jours puis transférées et maintenues à basses températures (4° ou 10°C) demeurent vivantes, comme en témoigne leur capacité de réduire le sel de tétrazolium (INT), même après 8 semaines de séjour à ces températures défavorables (Tableau 3).

De même lorsque l'on maintient pendant plusieurs semaines un inoculum à basses températures (4°C ou 10°C), celui-ci demeure capable de reconstituer une colonie importante après retour dans des conditions favorables de température (Tableau 3).

Discussion

Les résultats obtenus montrent que les souches de *Frankia* isolées de stations à climat froid sont incapables de se développer à des températures inférieures à 15°C. Elles ne présentent aucune adaptation particulière vis à vis du facteur température, contrairement à ce que l'on connaît pour d'autres microorganismes fixateurs d'azote[1,10,13] ou non[16] isolés de sols froids. *Frankia* doit donc être considéré comme un microorganisme mésophile et non pas psychrotrophe[18]. Des résultats similaires avaient déjà été obtenus par Burggraaf et Shipton[6] pour trois autres souches de *Frankia*. Ce défaut d'adaptation aux basses températures paraît confirmé par l'absence de croissance lorsque

Tableau 2. Influence de températures alternées sur la croissance de souches pures de *Frankia*

Souches	Températures d'incubation							
	4°C–10°C		10°C–15°C		4°C–28°C		10°C–28°C	
	a*	b*	a	b	a	b	a	b
Ag N21	0.05	0.00	0.04	0.01	0.50	0.43	0.37	0.21
Ag N10 ai	0.09	0.03	0.05	0.04	–	–	0.41	0.50
Ag N12a	0.02	0.01	0.03	0.01	0.33	0.42	0.26	0.27
Ai N15a	0.05	0.00	0.04	0.02	0.44	0.50	0.37	0.46

* a – Valeur des densités optiques mesurées après 3 semaines d'incubation

  b – Valeur des densités optiques mesurées après 6 semaines d'incubation.

Table 2. Influence of alternated temperatures on the growth of pure Frankia *strains*

Tableau 3. Viabilité de colonies ou d'inoculums de cultures pures de *Frankia* après séjour de plusieurs semaines à des températures constamment basses

Souches	Colonies développées						Inoculums					
	Réduction INT[1]						DO des cultures après transfert à 28°C[2]					
	Durée séjour à 4°C			Durée séjour à 10°C			Durée séjour à 4°C			Durée séjour à 10°C		
	1s	3s	8s	1s	3s	8s	1s	3s	6s	1s	3s	6s
Ag N21	+ + +	+ + +	+ + +	+ + +	+ + +	+ + +	0.27 0.35*	0.21	0.25	0.23	0.24	0.34
Ag N10 ai	+ + +	+ + +	+ + +	+ + +	+ + +	+ + +	0.19 0.28*	0.21	0.27	0.19	0.20	0.24
Ag N12a	+ + +	+ + +	+ + +	+ + +	+ + +	+ + +	0.15 0.19*	0.15	0.22	0.14	0.17	0.26
Ai N15a	+ + +	+ +	+ +	+ + +	+ +	+ +	0.09 0.16*	0.09	0.14	0.08	0.08	0.17

[1] – Viabilité des colonies appréciée par observation microscopique des colonies préalablement développées à 28°C et maintenues ensuite pendant des temps variés à basse température et évaluation du nombre d'hyphes présentant des cristaux d'INT réduits.

[2] – Les tubes de culture, après inoculation, ont été placés à 4°C ou 10°C. Après 1 semaine, 3 semaines et 6 semaines, une série de tubes a été transférée a 28°C et la DO de la culture obtenue mesurée après 21 jours d'incubation à cette température.

* DO des cultures témoins maintenues à 28°C.

Table 3. Colonies or inocula viability of pure Frankia *strains after several weeks of incubation at low temperatures*

les souches sont soumises à des températures basses alternées. En effet, il est connu depuis longtemps que chez certains organismes de stations froides, une alternance de températures basses et plus favorable qu'une température constante beaucoup plus élevée. Ainsi la germination de plantes alpines peut avoir lieu sous une alternance 2°C–13°C alors que la température constante permettant d'obtenir le même pourcentage de germination est de l'ordre de 20°C[4].

Lorsque, au cours de la thermopériode, la température atteint

28°C, la croissance de toutes les souches de *Frankia* est active et l'influence d'une température froide n'est plus apparente. Toutes les souches de *Frankia* étudiées présentent une bonne résistance aux températures défavorables. Il est également remarquable que les hyphes eux-mêmes demeurent vivants bien qu'apparemment ils soient incapables de croissance à basses températures. Cette capacité de survivre à basse température pendant des temps relativement longs et la faculté qu'a *Frankia* de se développer rapidement lorsque la température redevient favorable permettent de mieux comprendre la présence de ce microorganisme dans le sol, que celui-ci soit soumis à un climat tempéré ou froid. En effet, il semble simplement nécessaire que la température du sol atteigne pendant quelques heures chaque jour une valeur de 18°C à 20°C pour que *Frankia* puisse s'y développer. On sait en effet que la croissance *in vitro* de certaines souches de *Frankia* n'est pas négligeable à ces températures[6]. Une température de l'ordre de 20°C est fréquemment atteinte, en période estivale, dans la plupart des sols, même les lithosols morainiques[16,9].

Remerciements   Les auteurs remercient Madame N. Guillaumaud et Mademoiselle E. Juvin pour leur aide lors de la réalisation de ce travail.

Références

1   Alexander V 1974 A synthesis of the IBP Tundra Biome Circumpolar study of nitrogen fixation. *In* Soil Organisms and Decomposition in Tundra. Eds. A J Holding *et al.* pp 109–121, Tundra Biome Steering Committee (Stockholm).
2   Becking J H 1970 Frankiaceae fam. nov. (Actinomycetales) with one new combination and six new species of the genus *Frankia* Brunchorst 1886. Int. J. Syst. Bacteriol. 20, 201–220.
3   Becking J H 1977 Nitrogen fixation in higher plants other than legumes. *In* Dinitrogen Fixation. II. Eds. R W F Hardy and W S Silver, pp 185–275, Wiley and Sons Inc. Publ. New York.
4   Billing W O and Mooney H A 1968 The ecology of arctic and alpine plants. Biol. Rev. 43, 481–529.
5   Bitton G and Koopman B 1982 Tetrazolium reduction malachite green method for assessing the viability of filamentous bacteria in activated sludge. Appl. Environ. Microbiol. 43, 964–966.
6   Burggraaf A J P and Shipton W A 1982 Estimates of *Frankia* growth under various pH and temperature regimes. Plant and Soil 69, 135–147.
7   Cleve van K, Viereck L, Schlentner R 1971 Accumulation of nitrogen in an alder (Alnus) ecosystem near Fairbanks, Alaska. Arctic and Alpine Res. 3, 101–114.
8   Crocker R L and Major J 1955 Soil development in relation to vegetation and surface age at glacier Bay, Alaska. J. Ecol. 43, 427–448.
9   Debaud J C 1983 Recherches Ecophysiologiques sur les espèces alpines des genres *Clitocybe* et *Heleboma* associées à *Dryas octopetala*. Thèse, Univ. Lyon 206 p.
10   Ek-Jander J C and Fahraeus G 1971 Adaptation of Rhizobium to subarctic environment in Scandinavia. Plant and Soil Spéc Vol Eds. T A Lie and E G Mulder pp 129–137.

11   Faure-Raynaud M, Horrière F, Simonet P, Moiroud A 1982 Caractéristiques principales de quelques souches pures de *Frankia* isolées de nodules d'Aulnes de la flore française. *In* Energies renouvelables en milieu rural. Colloque Limoges, France pp 121–125.

12   Gounot A M 1960 Etude microbiologique de boues glaciaires arctiques. C.R. Acad. Sci. Paris D 226, pp 1437–1438.

13   Hardarson C and Jones D G 1979 Effect of temperature on competition amongst strains of *Rhizobium trifolii* for nodulation of two white clover varieties. Ann. Appl. Biol. 92, pp 229–236.

14   Lalonde M, Calvert H and Pine S 1981 Isolation and use of *Frankia* strains in actinorhizae formation. *In* Current Perspectives in Nitrogen Fixation. Eds. A H Gibson and W E Newton. pp 296–299 Australian Academy Sciences Canberra.

15   McDonald I J, Quadling C, Chambers A F 1963 Proteolytic activity of some cold-tolerant bacteria from arctic sediments. Can. J. Microbiol. 9, pp 301–315.

16   Moiroud A 1976 Etude Ecologique des marges glaciaires, en particulier de leur micropeuplement. Thèse, Univ. Lyon, 168 p.

17   Moiroud A et Capellano A 1979 Etude de la dynamique de l'azote à haute altitude. Fixation (réduction de C_2H_2) per *Alnus viridis* Chaix et étude ultrastructurale des nodules. Symposium Physiologie des Racines et Symbioses, Nancy, pp 365–371, Eds. A Riedacker I.N.R.A.-C.N.R.F. Nancy.

18   Morita R Y 1975 Psychrophilic bacteria. Bacteriol. Rev. 39, pp 144–167.

Plant and Soil 78, 99–104 (1984).
© 1984 Martinus Nijhoff/Dr W. Junk Publishers, The Hague.

The biosynthesis of indole-3-acetic acid by *Frankia*

C. T. WHEELER, A. CROZIER and G. SANDBERG*
Department of Botany, The University, Glasgow G12, Scotland

Key words Actinorhizas Biosynthesis *Frankia* Indole-3-acetic acid Nodules

Summary High perfomance liquid chromatography (HPLC) of the products of [5-³H]trypto-phan metabolism by *Frankia* sp. Avc I1 indicates that small amounts of [³H]indole-3-acetic acid (IAA) are excreted into the growth medium. *Frankia* has a limited capacity for the catabolism of [2-¹⁴C]IAA and the product that accumulates is different from that detected in *Rhizobium japonicum* cultures following inoculation with [2-¹⁴C]IAA. The data imply that the rate of turnover of IAA is much more rapid in *Rhizobium* than *Frankia* and that the two organisms employ different routes for the catabolism of IAA.

Introduction

It is almost fifty years since it was proposed that auxin secreted by the legume microsymbiont *Rhizobium,* may be a regulator of the development of nitrogen-fixing root nodules[7]. Confirmation of IAA excretion by cultured *Rhizobium* has been obtained subsequently[2,3,9], but there is still no direct evidence for the involvement of endogenous IAA from *Rhizobium* in nodule development. Indeed, research into the causation of root hair curling, which is the first obvious symptom of infection, suggests that IAA is not essential for this part of the nodulation process. Moderate root hair curling in clover is initiated by a poorly diffusible, heat stable product of *Rhizobium* which is unlikely to be IAA[11] and, in addition, mutant strains of *Rhizobium* which adhere to clover root hairs and which produce IAA do not cause root hair curling[3]. A lack of correlation between the ability of mutant *Rhizobium* to induce nodulation and synthesize IAA[3,9] also casts doubt on the primary importance of microsymbiont IAA in the control of nodulation, although involvement in aspects of nodule development other than the early stages of nodulation cannot be excluded.

Root hair curling also accompanies infection in actinorhizal plants, where the nodules develop as modified, highly branched lateral roots induced between the normal points of lateral root production in the pericycle[1]. A role for growth substances in the formation of these nodules is suggested by the induction of lateral root primordia,

*Present address: Department of Forest Genetics and Plant Physiology, The Swedish University of Agricultural Sciences, S-901 83 Umeå, Sweden.

following administration of auxins and kinetin to young plants[1]. Unlike the legume-*Rhizobium* symbiosis, there is currently no information concerning the ability of *Frankia* to synthesize and excrete phytohormones, which might contribute to the range of growth substances known to occur in actinorhizal nodules[10]. Clearly, such information is a necessary precursor to investigation of the role of phytohormones in the regulation of actinorhizal nodule formation. Because IAA normally is ascribed a major role in lateral root formation[8], this investigation was concerned to establish the capacity of cultured *Frankia* for biosynthesis and secretion of IAA.

Materials and methods

Culture of micro-organisms

Rhizobium phaseoli gRp1 was isolated from nodules harvested from *Phaseolus vulgaris* (L) cv. Canadian Wonder, grown in the Glasgow University Botany Department Experimental Gardens, and maintained on slopes of yeast-mannitol agar. For metabolism experiments, conical flasks each containing 125 ml culture medium (KH_2PO_4 1.0 g, K_2HPO_4 1.0 g, KNO_3 0.5 g, $MgSO_4.7H_2O$ 0.25 g, $CaCl_2.2H_2O$ 0.15 g, $FeCl_3.6H_2O$ 0.01 g, mannitol 10 g, biotin 1 mg, thiamin 1 mg, pantothenic acid 1 mg, trace elements[4], distilled water to 1 l pH 6.8) were inoculated with *Rhizobium* and shaken in the dark at 25°C on an orbital shaker for 10 days prior to use.

The *Frankia* strain used was the isolate Avc I1 of Dwight Baker, from *Alnus viridis* ssp. *crispa*. Cultures were maintained in Burggraaf's medium 10 with ammonium nitrate replaced by NH_4Cl, 0.01 gl^{-1}. For metabolism experiments, *Frankia* was grown in 125 ml culture medium in the dark at 25°C for 6 weeks prior to use.

Feeding of radio-labelled compounds

All radiochemicals were purchased from Amersham International, U.K. Tryptophan used was DL-[U-^{14}C]tryptophan, specific activity 84 mCi mmol^{-1} and [5-^{3}H]tryptophan, 29 Ci mmol^{-1}. IAA was [2-^{14}C]indole-3-acetic acid, specific activity 55 mCi mmol^{-1}. Radiolabelled compounds were fed to flasks maintained in the dark at 25°C. *Rhizobium* cultures were shaken on an orbital shaker during feeding but *Frankia* cultures were static with occasional manual shaking.

Extraction and partitioning

After incubation, cultures were filtered through Nalgene 0.2 μm filter units. Sodium diethyl dithiocarbamate (0.005 M) and 2 mg l^{-1} 'cold' IAA were added to the filtrate to inhibit breakdown of radiolabelled IAA. The filtrate was adjusted to pH 3.0 and partitioned five times against 2/5 volumes ethyl acetate. The ethyl acetate was stored overnight at -20°C, filtered to remove ice, dried with anhydrous sodium sulphate, refiltered and reduced to dryness *in vacuo*.

High performance liquid chromatography

Solvents were delivered at a flow rate of 1 ml min^{-1} by a Spectra Physics SP 8100 liquid chromatograph. Samples were introduced off-column via a Valco sample valve with a 250 μl sample loop. A 250 × 5 mm (i.d.) ODS-Hypersil (5 μm) column eluted over 25 min with gradient of 20–70% methanol in 20 mM pH 3.5 ammonium acetate buffer was used for reverse-phase HPLC. Normal-phase HPLC was carried out on a 250 × 5 mm (i.d.) column packed with CPS-Hypersil (5 μm) using a mobile phase of hexane:ethyl acetate (85:15) in 1% acetic acid. Column effluent was mixed at a 'T'-junction with scintillation cocktail being pumped by a

Table 1. Details of radioactive substrates fed to *Rhizobium* and *Frankia* cultures

Substrate	Incubation period	Radioactivity recovered in acidic ethyl acetate extract	
		dpm	% applied label
Rhizobium			
[^{14}C] tryptophan (1.32 × 10^6 dpm)	1 h	516 × 10^3	39
[2-^{14}C] IAA (9.6 × 10^6 dpm)	1 h	9.4 × 10^6	98
Frankia			
[5-^{3}H] tryptophan (12.8 × 10^6 dpm)	24 h	321 × 10^3	13
[2-^{14}C] IAA (9.6 × 10^6 dpm)	24 h	9.6 × 10^6	100

Reeve Analytical reagent delivery unit and directed to a Reeve Analytical radioactivity monitor with a 400 μl flow cell. A scintillant comprising 10 g PPO, 330 ml Triton X-100, 670 ml distilled xylene and 150 ml methanol was used for reverse-phase analyses at a 3:1 scintillant-eluant ratio. The solvent from the normal-phase HPLC column was mixed with a scintillant containing 12 g PPO, 150 g napthalene, 50 ml Triton X-100 and 1.0% distilled toluene to give a 2:1 scintillant-eluant ratio.

Results and discussion

The data obtained are summarized in Table 1 and Fig. 1. After 1 h incubation of [^{14}C] tryptophan with *Rhizobium*, 39% of the applied label was recovered in an acidic, ethyl acetate soluble extract (Table 1). As tryptophan does not migrate into ethyl acetate from an acidic aqueous phase it is evident that a considerable amount of metabolism occurred during a relatively short incubation period. Analysis of the ethyl acetate extract by reverse-phase HPLC revealed the presence of a number of components (Fig. 1A). One of the peaks, corresponding to 4.3% of the applied label, co-chromatographed with IAA in both reverse- and normal-phase HPLC systems. The peak labelled X was the sole product when [2-^{14}C] IAA was incubated with *Rhizobium* (Fig. 1B). Thus, with the exception of X, IAA does not appear to act as an intermediate in the production of the [^{14}C] tryptophan metabolites shown in Fig. 1A. More than 30% of a [2-^{14}C] IAA substrate was converted to X by *Rhizobium* during a 1 h incubation (Fig. 1B). The biosynthesis and catabolism data imply that there is a rapid rate of turnover of IAA in gRp1 cultures. This clearly contrasts with some other *Rhizobium* strains which it is claimed do not degrade IAA[11].

When *Frankia* cultures were incubated with [^{14}C] tryptophan, under similar conditions to those used for *Rhizobium*, metabolite [^{14}C] IAA was barely detectable. *Frankia* cultures were therefore fed high specific activity [^{3}H] tryptophan and incubated for a period of 24 h. Under these conditions 13% of the applied label was extracted into ethyl

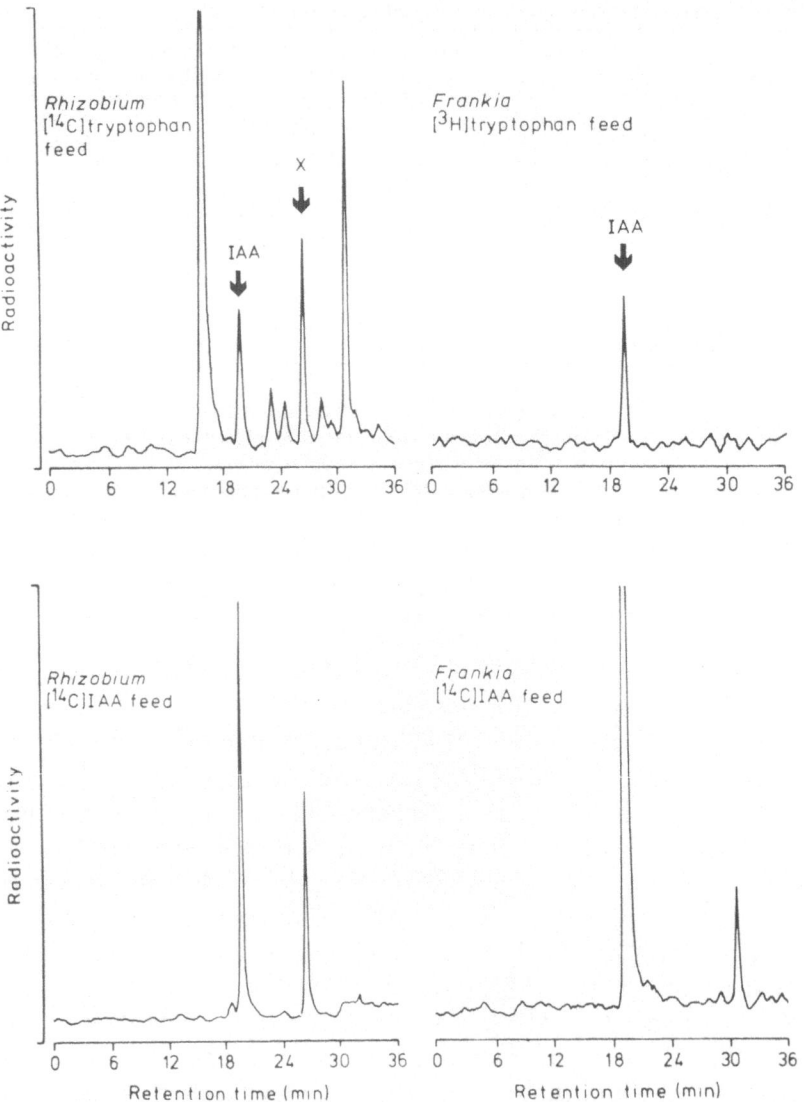

Fig. 1. Reverse-phase HPLC analyses of the acidic ethyl acetate-soluble fraction from *Rhizobium* and *Frankia* cultures following incubation with radiolabelled tryptophan and IAA. Column: 250×5 mm (id) ODS-Hypersil (5 μm). Mobile phase: 25 min gradient, 20–70% methanol in 20 mM, pH 3.5 ammonium acetate buffer. Flow rate: 1 ml min^{-1}. Samples: A 1/20 aliquot of an acidic ethyl acetate-soluble fraction from a *Rhizobium* culture (125 ml, 3.08 $\times$ 10^9 cells ml^{-1}) incubated for 1 h with [^{14}C]tryptophan, B 1/400 aliquot of an acidic ethyl acetate-soluble fraction from a *Rhizobium* culture (125 ml, 2.84 $\times$ 10^9 cells ml^{-1}) incubated for 1 h with [2-^{14}C]IAA; C 1/100 aliquot of an acidic ethyl acetate-soluble extract from a *Frankia* culture (27.2 mg dry weight) incubated for 24 h with (5-^{3}H)tryptophan; D 1/200 aliquot of an acidic ethyl acetate soluble extract from a *Frankia* culture (24.8 mg dry weight) incubated for 24 h with [2-^{14}C]IAA. Detector: Homogenous radioactivity monitor.

acetate (Table 1) indicating that tryptophan is metabolised at a much slower rate in *Frankia* than in *Rhizobium*. The acidic, ethyl acetate extract from *Frankia* was less complex than that from *Rhizobium* (Fig. 1A) as when analysed by reverse-phase HPLC it was found to contain only one major product (Fig. 1C). This constituent, which corresponded to 2.6% of the applied [^{14}C] tryptophan substrate, co-chromatographed with IAA in both the normal- and reverse-phase HPLC systems. [2-^{14}C] IAA was catabolised at a slow rate by *Frankia*. After 24 h there was only a 3.7% conversion to a less polar catabolite (Fig. 1D) that had chromatographic properties different to those of compound X produced by *Rhizobium*. The identity of the two IAA-derived products is as yet unknown although their HPLC properties suggest it is unlikely that they are sugar or amino acid conjugates.

Although IAA is a product of tryptophan metabolism in *Frankia*, its rates of synthesis and catabolism are apparently much slower than in *Rhizobium*. Neverthless, it is evident that, as in *Rhizobium*-legume symbiosis, investigations of actinorhizal nodule development must consider the possibility that IAA, which may be involved in the regulation of nodule development, could originate from the host plant cells and/or the invading *Frankia*, even though the contribution of the actinomycete to the IAA content of the nodule may be small.

Acknowledgement The authors wish to thank Mrs Alison Sutcliffe for technical assistance.

References

1   Angulo Carmona A F 1973 La formation des nodules fixateurs d'azote chez *Alnus glutinosa* (L.) Vill. Acta Bot. Neerl. 23, 257–303.
2   Badenoch-Jones J, Summons P E, Entsch B, Rolfe P G, Parker C W and Letham D S 1982 Mass spectrometric identification of indole compounds produced by *Rhizobium* strains. Biomed. Mass Spectrom. 9, 429–437.
3   Badenoch-Jones J, Summons R E, Djordjevic M A, Shine J, Letham D S and Rolfe B G 1982 Mass spectrometric quantification of indole-3-acetic acid in *Rhizobium* culture supernatants; relation to root hair curling and nodule initiation. Appl. Environ. Microbiol. 44, 275–280.
4   Cannon F C 1981 *In* Methods for evaluating biological Nitrogen Fixation Ed. F J Bergersen, Wiley, Chichester, p 395.
5   Dullaart J 1970 The bioproduction of indole-3-acetic acid and related compounds in root nodules and roots of *Lupinus luteus* L. and by its rhizobial symbiont. Acta Bot. Neerl. 19, 573–615.
6   Shipton W A and Burggraaf A J P 1982 A comparison of the requirements for various C and N sources and vitamins in some *Frankia* isolates. Plant and Soil 69, 149–161.
7   Thimann K V 1936 On the physiology of the formation of nodules on legume roots. Proc. Natl. Acad. Sci. USA 22, 511–514.

8   Torrey J G 1976 Root hormones and plant growth. Annu Rev. Plant. Physiol 27, 435 – 459.
9   Wang T L, Wood E A and Brewin N J 1982 Growth regulators, *Rhizobium* and nodulation in peas. Planta Berlin 155, 345 – 349.
10 Wheeler C T, Henson I E and McLaughlin M E 1979 Hormones in plants bearing actinomycete nodules. Bot. Gaz. 140 (Suppl.), 552 – 557.
11 Yao P Y and Vincent J A 1976 Factors responsible for the curling and branching of clover root hairs by *Rhizobium*. Plant and Soil 45, 1 – 16.

Plant and Soil 78, 105–128 (1984).
© 1984 *Martinus Nijhoff/Dr W. Junk Publishers, The Hague.*

Identification of the endophypte of *Dryas* and *Rubus* (Rosaceae)

J.H. BECKING
Research Institute ITAL, P.O. Box 48, 6700 AA Wageningen, The Netherlands

Key words Cytology of root nodules *Dryas drummondii Frankia* Geographic distribution
Nitrogen fixation *Rubus ellipticus* Ultrastructure of endophyte

Summary Root nodules of *Dryas drummondii* are of the coralloid type (*Alnus* type). The endophyte is present in the middle cortical cells of the root-nodule tissue. Transmission electron micrographs revealed an actinorhizal endophyte with septate hyphae and non-septate spherical or ovoid vesicles. Vesicles always possess at the base a septum; septa formation in the endophyte is always associated with the presence of mesosomes. Branching of the endophyte is not necessarily correlated with septum formation. Hyphal structures are more prominent in the apical part of the root nodule and vesicles are more numerous in a broad zone below this. In the middle and towards the base of the root nodule the endophytic structures appear in a stage of disintegration. Vesicles appear in a broad region near the periphery of the host cell and regularly show no strict orientation towards the host-cell wall. In the center of the host cells only hyphae occur. In the intercellular spaces between the host cells the *Frankia* endophyte produces spore-like structures although the outline of the sporangia is often faint.

The coralloid root of *Rubus ellipticus* shows characteristically a basal rootlet initiated below the dichotomous branching of the nodular lobes, but extending beyond the root nodule. The endophyte is only present in the outer cortex of the root nodule in a 1-2 cell wide layer. This endophytic layer is bounded, internally as well as externally, with a 4-5 cell wide layer of tannin-filled host cells. The implications of this situation are discussed. Tannin-filled cells occur regularly in *Rubus* species and their arrangement has been used for taxonomic purposes within the genus. The *Rubus* endophyte is a *Frankia* species with septate hyphae and distinctly septate spherical vesicles. The ultrastructure of the vesicles of the *Rubus* endophyte is very similar to that of the *Alnus* endophyte.

Introduction

At present six genera of plants within the Rosaceae are known to bear root nodules and to have the capacity to fix atmospheric nitrogen. Five of these nodulating genera, namely *Dryas, Purshia, Cercocarpus, Chamaebatia*[16] and *Cowania*[26], fall into the tribe Dryadeae which, however also includes some other genera (for example *Geum, Coluria, Waldsteinia* and *Fallugia*) so far not reported to be nodule-bearing. The only other representative within the Rosaceae not belonging to the tribe Dryadeae observed to be nodule-bearing is found in the genus *Rubus* of the tribe Rubeae[27] or Rubieae[17]. Here only a single species of about 250 *Rubus* species described[32], namely *Rubus ellipticus* J.E. Smith, was observed to be root nodulated. It is therefore an apparent exception and as pointed out earlier[7], such a situation seems to imply that the nodule-forming capacity arose after speciation in this tribe occurred.

In this study the morphological features of the root nodules, the light microscopic appearance and fine structure of the root-nodule endophyte of *Dryas drummondii* Richards and *Rubus ellipticus* J.E. Smith are described. The morphology and fine structure of the root nodules of *Dryas drummondii* have already been studied by Newcomb[25]. The present contribution will give however some additional information on the ultrastructure of the *Dryas* endophyte and will describe spore-like structures in this *Frankia* symbiont. Root nodulation in *Rubus ellipticus* was first observed by Mrs. Sri Soemartono in Java as reported by Bond[5]. Becking[3] confirmed root nodulation in this plant species and by the acetylene reduction assay he proved the N_2-fixing capacity of the root nodules. However, the nodule structure and the fine structure of the endophyte of *Rubus ellipticus* were not yet studied. The present investigation will present information on the type of nodulation, the cytology of the root nodules and the fine structure of the endophyte, which proved to be a *Frankia* species.

Finally, the frequency or rareness of root nodulation in the two above-mentioned Rosaceae species is discussed and in relation to this, the peculiar geographic distribution of these species is evaluated.

Materials and methods

Plant material

Root-nodule material of *Dryas drummondii* was collected by Mr. R.E. Henderson near Haines at the Haines Highway, Mile 25 and 27, Alaska in September 1968 and 1969. The fresh nodule material was sent to me by air mail and fixed and embedded within 4 days after collecting. At the above-mentioned sites the plants grew on a glacier silt that may be a foot or more thick, but the roots extended down to a gravel bed that underlies the silt. All nodules occurred in the gravel — never in the silt (R.E. Henderson, pers. communication). At time of collecting the ground was covered with snow but was not frozen. The plants were in a dormant condition for winter.

The *Rubus ellipticus* root nodules were collected from a plant growing in the Cibodas Mountain Gardens, Mt. Pangrango-Gedeh, 1450 m altitude, where the plant is introduced. This is the same site, where the root nodules used for the previous physiological experiments[3] were obtained. In 1974, 1977, and 1978 root nodules were observed, but later examinations (1982, 1983) showed that this plant was removed or had died, and no further root nodules on other *Rubus ellipticus* plants growing in these Botanic Gardens could be found. This indicated an extreme rareness of the occurrence of root nodulation and although the same plant species was growing in identical soil (an iron-rich latosol) at the same site, root nodulation in these plants was absent. Fresh root nodules of *Rubus ellipticus* collected in July 1977 and September 1978, were fixed in the field in Karpechenko and FAA fixatives (see below). Moreover, the external morphology of the root nodules was investigated in the field and documented by photographs.

Structural studies

For bright-field light microscopy nodule slices of fresh root nodules were fixed in Karpechenko (chromic-acetic-formic acid mixture) and FAA (formaldehyde-acetic acid mixture) and after dehydration embedded in paraffin-wax[15]. Sections $3.0-6.0\,\mu m$ thick were stained with a number of stains: toluidine blue (0.1% aqueous), toluidine blue and subsequently with Orange G (0.1% in clove oil), safranin (0.5% aqueous) and cystal violet (1.0% aqueous), orcein (conc. solution in 1.0% acetic acid)[15], or with erythrosine (1.0% aqueous) and counterstained with methylene green (0.25% in ethanol)[22].

For transmission electron microscopy (TEM) the root nodules were dissected in small pieces of 1–2 mm in diameter. Subsequently, these nodule-lobe slices were fixed in glutaraldehyde (2.5%) in 0.025 M potassium phosphate buffer (or Kellenberger buffer), pH 6.8 for 2–4 h at room temperature, washed in buffer, postfixed in OsO_4 (1.0% in buffer), and again washed in buffer. The slices were stained with uranylacetate (0.5 or 1.0%) in phosphate buffer or Kellenberger buffer (pH 5.3 or 4.7) *in vacuo* for 1 h at room temperature. Dehydration occurred with an ethanol series with c. 20% increments. Via propyleneoxide, the nodule slices were finally brought into an Epon-Araldite embedding medium.

Results

The general morphology of *Dryas* root nodules is illustrated in Fig. 1. The root nodules were very much enlarged and thickened as compared with the very fine rootlets to which they were attached. The nodular lobes showed arrested growth of the apical meristem and a characteristic dichotomous branching. In this respect they are very similar to *Alnus* root nodules and therefore can be classified as the *Alnus*-type of root nodulation. The root nodules were much smaller and far less compact than those depicted by Newcomb[25]. In the root nodules examined by me, the nodular lobes were approximately 5–6 mm in length and approximately 1.2 mm in diameter, while the supporting rootlet had a diameter of only 0.2–0.3 mm. Young root nodules show generally thicker nodular lobes and are less branched (Fig. 1A) than older root nodules (Fig. 1B). The latter nodules show particularly at the base a reduction in diameter of the nodular lobes probably by partial necrosis of the tissue.

Light microscopy (not illustrated here) and transmission electron microscopy revealed within the host cells two endophytic forms, *i.e.* hyphae and vesicles. The hyphae (0.3–0.7 μm in diameter) were septate like *Frankia* spp. in other symbioses, *e.g.* the alder endophyte[4]. Septum formation was always associated with distinct plasmalemmosomes or mesosomes[9, 10, 11] (Fig. 2A). Branching of the hyphae was not necessarily correlated with crosswall formation (Fig. 2B). Cortical cells containing the endophyte contained either hyphae or hyphae bearing at the tip non-septate vesicles. Near the nodule apex the infected host cells revealed only hyphae, followed by a broad zone in which the cortical cells contained hyphae and vesicles, whereas in the middle to the base of the older root nodules the infected cells show endophytic forms apparently in the process of lysis or disintegration (see Fig. 10).

The vesicles are 2.5–3.6 μm in diameter and ovoid or spherical in

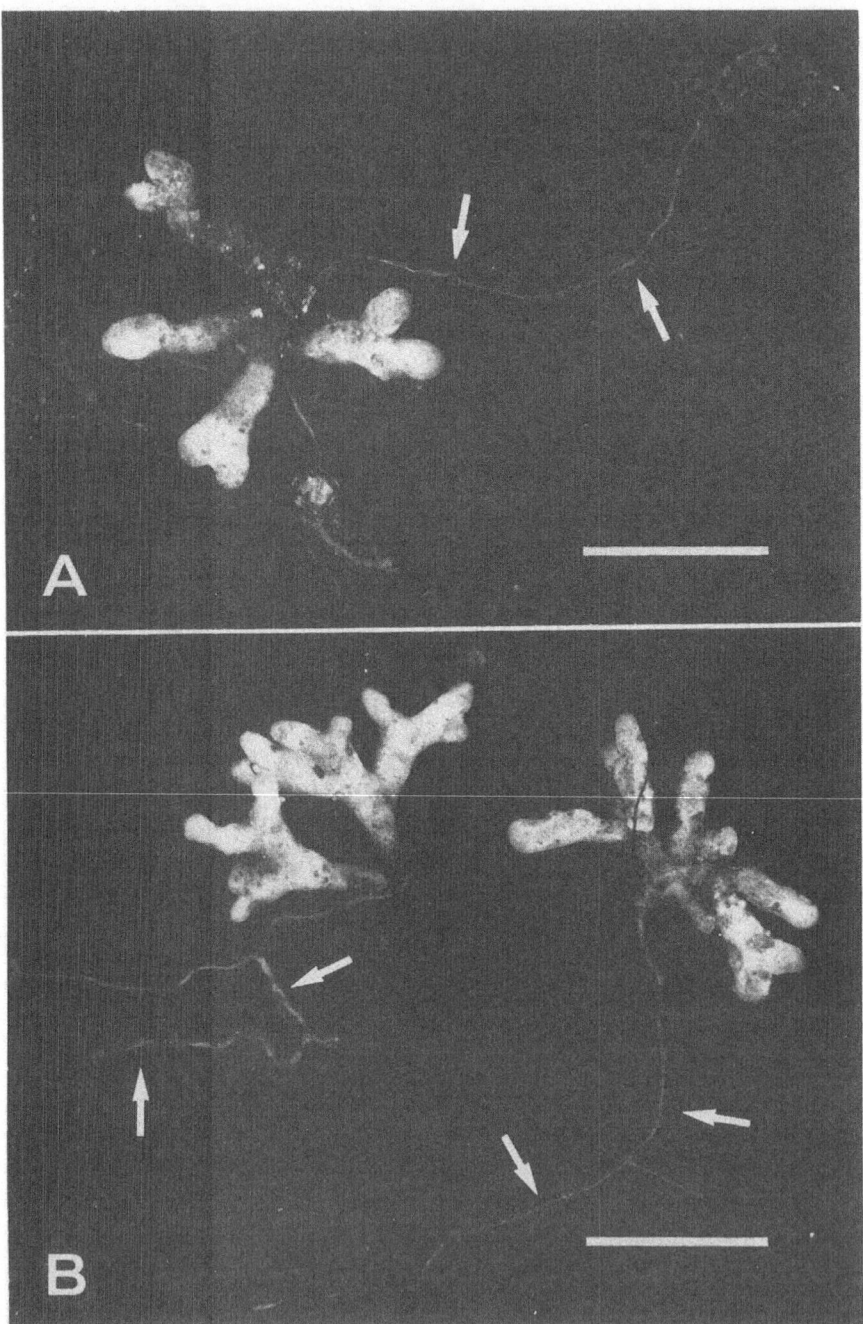

Fig. 1. Morphology of *Dryas drummondii* root nodules. Note the radially thickened root nodules compared to the very fine rootlets (see arrows) to which the root nodules are attached. Bar scale = 0.5 cm.

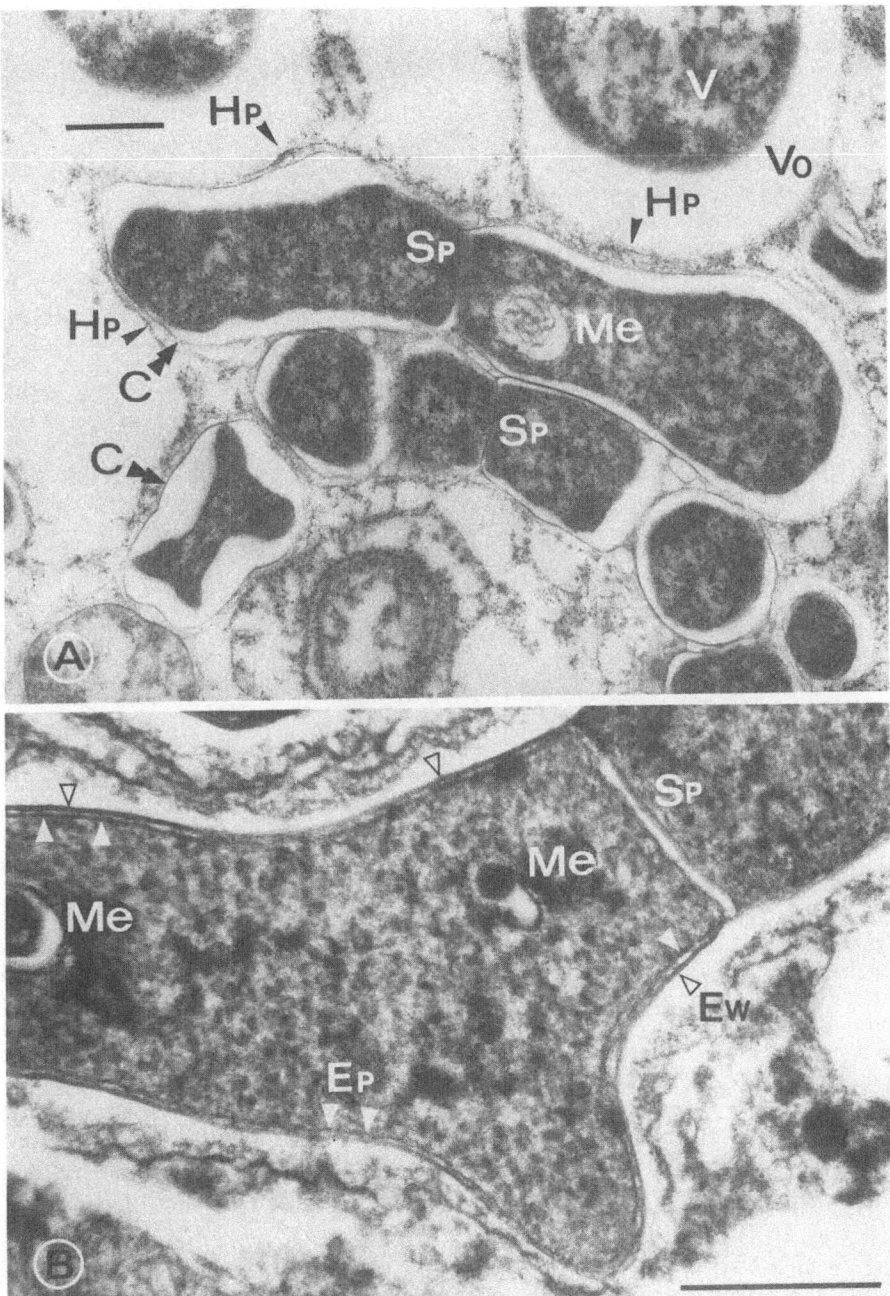

Fig. 2. Transmission electron micrograph of the hypal form of the *Dryas drummondii* endophyte. Notable are septa (**Sp**) in the hyphae and the presence of prominent mesosomes (**Me**) associated with the septum formation. The endophyte is separated from the host cytoplasm by the endophytic cytoplasmic membrane (**Ep**), the endophytic cell wall (**Ew**), a capsule layer (**C**), and the host cytoplasmic membrane (**Hp**). In the upper portion of Fig. 2A sections through vesicles (**V**) are visible and a 'void space' (**Vo**) between endophyte and capsule. Bar scale = 0.5 μm.

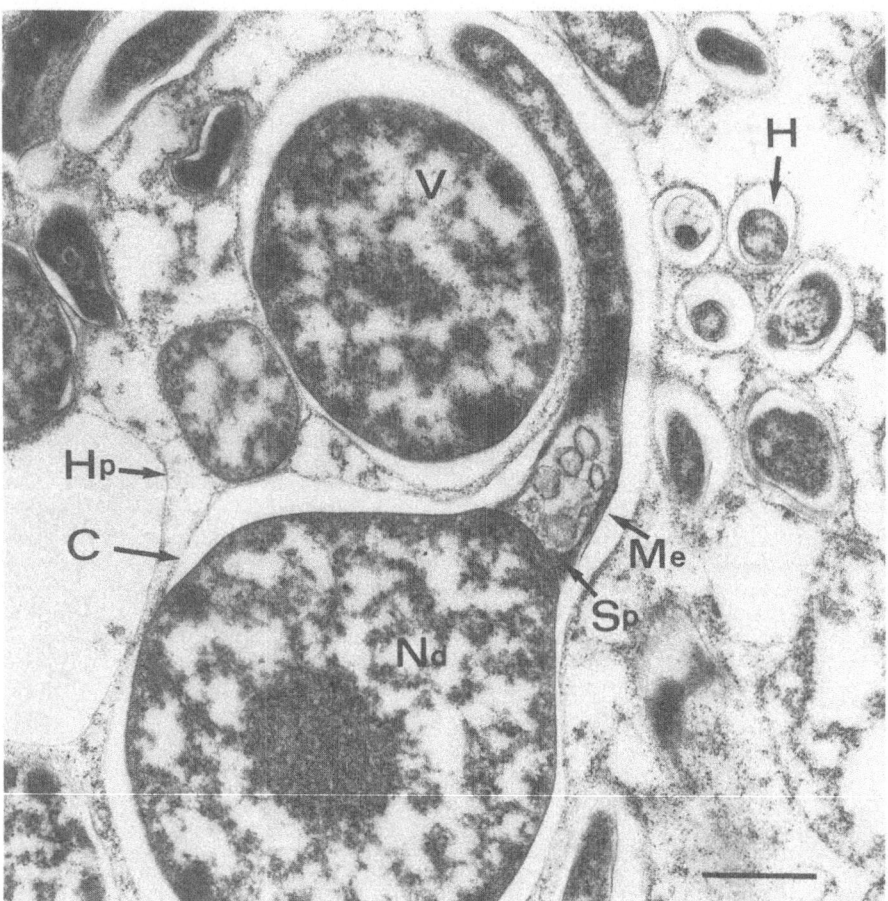

Fig. 3. Transmission electron micrograph of vesicles (**V**) of the *Dryas* endophyte. Note the basal septum (**Sp**) in the non-septate vesicle. The latter septum is associated with distinct mesosomes (**Me**). The nucleoid region (**Nd**) in the vesicle characterized by a network of electron-dense fibrils shows in the centre electron-dense material. To the right hyphae (**H**) in transverse section. The capsule layer (**C**) and the host cytoplasmic membrane (**Hp**) are also visible. Bar scale = 0.5 μm.

shape (Figs 3 and 4). The vesicles always show a well-defined septum at their base forming the connection with the hyphae. Again septum formation is associated with the distinct presence of mesosomes (Fig. 3, see Me). In the *Alnus* endophyte and in some other *Frankia* associations the vesicles are situated near the periphery of the host cell being oriented to the host-cell wall. Likewise, in *Dryas,* hyphae occur only in the centre portion of the host cell, but vesicle formation appears to be in a broader zone near the outside. Moreover, the vesicles regularly show no defined orientation towards the periphery. For instance one vesicle may be oriented towards the peripheral region of the host cell,

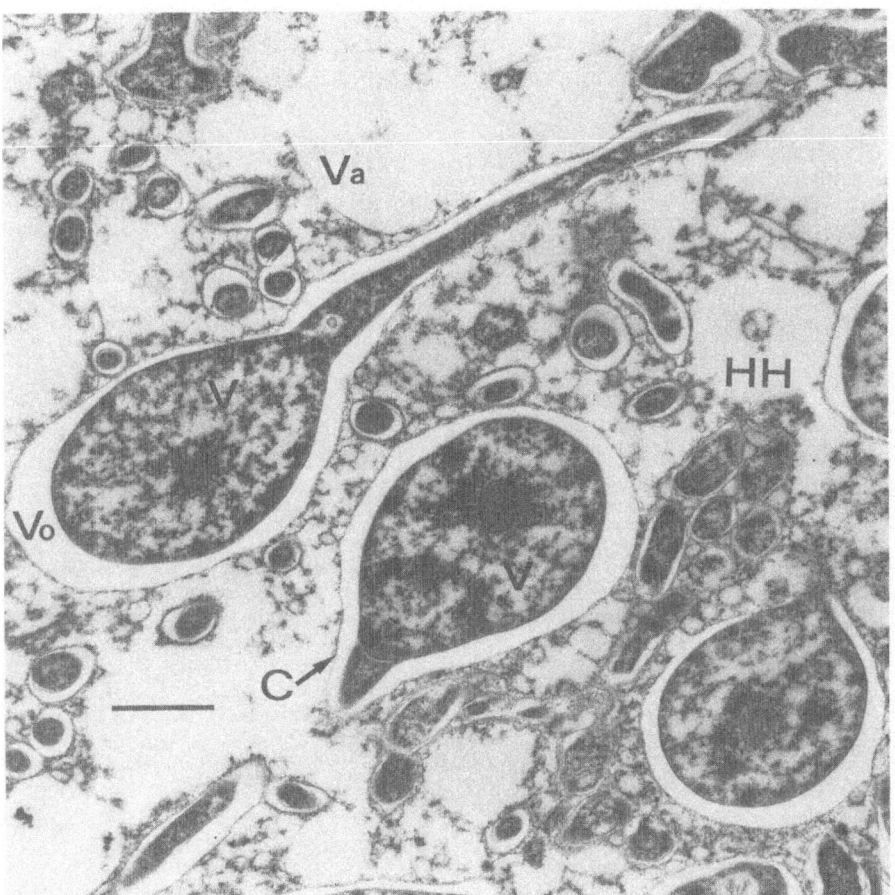

Fig. 4. Transmission electron micrograph of the *Dryas* endophyte showing vesicles (**V**) and hyphae (**HH**). The vesicles have no strict orientation towards the peripheral region of the host cell as shown by the opposite direction of two adjacent vesicles. The hyphae (**HH**) sometimes form strands, but each hypha is enclosed by a capsule (**C**). Note the 'void space' (**Vo**) between vesicle (**V**) and capsule layer (**C**), and the numerous vacuoles (**Va**). Bar scale = 1.0 μm.

but an adjacent vesicle may be oriented in the opposite direction (Fig. 4). Transmission electron micrograph Fig. 5 shows probably a nearly tangential section through a host cell containing vesicles near the periphery of the cell.

The non-septate vesicles have a clear nucleoid region with an electron-dense centre (Figs. 3 and 4). In contrast to electron micrographs by Newcomb[25] the hyphae and vesicles in our preparations showed about the same content of ribosomes as evident from the approximately same electron densities of both cell types (*e.g.* Figs. 3 and 4). The electron micrographs demonstrated that the hyphae as well as the vesicles are surrounded by a common capsule and a 'void space'

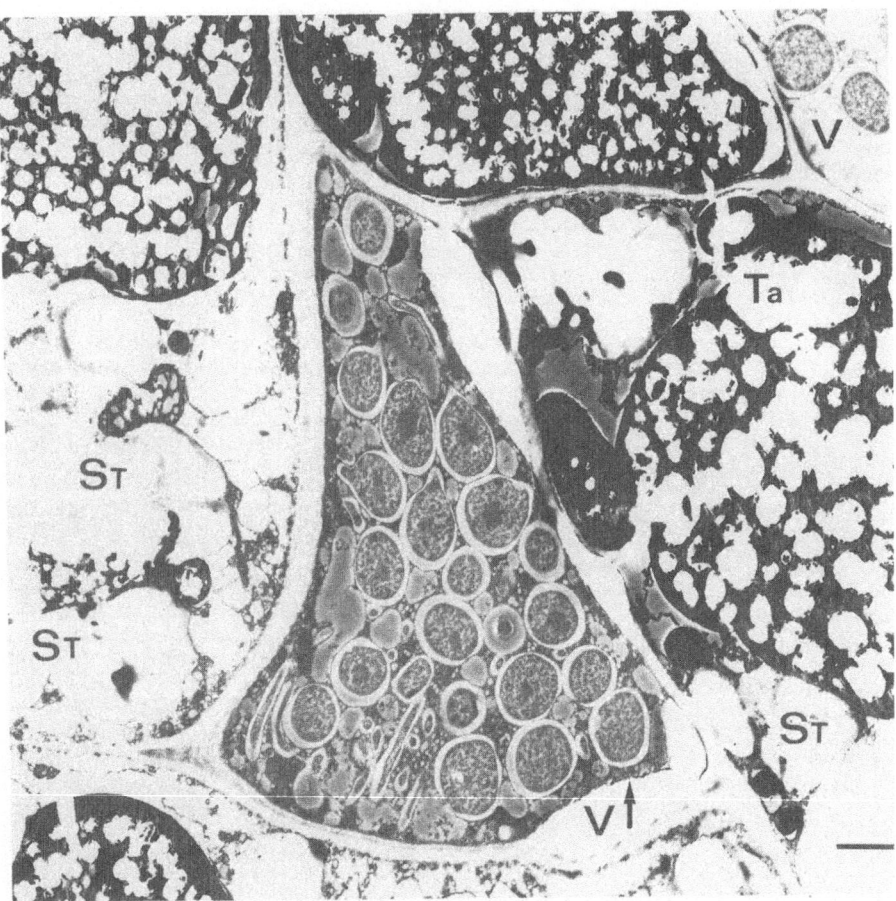

Fig. 5. Transmission electron micrograph of a host cell of *Dryas* containing vesicles (**V**) at low magnification. The random distribution of the vesicles throughout the host cell is probably caused by the nearly tangential section close to the periphery of the host cell. Note the absence of starch (**ST**) and tannin (**Ta**) reserve substance in the infected host cell in contrast to the adjacent host cells. Bar scale = 2.0 μm.

between capsule and endophyte. The 'void' space' is probably an artifact arising from fixation or other preparatory steps[20]. The endophyte cells show a distinct cell wall and an endophytic cytoplasmic membrane. The host cytoplasmic membrane separates the capsule from the host cytoplasm (Figs. 2, 3, and 4). Within the host cells hyphae may occur in clusters or strands, but every individual hypha seems to be surrounded by a capsule (see HH, Fig. 4). Mitochondria of about the same dimensions as the cross sections through the hyphae are often associated with the vesicles (see Mi, Fig. 6). Numerous vacuoles are randomly distributed in host cells containing endophytic structures (Figs. 4 and 6). Host cells containing the endophyte are devoid of

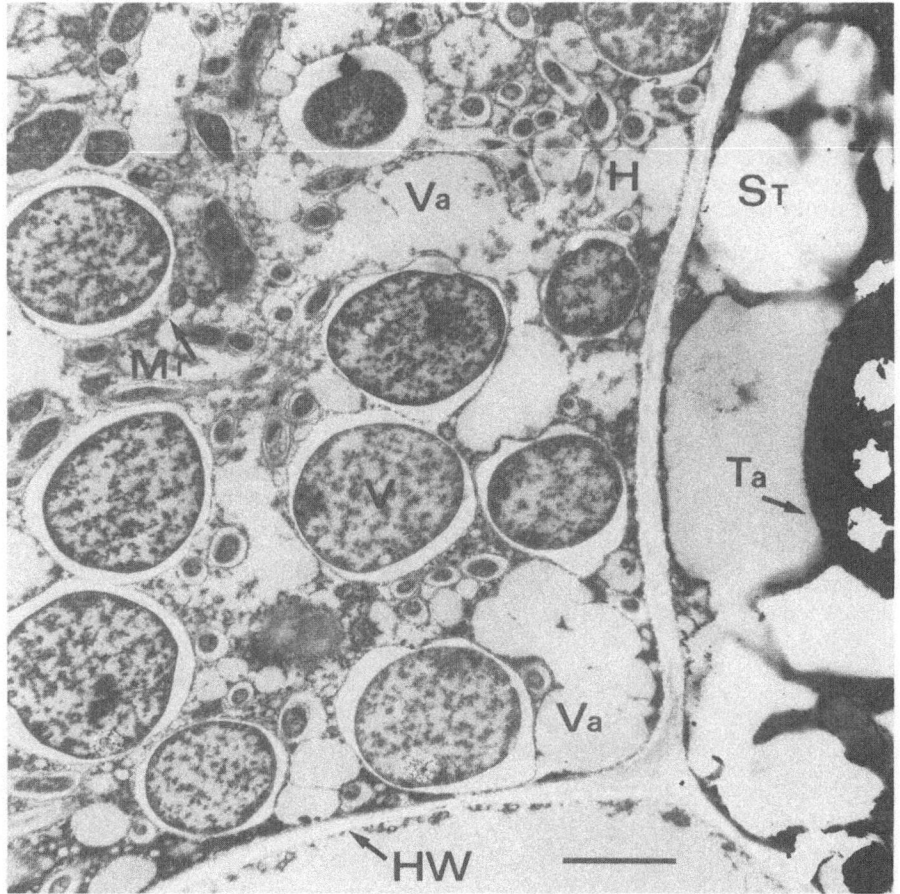

Fig. 6. Transmission electron micrograph of the *Dryas* endophyte at higher magnification. Note the host cell filled with vesicles (**V**) forming a broad zone near the periphery of the cell close to the host-cell wall (**HW**). Hyphae (**H**) are more prominent to the centre of the host cell, but also occur near the periphery. Mitochondria (**Mi**) of about the same size as the cross sections through the hyphae are often associated with the vesicles. Also numerous vacuoles (**Va**) are present in vesicle-containing host cells. Starch (**ST**) granules and tannin (**Ta**) deposits are numerous in the adjacent endophyte-free host cell. Bar scale = 2.0 μm.

starch and phenolic or tannin deposits, in contrast to adjacent non-infected host cells, where starch granules in amyloplasts and tannin deposits are frequent (see Figs. 5 and 6). The endophyte may induce these differences by the utilization of these compounds itself or by causing an increased metabolic activity of the host cell.

In the intercellular spaces between the host cells endophytic cells with irregular forms occur suggesting distorted hyphae. In many of these structures mesosomes are prominent (see Me, Fig. 7). In some of the intercellular spaces these endophytic cells possess rather thick cell walls suggestive of spores (Fig. 8). In other electron micrographs, groups

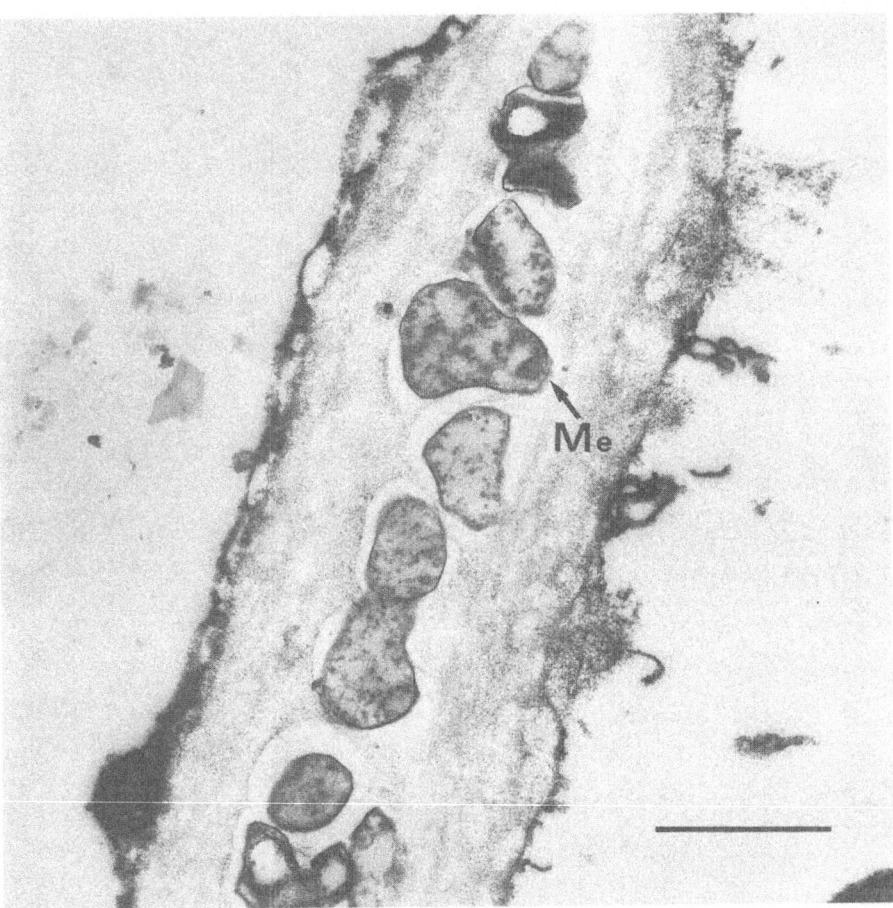

Fig. 7. Transmission electron micrograph of the *Dryas* endophyte. Irregular endophytic cells suggesting distorted hyphae occur in the intercellular spaces between the host cells. In some of these cells mesosomes (**Me**) are prominent. The endophytic cells have usually rather distinct cell walls and are embedded in a matrix of probably polysaccharide nature. Bar scale = 1.0 μm.

of cells give the impression to be spore-like structures possibly within a sporangium (Fig. 9). Spore formation in the *Dryas* endophyte and in other Rosaceae symbioses has so far never been observed. However, these groups of polyhedral cells are very reminiscent of a similar type of endophytic cells in *Alnus glutinosa* root nodules, which could be identified as spores[4]. In agreement with the latter observation, these structures only occur in dead host cells devoid of cytoplasm and cytoplasmic organelles. In the case of *Dryas* the intercellular spaces are also free of cytoplasm.

From the middle towards the base of mature root nodules the hyphae and vesicles appear in the process of lysis or disintegration (Fig. 10). It might be that by such a process nitrogenous compounds

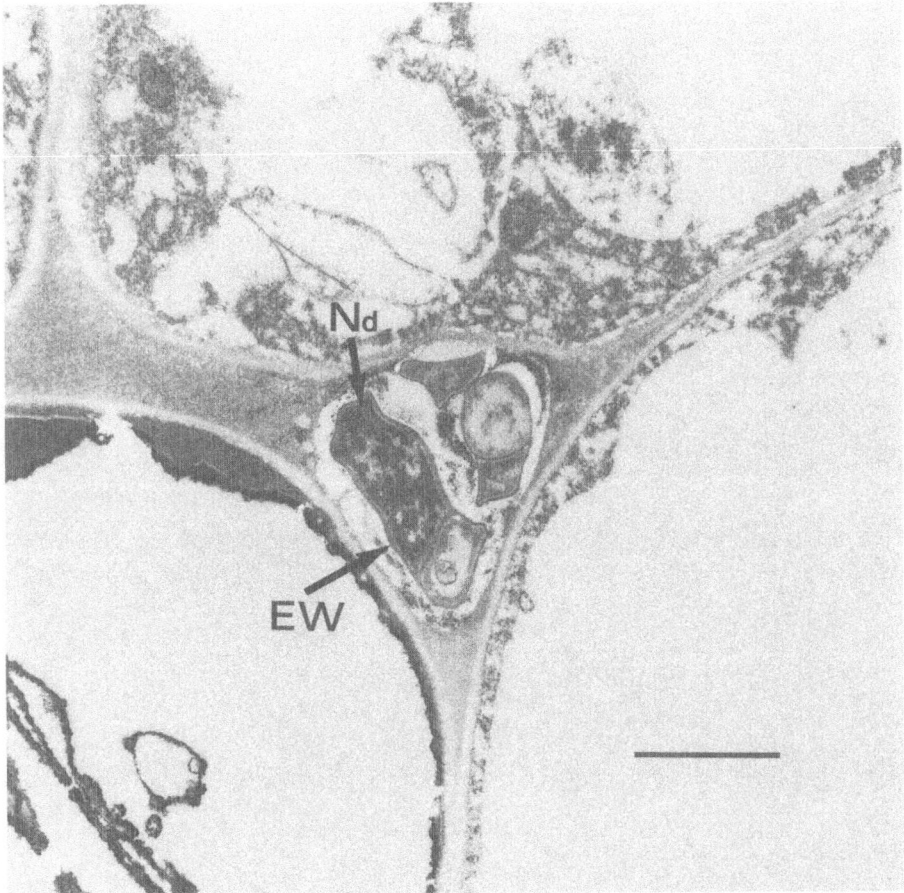

Fig. 8. Transmission electron micrograph. Possible spores of the *Dryas* endophyte in the intercellular space between living host cells. Note the very thick cell walls of the endophytic cells
(EW), their polyhedral shape and the presence of dense nucleoid material (Nd) within these
spore-like cells. Bar scale = 1.0 μm.

released from the endophytic structures become available to the plant.
Notable is that in this stage many plasmodesmata (see Pd, Fig. 10),
which otherwise are rather indistinct or infrequent in host cell walls,
connect such host cells containing disintegrating endophytic structures
with adjacent active endophyte-free host cells.

Rubus ellipticus

Root nodules of *Rubus ellipticus* have a peculiar morphology. They
resemble superficially those of the *Alnus*-type by a dichotomous
branching of nodular lobes and arrested growth of the apical meristem
(Fig. 11B). Sometimes, the nodular lobes appear to be trifurcate
branched, however always one of these nodular lobes arises below the

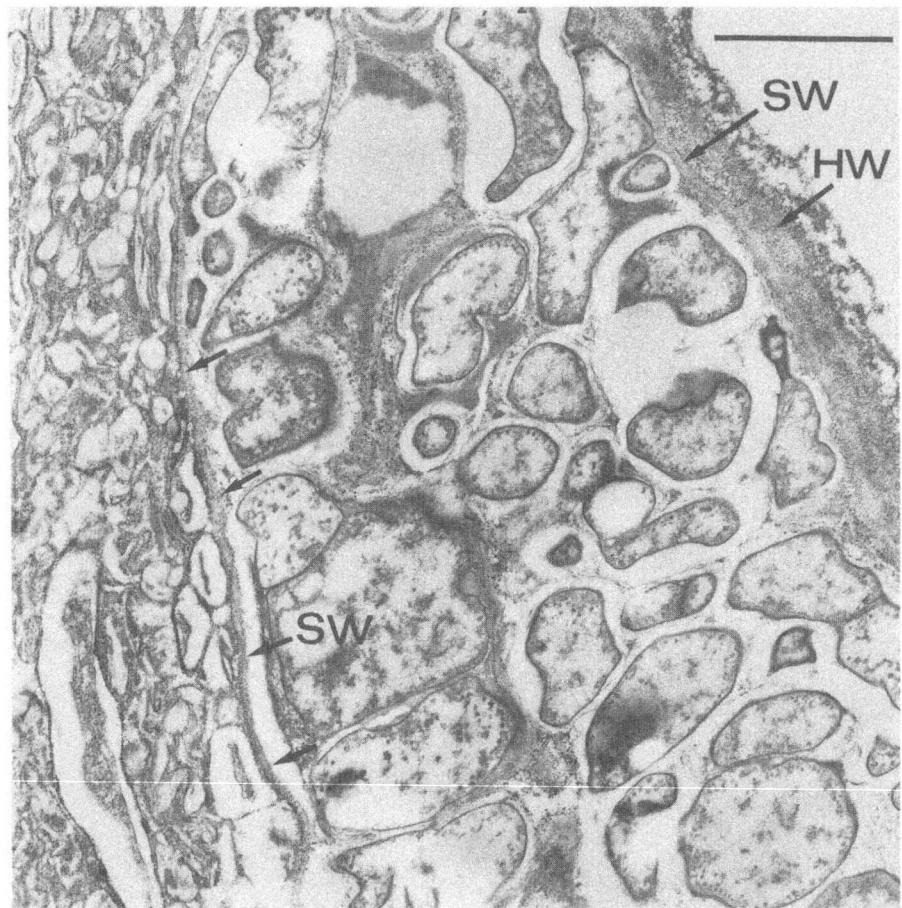

Fig. 9. Transmission electron micrograph. Possible spores of the *Dryas* endophyte in the intercellular space between host cells. The spore-like structures occur in groups and the possible outline of a sporangial wall (**SW**) is visible. In the upper right corner of the micrograph this sporangial wall is close to the host-cell wall (**HW**). Bar scale = 1.0 μm.

insertion of the two other lobes. Hence the branching is still dichotomous (Fig. 11A). A special feature of *R. ellipticus* root nodules is the phenomenon that the uninfected rootlet accompanying the nodular lobes, does not sprout from the nodular tip being an unarrested nodule-lobe meristem like in *Myrica* or *Casuarina* root nodules, but the rootlet arises below the other branchings. It is therefore in fact a basal rootlet as illustrated in the line drawing (Fig. 11C). So far, in other actinorhizal plant species no root nodules having such basal nodular rootlets have been observed.

Transverse sections (Fig. 12) of the nodular lobes of *Rubus ellipticus* root nodules stained with toluidine blue and other stains showed a central stele with xylem and phloem vascular bundles. The central stele

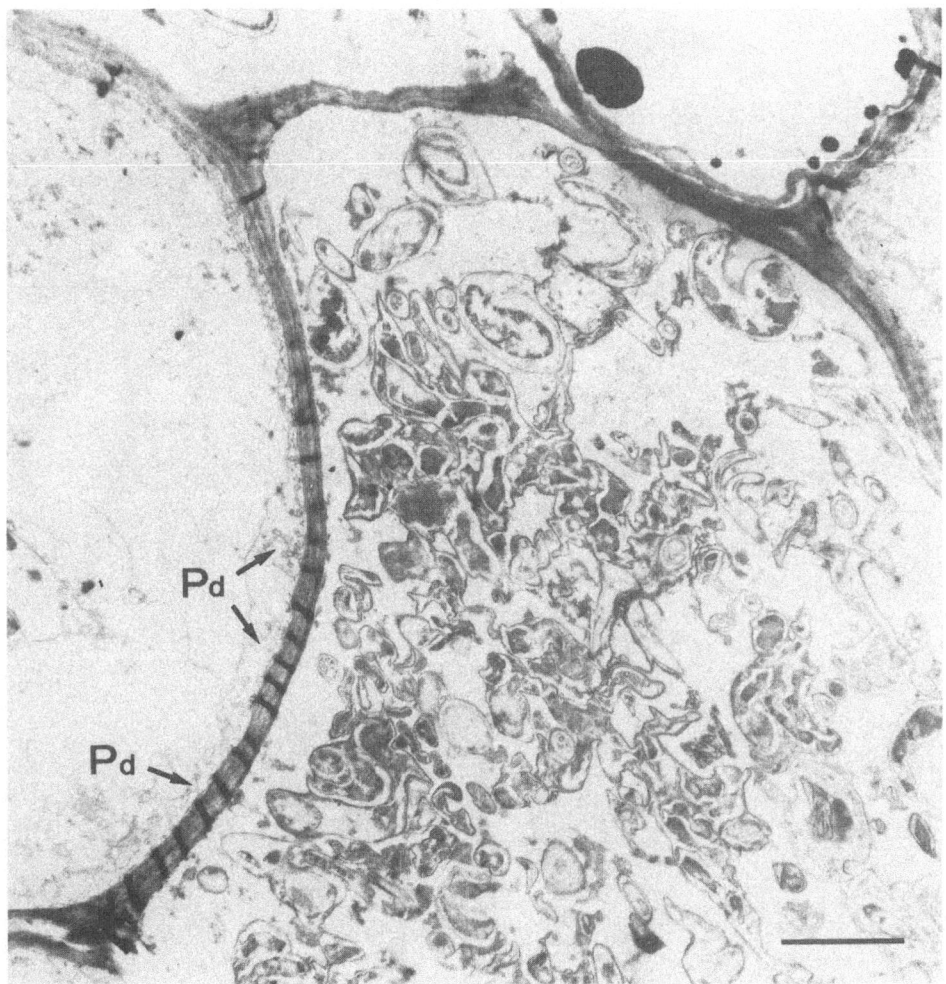

Fig. 10. Transmission electron micrograph. Host cell of *Dryas drummondii* containing vesicles and hyphae near the base of the root nodule. In this region resorbtion of the endophyte by the host occurs. Notable are the many plasmodesmata (**Pd**), which connect this host cell with the adjacent endophyte-free host cell. Bar scale = 2.0 μm.

is surrounded by a distinct endodermis, which in radial direction is followed by a 2–3 cells deep layer of cortical parenchyma completely filled with tannin deposits. Subsequently, there is a layer of radially enlarged parenchyma cells of the inner cortex, which appears to be empty as only a few of these cells contain large crystal inclusions. These crystals are probably silica crystals, since crystalline silica deposits have regularly been observed in cortical cells of other Rosaceae, including the genus *Rubus*[24] (l.c. p. 92). In peripheral direction this layer is followed by other layers containing 4–5 host cells completely filled with tannin deposits (Fig. 12A). Then, further to the

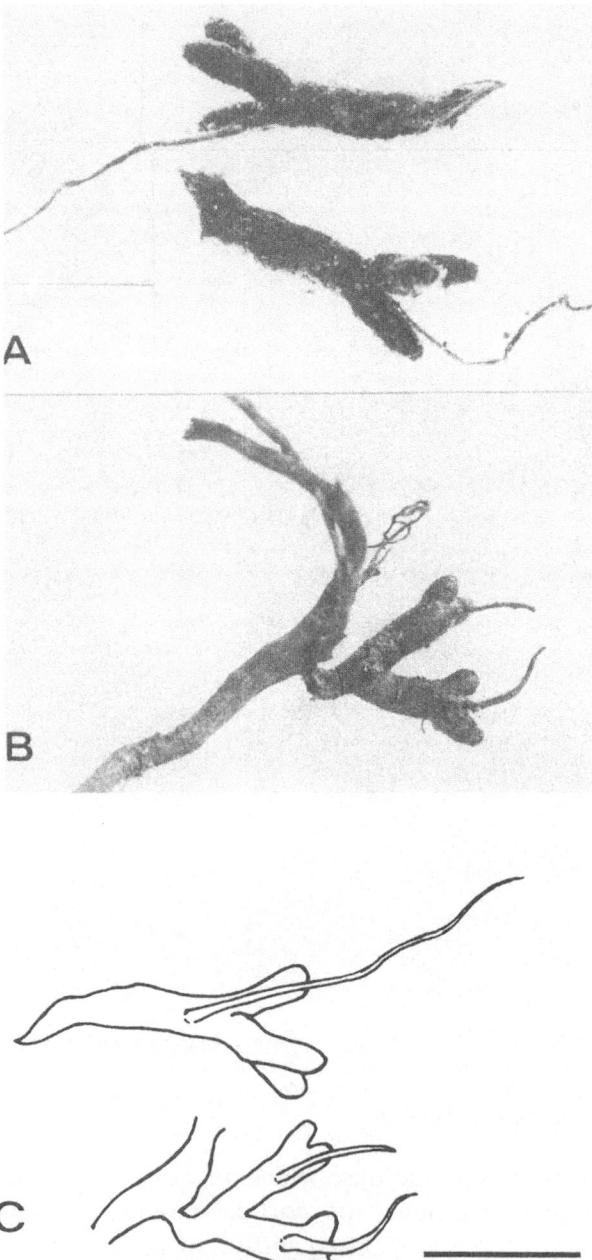

Fig. 11. External norphology of *Rubus ellipticus* root nodules.
(**A**) – Trifurcate root nodule with the basal rootlet.
(**B**) – Dichotomous branched root nodule with basal rootlet and showing the attachment of the root nodule to the root system.
(**C**) – Line-drawing of the root nodules showing the mode of insertion of the basal rootlet. Bar scale = 0.5 cm.

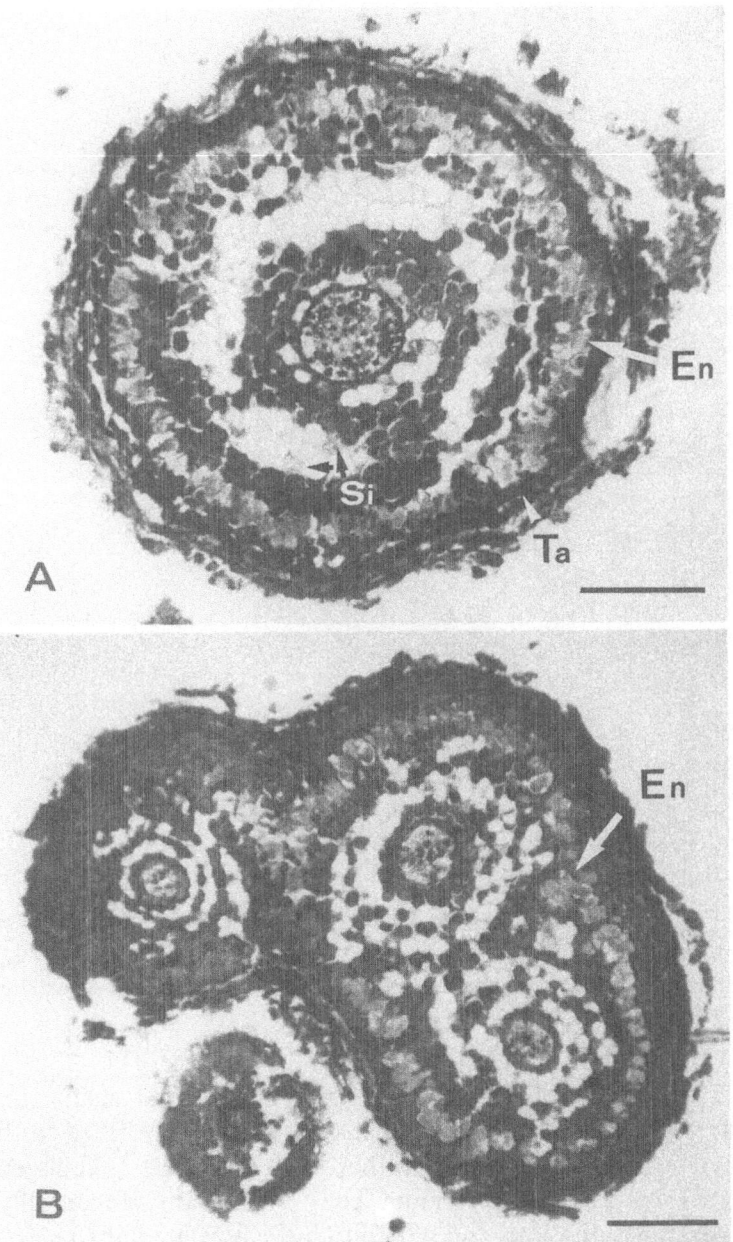

Fig. 12. Light micrographs of transverse paraffin-wax sections of *Rubus ellipticus* root nodules. (A) − Section through a nodular lobe of a dichotomous branched root nodule. (B) − Section through a trifurcate root nodule showing the three steles of the nodular lobes and the endophyte-free rootlet. From the inner to the outer side the following tissues can be distinguished: (1) vascular cylinder with phloem and xylem elements, (2) endodermis, (3) radially enlarged inner cortical cells with some containing silica (Si) crystals, (4) cell layer with tannin deposits, (5) middle cortical cells, (6) tannin (Ta) layer, (7) endophytic layer (En), (8) tannin layer (Ta), (9) periderm, and (10) epidermis. Bar scale = 200 μm.

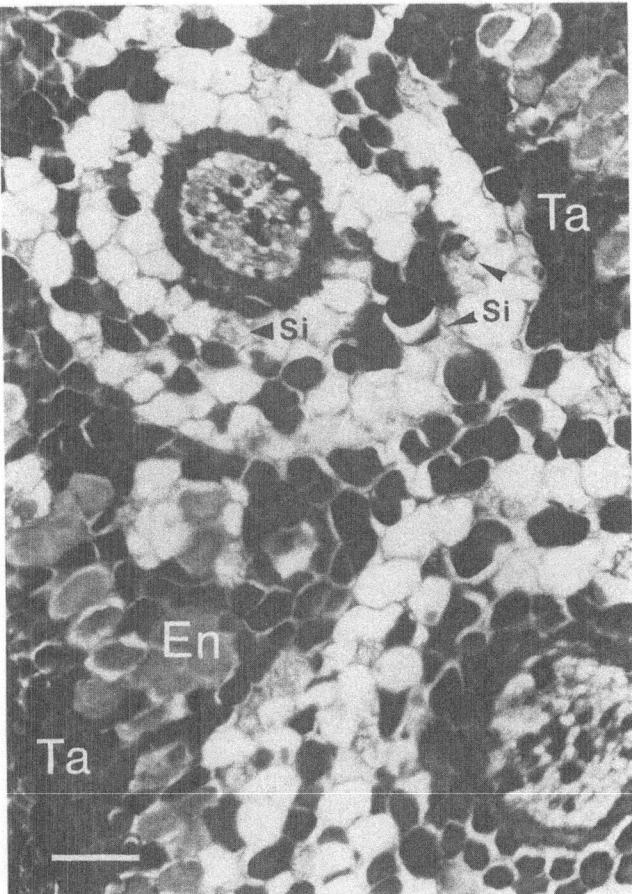

Fig. 13. Light micrograph. Higher magnification of a transverse section of a *Rubus* root nodule. Notable are the central stele with phloem and xylem elements, the cell layer containing the presumed silica (**Si**) crystals intermingled with tannin-filled cells (**Ta**) enclosing the endophyte-containing cells (**En**) on both sides. Bar scale = 50 μm.

outside in the outer cortex, there is a layer consisting of 1–2 cells which contains the endophyte. This endophytic layer is again surrounded by a 3–4 cell wide layer of host cells completely filled with tannin. Finally, near the periphery a rather loose periderm (phellogen and phelloderm) is present containing cells which are tangentially elongated by periclinal divisions and which tissue is bordered at the outside by an epidermis. Hence, in *Rubus* the endophytic layer is only 1–2 cells wide and on both sides, external as well as internal, completely surrounded by host cells filled with tannin. A trifurcate root nodule shows essentially the same histological structure as above. Here also the endophyte-containing host cells are only present in the outer cortex and in the same way completely enclosed by two tannin-containing host cell layers (Figs. 12B and 13).

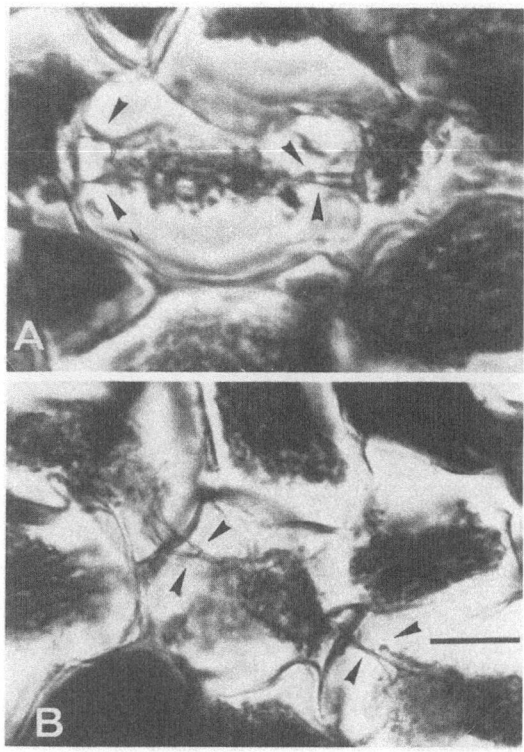

Fig. 14. Light micrographs. (A) and (B) – Enlarged views of a transverse section of a *Rubus* root nodule showing the endophytic zone. The outer cortical cells show massive occupation by the *Frankia* endophyte mycelia and cell-to-cell connections of the filaments (see arrows). Bar scale = 10 μm.

At higher magnification in the outer cortex *Frankia* hyphae and vesicles could be observed in the endophyte-containing host cells (Fig. 14). Also hyphal strands could be seen which apparently had hydrolyzed the primary cell wall of the host and in this way invaded one cortical cell from an adjacent cell (Fig. 14). The vesicle structures of the endophyte appeared in these sections to be mainly present in the peripheral region of the host cell. Its distribution limited to the periphery of the host cell could be confirmed by transmission electron microscopy. Unfortunately, fresh root-nodule material of *Rubus ellipticus* for the preparation of ultra-thin sections for transmission electron microscopy was not available and only preparations originally fixed for light microscopical purposes could be used. Further ultra-thin sectioning proved to be extremely difficult, because of the frequency of tannin-filled cell layers in the nodule tissue hampered considerably the penetration of the Epon-Araldite resin. This difficulty could only be partly overcome by making first small radial paraffin-wax sections and

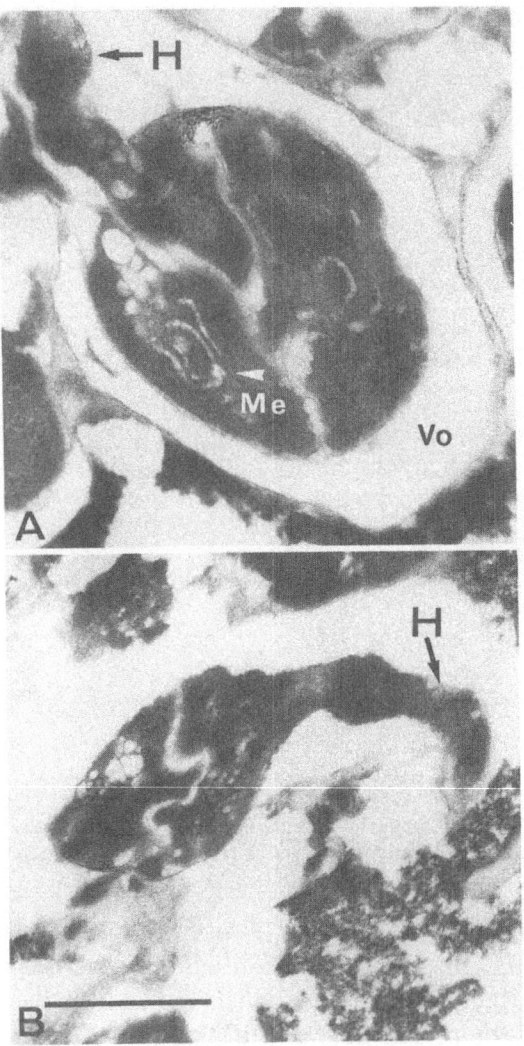

Fig. 15. Transmission electron micrographs (**A** and **B**) of vesicles of the *Rubus* endophyte showing prominent compartmentation of the vesicles with numerous imcomplete cross-walls. Note the 'void space' between the endophyte and capsule and the hypha (**H**) bearing the vesicle. Within the vesicles also many mesosome (**Me**) structures are visible. Bar scale = 2.0 μm.

after deparaffinization embedding them in the Epon-Araldite resin. The transmission electron micrographs finally obtained showed very clearly that the endophyte of *Rubus* is a *Frankia* species. The hyphae 0.4–0.6 μm in diameter bear at their tip roughly spherical vesicles with numerous septa and mesosomes very similar to those of the alder endophyte. The vesicles measuring 3.6–4.6 μm in diameter show, similar to the alder endophyte[4], many internal incomplete cells walls (Fig. 15).

Discussion

The examination of the two Rosaceae species reported above provided evidence that the endophyte of both species is actinorhizal and therefore is a *Frankia* species. In *Dryas drummondii* root nodules the endophyte represents itself in the living host cell in the form of hyphae and non-septate vesicles. Like in the other *Frankia* symbioses the endophytic structures are separated from the host cytoplasm by the endophytic cytoplasmic membrane, the endophytic cell wall, a capsule layer and the host cytoplasmic membrane. The non-septate vesicles are always at the tip of the hyphae and invariably they have at the base a septum, associated with mesosomes, which delimits it from the septate hyphae. As observed in other *Frankia* species[4] septa formation in the hyphae is not necessarily correlated with branching. Also the occurrence of mesosomes or plasmalemmosomes in hyphae and vesicles in association or in close proximity of septa is a prominent feature in other *Frankia* species[4,6,13,14,29]. In the host cells near the apex of the nodular lobes, solely hyphae occur as can be expected in very young host cells. It can be argued that the non-septate vesicles are merely young vesicles, which in an older stage may produce septa. This is, however, unlikely as in perennial root nodules at the end of the growing season mature vesicles certainly must be present. Moreover, Newcomb[25] probably investigating summer root nodules of *Dryas drummondii* also found only non-septate vesicles within the host cells.

In the *Dryas* endophyte the vesicles appear in a broader zone along the host-cell wall than usually is found in the *Alnus* endophyte[4]. Moreover, the vesicles are often not strictly oriented to the host-cell wall, but also vesicles may occur with an opposite orientation. In contrast to Newcomb's[25] observations, we observed spore-like structures in the *Dryas* endophyte. From all Rosaceae root nodules so far investigated, it may be the first suggestion of spore formation in this group. Spore-like structures solely occur in the intercellular spaces between the host cells. In the *Alnus glutinosa* endophyte spore formation only occured in dead or empty host cells[4]. Therefore, it is conceivable that spores in *Dryas* root nodules only occur in the intercellular spaces which are devoid of cytoplasm. Moreover, in free-living (*i.e. in vitro* cultivated) *Frankia* species spore formation invariably occurs. This is so even in those endophytes inhabiting the American *Alnus* species, where within the root nodules spore formation is absent or very rare.

On the basis of fine structure the *Frankia* symbioses can be divided in those with septate vesicles and those with non-septate vesicles. The presence of septa in the vesicles had been observed in the more or less spherical vesicles of the *Frankia* endophyte of *Alnus, Elaeagnus,*

Shepherdia Hippophaë, Colletia, and *Discaria.* Non-septate vesicles as found in the *Dryas* endophyte are present in the endophytes of *Ceanothus* species[13,29,30] and in *Purshia tridentata*[6]. The relationship of the *Dryas* endophyte with the two above mentioned endophytes merely on basis of morphology is however difficult to form a judgement. Lalonde[18,19] has shown that the endophytic structures within the root nodules change with the host species concerned. On the other hand it is striking, that all non-septate vesicle producing *Frankia* species and their host plants are restricted to North America. Hence it is not inconceivable that such a relationship might exist.

Noteworthy is the peculiar distribution of *Dryas* and its endophyte in Northern America. Nodulation of *Dryas drummondii* has only been reported in a few restricted areas in Alaska (Palmer area[1], Glacier Bay[21] and in the environments of Haines as shown in this study) and in Canada (Mt. Robson, British Columbia[21], Kluane Lake, Yukon (K. Huss-Danell, pers. comm.), St. Lawrence river area[21] and Gaspé Peninsula[25], Quebec, and near Calgary, Alberta[25]). On many other places where *Dryas drummondii* was examined for nodulation, it was non-nodulated. A similar situation occurs with some *Ceanothus* species[5], but in *Dryas drummondii* also at sites where root nodules were found, the nodulation of the plants sometimes appears to be scarce and in many plants even to be absent (R.E. Henderson, pers. comm.). The proper assessment of the frequency of nodulation in natural habitats is usually difficult and often depends on the persistence and patience of the observer and careful excavations. As reported for *Ceanothus*, it may be however, that the correct nodule organism is sometimes absent from the soil or soil conditions inhibit nodulation.

It is moreover strange, that although the natural distribution of *Dryas drummondii* is from Central Alaska south to Oregon, U.S.A., nodulation in the southern localities has never been reported. Further, Sprague (cited by Lawrence[21], l.c. p.800) observed at a *Dryas drummondii* site with root nodulation, also nodulation in the two other *Dryas* species occurring sympatrically, *i.e., Dryas octopetala* and *D. integrifolia*. Now, in contrast to the yellow-flowering *Dryas drummondii*, the white-flowering *Dryas octopetala* has a very extensive distribution in the Northern Hemisphere. Its distribution includes the arctic and subarctic of North America (Alaska and Canada) to northern Europe (Northern Ireland, Middle and North England, Scotland, Scandinavia and northern Russia), northern Central Asia to Japan, with outcrops in the Alps in Central Europe, the Pyrenees in south-west mountains of western North America as far south as Colorado. In all these localities nodulation of *Dryas octopetala* has never been reported (see IBP survey[5]). In addition, *Dryas octopetala* had in the Glacial times of the Dilivium a very extensive distribution in Europe and Central Asia and the plant is commonly used as a guide fossil of this geological time (Dryas Period). One may speculate that the nodule-forming habit in *Dryas drummondii* is a rather recent acquisition of this species, which occurred after speciation within this genus. This has developed in the centre of the distribution area of the genus containing three sympatric species. Moreover, *D. drummondii* had at the same time evolved some genetic features which favour the development of a symbiosis. Such a genetic development and ability might be absent in the other *Dryas* species, including *Dryas octopetala*. For this reason the latter species might be non-nodulated, in spite of ecological advantages as demonstrated by its very wide distribution range. In this context it would be highly desirable that Sprague's observation[21] of root nodulation in *Dryas octopetala* in Alaska will be confirmed by other workers in this field.

The preceding experiments have produced evidence that the root-nodule endophyte of *Rubus ellipticus* is actinorhizal. It was shown that the root nodules of *R. ellipticus* differ from all other non-leguminous root nodules so far described by the presence of an endophyte-free basal rootlet initiated below the dichotomous branching of the nodular lobes, but extending far beyond the root nodule. The dichotomous branching of the nodular lobes and the arrested growth of the apical meristem would classify this type of root nodulation as the *Alnus*-type[2]. *Myrica* and *Casuarina* root nodules show the feature that the apical meristem of the nodular lobes after a period of arrested growth, resumes its growth and produces a normal but negative geotropic rootlet. Therefore, these root nodules finally become covered with upward growing rootlets (*Myrica/Casuarina*-type of root nodulation)[2]. The *Rubus*-type of root nodulation is fundamentally different from the *Myrica/Casuarina*-type of root nodulation since the rootlet is not derived from the apical meristem of the nodular lobe, but from a basal lateral root meristem.

Transmission electron micrographs of the *Rubus* endophyte have demonstrated that the *Frankia* endophyte possesses septate hyphae which bear at their tips nearly spherical vesicles with many internal cross walls. Like in the alder endophyte[4], vesicle formation in *Rubus* is mainly restricted to the peripheral region of the host cell close to the host-cell wall. Also in other aspects of fine structure (*e.g.* the number of cross walls and the presence of mesosomes), the vesicles of the *Rubus* endophyte are very similar to those of the *Alnus* endophyte.

Anatomically *Rubus ellipticus* root nodules revealed a structure characteristic for the genus *Rubus*. Layers of tannin-filled host cells are distributed in the cortical parenchyma cells. This is a typical feature of the genus *Rubus* and the arrangement of the tannin cells in the pith of *Rubus* has been used for taxonomic purposes[12, 23, 28]. It is therefore very unlikely, that these roots nodules of *Rubus* are confused in the field with the root nodules of another non-leguminous plant species. Moreover, in contrast to the actinorhizal symbioses of other non-legumes, the endophyte-infected cells are only present in the outer cortex. The endophyte-containing cell layer is only $1-2$ cells wide and completely surrounded on both sides by $3-4$ cell wide layers of tannin-filled cells. The impression is gained (especially because the difficult penetration of reagents, fluids and embedding resins) that the tannin-filled host cells form a structural barrier for O_2 diffusion to the endophyte, thus protecting the nitrogenase for O_2 inhibition in the same way as found by Tjepkema and Cartica[31] in *Parasponia* root nodules. In this region of the *Rubus* root nodule, intercellular spaces between the host cells are rare or even absent.

Many questions remain to be solved with respect to root nodulation in *Rubus*. For instance, why in this very large genus with 250 or more species[32], has at present root nodulation only been observed in *Rubus ellipticus*. Moreover, the distribution area of *Rubus ellipticus* shows a native distribution of this species in the Himalayas from East Pakistan and India to Nepal and northern Burma, Thailand, Laos and Vietnam, and China (Yunnan, Szechuan, Kweichow) and secondary distributions in Sri Lanka and on Luzon in the Philippines. Chaudhary *et al.*[8] observed, however, the absence of root nodulation in native *Rubus ellipticus* populations in N.E. Pakistan (Karakar Pass and Murree Hills). One may ask why *R. ellipticus* possesses root nodules on Java where it is introduced and not in the indigenous populations so far examined. Why are *Rubus ellipticus* root nodules so rare? As already mentioned, on Java (Cibodas Mountain Gardens), root nodulation in neighbouring *Rubus ellipticus* plants growing at the same site in identical soil was absent. In this context, one may also speculate on the origin of root nodulation in this species. It is likely that the nodule-forming capacity arose after speciation in this genus, because it is absent in other species of this genus. It might have been induced by some special genetic features of this species, which are the prerequisites for the evolvement of such a symbiosis, *e.g.* the development of species-specific tannin layers around the endophytic layer causing O_2 diffusion limitation. One may even speculate further, it may be a very recent achievement produced locally with a foreign endophyte, for instance one of the *Frankia* endophytes inhabiting *Alnus* root nodules to which the *Rubus* endophyte shows a striking resemblance.

Acknowledgements The author is greatly indebted to Mr. R.E. Henderson, Haines, Alaska for collecting and supplying me with the *Dryas drummondii* root nodules.

I would like to thank the Netherlands Foundation for the Advancement of Tropical Research (WOTRO) for financial support for my visits to Indonesia in 1971 and 1974 and likewise the authorities of the Lembaga Biologi Nasional (LBN) for giving me working facilities at the Treub Laboratory at Bogor, Java, Indonesia during that period.

References

1 Allen E K, Allen O N and Klebesadel L J 1963 An insight into symbiotic nitrogen-fixing plant associations in Alaska. Science in Alaska, Proc. 14th Alaskan Sci. Conf., Anchorage, Alaska, pp 54–63.

2 Becking J H 1977 Dinitrogen-fixing associations in higher plants other than Legumes. *In* A Treatise on Dinitrogen Fixation, Section III: Biology. Eds. R W F Hardy and W S Silver, John Wiley & Sons, New York, 185–275.

3 Becking J H 1979 Nitrogen fixation by *Rubus ellipticus* J E Smith. Plant and Soil 53, 541–545.

4 Becking J H, De Boer, W E and Houwink A L 1964 Electron microscopy of the endophyte of *Alnus glutinosa*. Antonie van Leeuwenhoek, J. Microbiol. Serol. 30, 343–376.

5 Bond G 1976 The results of the IBP survey of root-nodule formation in non-leguminous angiosperms. *In* Symbiotic Nitrogen Fixation in Plants. Ed. P S Nutman, Int. Biol. Programme, vol. 7, Cambridge Univ. Press, pp 443–474.

6 Bond G 1976 Observations on the root nodules of *Purshia tridentata*. Proc. R. Soc. Lond. B, 193, 127–135.

7 Bond G and Becking J H 1982 Root nodules in the genus *Colletia*. New Phytol. 90, 57–65.

8 Chaudhary A H, Khokhar, S N, Zafar, Y and Hafèez, F 1981 Actinomycetous root nodules in Angiosperms of Pakistan. Plant and Soil 60, 341–348.

9 Edwards M R and Gordon M A 1962 Membrane systems of *Actinomyces bovis, In* Electron Microscopy. Ed. S S Breese Jr., 5th Int. Congr. Electron Microscopy, Academic Press, New York, p. UU-3.

10 Edwards M R and Stevens, R W 1963 Fine structure of *Listeria monocytogenes*. J. Bacteriol. 86, 414–428.

11 Fitz-James P C 1960 Participation of the cytoplasmic membrane in the growth and spore formation of Bacilli. J. Biophys. Biochem. Cytol. 8, 507–528.

12 Fritsch F E 1887 Anatomische systematische Studien über die Gattung *Rubus*. Sitz.-Ber. Wiener Akad. 95, 28 pp and 2 Plates.

13 Gardner I C 1976 Ultrastructural studies of non-leguminous root nodules. *In* Symbiotic Nitrogen Fixation in Plants. Ed. P S Nutman. Int. Biol. Programme, vol 7, Cambridge Univ. Press, pp 485–495.

14 Gardner I C and Gatner E M S 1973 The formation of vesicles in the developmental cycle of the nodular endophyte of *Hippophaë rhamnoides* L. Arch. Mikrobiol. 89, 233–240.

15 Gerlach D 1969 Botanische Mikrotechnik. Verlag Georg Thieme, Stuttgart, 298 p.

16 Heisey R M, Delwiche C C, Virginia R A, Wrona, A F and Bryan, B A 1980 A new nitrogen-fixing non-legume: *Chamaebatia foliosa* (Rosaceae). Am. J. Bot. 67, 429–431.

17 Hutchinson J 1964 The Genera of Flowering Plants (Angiospermae), vol. 1, Dicotyledones. Clarendon Press, Oxford, 516 p.

18 Lalonde M 1978 Confirmation of the infectivity of a free-living actinomycete isolated from *Comptonia peregrina* (L.) Coult. root nodules by immunological and ultrastructural studies. Can J. Bot. 56, 2621–2635.

19 Lalonde M 1979 Immunological and ultrastructural demonstration of nodulation of the European *Alnus glutinosa* (L.) Gaertn. host plant by an actinomycetal isolate from the North American *Comptonia peregrina* (L.) Coult. root nodule. Bot. Gaz. (Chicago) 140 (Suppl.), S35–43.

20 Lalonde M, Knowles R and Devoe I W 1976 Absence of 'void area' in freeze-etched

vesicles of the *Alnus crispa* var. *mollis* Fern. root nodule endophyte. Arch. Microbiol. 107, 263–267.

21   Lawrence D B, Schoenike R E, Quispel A and Bond G 1967 The role of *Dryas drummondii* in vegetation development following ice recession at Glacier Bay, Alaska, with special reference to its nitrogen fixation by root nodules. J. Ecol. 55, 793–813.

22   Löhnis M P 1930 Investigations upon the ineffectiveness of root-nodules on leguminosae. Zentr. Bakteriol. Parasitenkd. II Abt. 80, 342–368.

23   Metcalfe C R and Chalk, L 1950 Anatomy of the Dicotyledons, vol. 1, Clarendon Press, Oxford, 724 p.

24   Metcalfe C R and Chalk L 1983 Anatomy of the Dicotyledons, vol. 2, 2nd Ed., Wood structure and Conclusion of the General Introduction. Clarendon Press, Oxford, 297 p.

25   Newcomb W 1981 Fine structure of the root nodules of *Dryas drummondii* Richards (Rosaceae). Can. J. Bot. 59, 2500–2514.

26   Righetti T L and Munns D N 1980 Nodulation and nitrogen fixation in Cliffrose (*Cowania mexicana* var. *stansburiana* (Torr.) Jeps.). Plant Physiol. 65, 411–412.

27   Schulze-Menz G K 1964 Rosales. *In* A. Engler's Syllabus der Pflanzenfamilien, 12th ed. Ed. H Melchior, Gebr. Borntraeger, Berlin, 193–242.

28   Solereder H 1899 Systematische Anatomie der Dicotylendonen. Verlag Ferd. Enke, Stuttgart, 984 pp.

29   Strand R and Laetsch W M 1977 Cell and endophyte structure of the nitrogen-fixing root nodules of *Ceanothus integerrimus* H. and A, I. Fine structure of the nodule and its endosymbiont. Protoplasma 93, 165–178.

30   Strand R and Laetsch W M 1977 Cell and endophyte structure of the nitrogen-fixing root nodules of *Ceanothus integerrimus* H. and A. II. Progress of the endophyte into young cells of the growing nodule. Protoplasma 93, 179–190.

31   Tjepkema J D and Cartica R J 1982 Diffusion limitation of oxygen uptake and nitrogenase activity in the root nodules of *Parasponia rigida*. Plant Physiol. 69, 728–733.

32   Willis J C 1973 A Dictionary of the Flowering Plants and Ferns, 8th ed. (Revised by H K Airy Shaw), Cambridge Univ. Press, Cambridge, 1245 p.

Plant and Soil 78, 129–146 (1984).
Ms. Fr 27
© 1984 *Martinus Nijhoff/Dr W. Junk Publishers, The Hague.*

Physiological studies on N_2-fixing root nodules of *Datisca cannabina* L. and *Alnus nitida* Endl. from Himalaya region in Pakistan

FAUZIA HAFEEZ, ASHRAF H. CHAUDHARY and ANTOON D. L. AKKERMANS*
Department of Biological Sciences Quaid-i-Azam University, Islamabad, Pakistan

Key words Actinorhizas *Alnus nitida* Amino acid composition *Datisca cannabina* Hydrogen uptake Inoculation Nitrogen fixation Root nodules

Summary The nodulation and the morphology and physiology of the nodules were studied on *Datisca cannabina*, a perennial herb from northern Pakistan and *Alnus nitida*, a nodulated tree in the same locality. Both species bear coralloid clusters of actinorhizal nodules. The main free amino acid in *D. cannabina* nodules was arginine while the predominant free amino acid in *A. nitida* nodules was citrulline. The infectivity of crushed nodules of both types of plants on their respective host was about 10^6 infective particles per gram of nodule fresh wt. In cross-inoculation experiments crushed nodule inoculum from *A. nitida* failed to induce nodulation on *D. cannabina* seedlings but the crushed nodule inoculum from *D. cannabina* caused low nodulation on seedlings of *A. nitida* (10^3 infective particles. g nodule fresh wt.).

The activity of nitrogenase, hydrogenase and respiration (O_2 uptake) were measured in detached nodules, nodule homogenates and the 20 μm residue and 20 μm filtrate preparations from the nodules of both species. Both species showed similar patterns of activities except that only the nodule homogenate and 20 μm residue preparations from *D. cannabina* showed pronounced enhancement of the O_2 uptake by succinate which was further stimulated by ADP. This has in part been explained by the presence of mitochondria in close connection with the endophyte.

Introduction

In the mountainous region in the northern part of Pakistan, several woody N_2-fixing actinorhizal plants and legumes are important nitrogen suppliers to the ecosystem and act as prominent soil stabilizers. With exception of *Elaeagnus* spp., most actinorhizal plants prefer relatively moist sites and soil temperatures below 30°C, while certain legumes are also found on drier and warmer sites. Irrespective of these limitations, actinorhizal plants usually are much less attractive to the cattle than woody legumes and therefore can be planted in eroded areas of northern Pakistan which are often overgrazed.

During a field study of the mountainous flora of northern Pakistan, root nodules (actinorhizas) with *Frankia*-like actinomycetes as microsymbiont were found on woody species of *Alnus, Coriaria, Hippophaë, Elaeagnus* and on a perennial herb, *Datisca cannabina* L.[11,12,13,14]. Although nodulation of the latter species grown in a Botanical Garden

* Department of Microbiology, Agricultural University, Wageningen, The Netherlands

in Europe, has been reported in 1922 by Severini[33], it had remained unnoticed[11] and no attention had been paid to the nature of the endophyte and the function of the nodules until recently[14].

Preliminary observations had shown that the pattern of the endophyte distribution within *Datisca* nodules is reversed from other actinorhizas, *e.g. Alnus* and *Comptonia* spp.[10]. The vesicles are club-shaped or filamentous, compactly packed and arranged parallel to one another and perpendicular to the centre of the host cell[18]. Since no information was available on the physiology of this aberrant type of symbiosis, the nitrogenase and hydrogenase activities of both field-grown and greenhouse-cultivated *D. cannabina* plants were investigated and compared with the activities of nodules of *Alnus nitida* Endl., a native alder tree in northern Pakistan. In addition a study of the nodulation of both types of actinorhizal plants and the determination of the amino acid contents of their nodules was made.

Materials and methods

Description of the sites
D. cannabina plants used for the field studies were growing on steep hills in Kulali, Swat valley, northern Pakistan, at 1450 m elevation. The soil was loamy, mixed with rocks and gravel (pH 5–6). Plants grew preferentially on sites where water was percolating through the upper soil layer. Precipitation was restricted to December–February and June–August. In summer the soil temperature reached values of 26–37°C. In winter the temperature was 0–7°C and plants were covered with snow for about 20 weeks.

A. nitida was found along small streams at 1240 m elevation. Plants were collected in Madian, Swat. Seedlings were cultivated in plastic bags in a nursery in Abottabad at 1220 m elevation and one year old seedlings were transported intact to the Department of Biological Sciences, Islamabad. In summer the soil temperature in the field remained below 37°C and the soil remained permanently moist. In winter temperature may fall below freezing point. The pH of the soil was 6–7.

Growth and nodulation of D. cannabina
Nodulation of field-grown plants was studied by digging out plants and measuring the dry weight of roots, shoots, leaves and nodules.

Seeds were surface-sterilized for 20 min in a hypochlorite (1%) solution and rinsed with sterile water. The seeds were collected on a sterile 100 μm filter and subsequently transferred to the surface of Leonard jars (Fig. 3c) filled with coarse sand and moistened with Hoagland solution (half strength) with only 10 mg.1^{-1} of KNO_3. The plants were inoculated with a suspension of crushed nodules of either *D. cannabina* or *A. nitida*. The surface of the jars was covered with gravel (size 3–5 mm) to prevent desiccation and contamination. During the first 2 weeks the jars were covered with petri dishes.

Plants were grown in a growth chamber at 22°C and a light regime of 16 h light and 8 h dark. Plants were harvested 8 weeks after inoculation.

Growth and nodulation of A. nitida
Plants were raised from surface-sterilized seeds (procedure see above). Seeds were germinated on glass beads (size 2–3 mm) supplied with N-free Hoagland solution, half strength. After 4 weeks the seedlings were transferred to 330 ml pots with N-free Hoagland solution.

Spore type of nodules

Presence of spores in the nodules was determined microscopically in free hand sections, stained with Fabil reagent[28] according to Van Dijk and Merkus[17].

Measurement of nitrogenase activity and nitrogen content

Nitrogenase activity was measured by incubating whole plants (young seedlings) or excised nodulated root systems (older plants) in 200 ml vials with air and 10% C_2H_2 at ambient soil temperatures, unless otherwise mentioned. Ethylene production was measured by collecting 1 ml samples every 15 min over a period of one hour. The gas samples were stored in 5.6 ml tubes (Vacutainers) and analysed gas chromatographically within one week after the field experiment.

Total N content was determined as Kjeldahl-N[22].

Free amino acids in root nodules

Samples of 0.5 g fresh weight nodules were homogenized in a mortar at 4°C and suspended in 1.5 ml of sulphosalicylic acid (2% w/v). After 30 min of incubation at 4°C the suspension was centrifuged for 10 min at 40,000 x*g*. The supernatant was diluted 1:1 with lithium-citrate buffer (0.06 *M*) and the pH was adjusted to 1.9 with concentrated LiOH solution. The free amino acids were determined on an amino acid analyzer (Biotronic LC 6000E) connnected to an integrator (Spectra-Physics System 1).

Diaminopimelic acid (DAP) content of nodules and roots

Samples of 1.0 g fresh weight nodules and 2.1 g fresh weight of roots were collected from young *D. cannabina* seedlings (height 25 cm) cultivated in perlite with N-free Hoagland solution in a greenhouse at 20°C in Wageningen. The tissue was homogenized in a Virtis mixer (45 Hi speed) for 2 min at 5,000 rpm and the suspension was centrifuged. The pellet was washed with water two times and subsequently hydrolyzed overnight at 100°C in HCl (6 *N*). The HCl was removed by vacuum film evaporation. The residue was dissolved in sodium-citrate buffer (0.06 *M*, pH 2.0) and the insoluble material was removed by centrifugation. DAP was measured on an amino acid analyzer (Biotronic LC 6000E).

Hydrogen uptake and respiration by detached nodules

Samples of 1.5 g fresh nodules were incubated in 16.6 ml vials filled with air containing 1.0% H_2. After preincubation for 10 min the rates of respiration and net H_2 uptake were measured on a gas chromatograph equipped with a 1 m Molecular Sieve column and a thermal conductivity detector[31].

Hydrogen uptake by nodule homogenates

Nodule homogenates were prepared as described previously[31]. Samples of 1.8 ml nodule homogenate, containing 2.16 g original fresh weight nodules, were injected in 16.6 ml vials which were previously filled with a gas mixture of N_2 and H_2 (0.05%) and 0.2 ml of a solution of an electron acceptor. The following electron acceptors were tested at final concentration of 10 m*M*: 2,6-dichlorophenolindophenol (DCPIP), methylene blue (MB) and phenazine metasulphate (PMS). Hydrogenase activity was assayed at 25°C by measuring the H_2 uptake gas chromatographically after preincubation for 15 min. All values were expressed as μmoles H_2 consumed per g original fresh weight nodules.

Respiration by nodule homogenates

Nodules of *D. cannabina* and *A. nitida* which were used for respiration studies had a nitrogenase activity of 2.0 and 4.6 μmoles C_2H_4 .g^{-1} nodule fresh weight.h^{-1}, respectively. Samples of about 5 g fresh weight were homogenized in a mixer (Virtis, 45 Hi-speed) for 2 min at 5,000 rpm in HEPES buffer, pH 7.4 (50 m*M*) containing sucrose (1.0 *M*), $MgCl_2$ (2 m*M*), KCl (1 m*M*), polyvinylpyrrolidone, PVP (K25; 4% w/v), defatted bovine serum albumin (0.1% w/v), EDTA (2 m*M*), dithioerythritol (5 m*M*) and $Na_2S_2O_4$ (20 m*M*). The homogenate was kept anaerobic by flushing with N_2. The homogenate was filtered through a 100 μm filter

Fig. 1. *D. cannabina* plants growing in natural habitat **(a)**, a typical female plant **(b)** and male plant **(c)**.

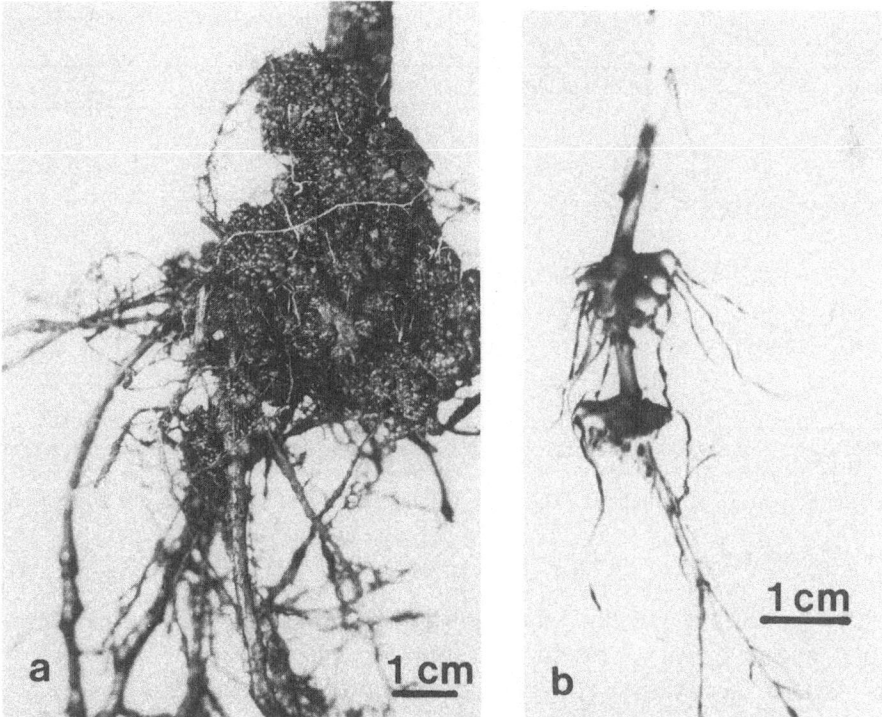

Fig. 2. Root nodules of *D. cannabina*. Nodule clusters (a) with positive geotropic roots (b).

to remove large plant cell debris. The filtrate was subsequently passed through a 20 μm filter and the 20 μm residue was washed and resuspended in HEPES buffer (pH 7.4; 50 mM) without dithionite and dithioerythritol. The O_2 consumption by the 20 μm residue was measured with an Oxygen Monitor (Yellow Springs Instruments, Model No. 53) after addition of NADH or different carbon substrates, viz. acetate, glucose, glutamate, malate, propionate and succinate (final concentration 10 mM).

Succinate dehydrogenase and NADH dehydrogenase were also determined qualitatively by incubating the 20 μm residue for 30 min in a Tris-HCl buffer (pH 7.4; 50 mM) with 0.1% 2,3,5-triphenyltetrazolium chloride (TTC) and NADH (50 mM) or sodium succinate (10 mM).

Results

Growth and nodulation of D. cannabina

D. cannabina is a dioecious perennial herb from the mountainous region in northern Pakistan. Female plants may reach a height of 150–180 cm at maximum maturity, while male plants remain smaller to a height of 120–150 cm (Fig. 1). All plants which were searched in Kulali (Swat), Naran (Khagan), Sharda (Neelam valley in Azad Kashmir) and Chitral valley were profusely nodulated, irrespective of the age of the plant (Fig. 2).

Table 1. Effect of inoculation with crushed nodules on the nodulation of *D. cannabina* and *A. nitida**

Nodule inoculum (g fresh wt.g⁻¹)	*D. cannabina*		*A. nitida*			
	D.c.	A.i.	D.c.		A.i.	
	%	%	%	N	%	N
0	0	0	0	0	0	0
10^{-7}	100	0	0	0	63	1.3
10^{-6}	100	0	0	0	44	1.4
10^{-5}	100	0	0	0	100	ND
10^{-4}	100	0	17	0.2	100	10
10^{-3}	100	0	50	0.9	100	ND
10^{-2}	100	0	100	5	100	> 20
10^{-1}	100	0	100	ND	100	> 20

* Plants were inoculated with crushed nodule suspensions of *D. cannabina* (D.c.) and *A. nitida* (A.i.) at different concentrations. Two months after inoculation the plants were harvested and the percentage of nodulated plants (%) and the number of nodules per plants (N) were counted.
ND: not determined.

The nodules were not uniformly distributed on the root system. Most nodules were present on superficial roots, 10–12 cm below the soil surface. Other roots were less nodulated and grew *ca* 20 cm deep in soil.

Root nodules of *D. cannabina* are superficially of the *Alnus*-type. They are perennial and show repeated closely set dichotomous branching, resulting in the formation of coralloid clusters with a diameter up to 6 cm (Fig. 2a). The apex of each lobe gives rise to a positively geotropic nodule rootlet up to 8 cm in length (Fig. 2b). The nodules occur laterally on primary as well as on secondary roots. They develop in an axil of very young lateral root and form 2–3 orders of dichotomy within one year. The apical part of the lobe is more or less swollen and contains most of the living *Frankia*-tissue. No spores were detected in nodule sections, indicating the spore negative character of the endophyte.

The colour of the nodules is yellow, due to the presence of the glycoside datiscin (= datiscosid) and some derivates[20, 21]. The colour turns yellowish brown when the nodules become older and when a periderm is formed.

Seedlings inoculated with crushed nodules of *D. cannabina* and grown in sand in Leonard jars formed nodules within three weeks after inoculation. With increasing amount of inoculum per pot, an increase was found in the percentage of plants nodulated and the amount of nodules per plant (Table 1). The yields of the plants was positively correlated with the amount of nodules formed (Fig. 3b). Nodulated plants turned green, grew rapidly on nitrogen-free medium

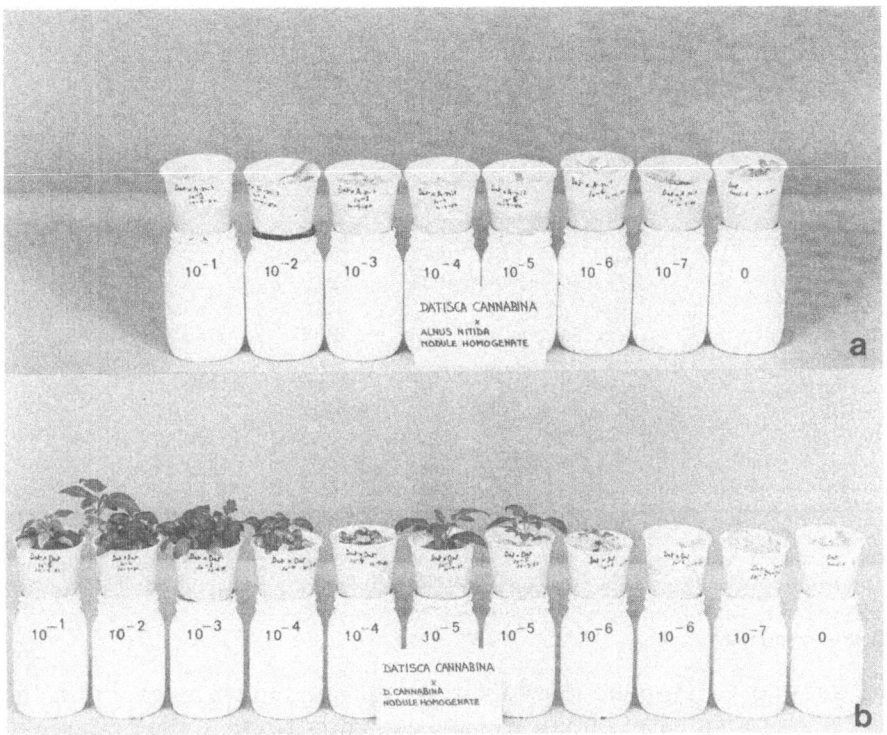

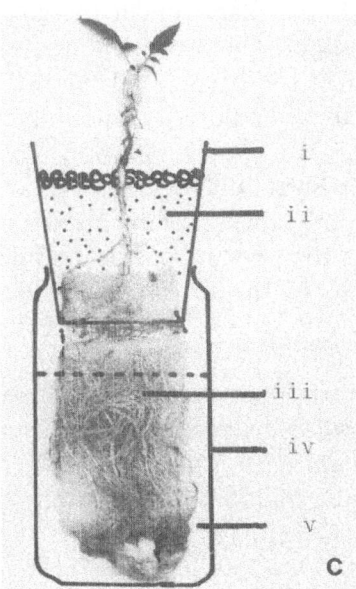

Fig. 3. Inoculation of *D. cannabina* seedlings with different concentrations of crushed nodules of *A. nitida* (a) and *D. cannabina* (b). Values on the pots denote the amount of inoculum (g fresh weight nodules. pot^{-1}).

Cross section of Leonard jar (c) with plastic coffee beaker (i), sand (ii), foam plastic strip (iii), glass bottle 330 ml (iv) and Hoagland solution (v).

Table 2. Nodulation of field-grown *D. cannabina* and *A. nitida*

Species	Number of plants ***	Number of nodules plant^{-1} *	g dry wet. plant^{-1}			
			nodules	roots	total	% nodule wt.
D. cannabina	18	7 (2−28)	0.044	0.288	0.518	8.5
D. cannabina	17	11 (3−16)	0.163	0.401	1.930	8.4
D. cannabina	1	153	0.783	5.283	12.066	6.5
D. cannabina	5	208 (140−260)	0.839	10.9	21.3	3.9
D. cannabina	4	78 (60−96)	0.450	ND	53.2	0.8
D. cannabina	3	63 (50−77)	1.366	19.6	60.366	2.2
D. cannabina	2	49 (40−57)	1.877	76.0	148.0	1.3
D. cannabina	3	194 (140−256)	1.610	432.0	600.0**	0.3**
A. nitida	10	11 (0−60)	0.086	ND	1.517	5.7
A. nitida	4	18 (5−32)	0.60	3.2	10.50	5.4
A. nitida	3	36 (21−66)	0.90	4.2	13.58	6.6
A. nitida	4	38 (19−57)	0.95	3.5	14.25	6.3

  * Extreme values are within brackets.
 ** Plants were partly cut at the time of harvest. Values presented in this table are minimum
    values.
*** Number of plants from the same locality and comparable size.
ND: Not determined.

and showed C_2H_2-reduction activity (data not shown here). From the results presented in Table 1 it can be computed that the amount of nodule mass needed to produce one nodule per plant, under the given experimental conditions, is 10^{-6} g fresh weight. One g fresh weight of nodules contained at least 1.4×10^6 infective particles. Under the experimental conditions *D. cannabina* seedlings could not be nodulated with crushed nodules of *A. nitida* (Fig. 3a).

Nodulation of field-grown plants from Swat of *D. cannabina* and *A. nitida* of different sizes is presented in Table 2. With increasing size of the plants both the number and dry weight of the nodules increased. The nodule dry weight, relative to the total plant weight increased during the first year up to 8.5% and subsequently decreased till 0.8% after *ca* 3 years when the plants had reached a size of 150 cm. Similar values for nodulation of older actinorhizas were also obtained for *Elaeagnus umbellata*, grown in somewhat drier sites in the same area. Three years old plants of *E. umbellata* with a height of 200 cm and an average total plant dry weight of 380 g had formed *ca* 60 nodules per plant which represented 0.7% of the total plant dry weight (unpublished results).

The nitrogen content of mature *D. cannabina* calculated on dry weight basis was somewhat lower than that of other plants, *e.g. Cannabis sativa*, grown in the same area (Table 3). This may be mainly due to the presence of a large proportion of shoots and roots with a

Table 3. Dry weight and nitrogen content of field-grown *Datisca cannabina* and *Cannabis sativa*

Plant part	D. cannabina		C. sativa	
	g dry wt. plant^{-1}	mg N. g dry wt^{-1}	g dry wt. plant^{-1}	mg N. g dry wt^{-1}
Leaves	2.10	17.0	1.23	25.0
Shoots	8.90	2.8	3.12	5.5
Roots	11.82	2.0	2.32	5.0
Nodules	0.95	7.5	–	–
Total	23.77	3.8	6.67	8.9

Plants grown in Kaghan, Pakistan and harvested in August 1980. The age of the *D. cannabina* plants was *ca* 2 years. Each value is the average of 3 replicates.

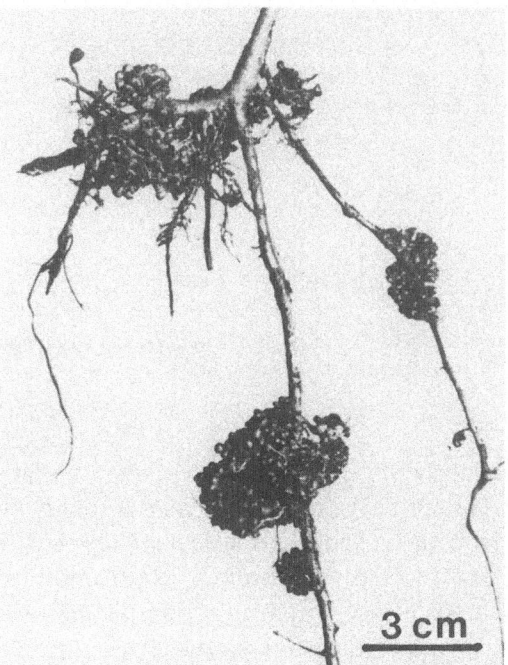

Fig. 4. Root nodules of *A. nitida*.

relatively low N content. The nodules had a significantly higher N content. *D. cannabina* produced a relatively large proportion of dry matter in the root system, *i.e.* about 50% of the total dry weight (Tables 2 and 3). This has an important function during the regeneration of the plants in springtime.

Growth and nodulation of A. nitida

A. nitida is a fast-growing tree along streams in the mountains of northern Pakistan. The trees were regularly nodulated (Fig. 4)

Fig. 5. Inoculation of *A. nitida* seedlings with crushed nodules of *A. nitida* (a) and *D. cannabina* (b) at different concentrations. Values on pots denote g fresh weight inoculum. pot^{-1}.

and the structure of the nodules and its endophyte was similar to other *Alnus* species (Hafeez, Chaudhary and Akkermans, to be published). Under light and electron microscope no spores were detected, indicating the spore-negative nature of the endophyte. Root nodules of *A. nitida* appear to have low amounts of oxidizable polyphenols since the tissue remained unstained for long time after sectioning.

Seedlings of *A. nitida* inoculated with crushed nodules of *A. nitida* (Fig. 5; Table 1) were rapidly nodulated. The amount of nodules formed was dependent on the concentration of the inoculum. Very high concentrations of inoculum usually resulted in a decrease of plant yield, due to introduction of plant pathogenic microorganisms that are present in low quantities in the inoculum. From the results presented in Table 1 it can be computed that crushed nodules of *A. nitida* were highly infective. Only 10^{-6} g fresh weight nodules was needed to produce one nodule per plant, and one g fresh weight nodules contained at least 0.7×10^6 infective particles.

A. nitida seedlings could also be nodulated by crushed nodules of

Table 4. Free amino acids in root nodules of *D. cannabina* (D) and *A. nitida* (A)

Compound	μmol g⁻¹ *		μmol %		% N**	
	D	A	D	A	D	A
Glutamic acid	2.299	0.948	34.4	16.6	20.4	8.1
Arginine	1.262	0.617	18.9	10.8	44.7	21.1
Citrulline	0	1.782	0	31.1	0	45.7
Aspartic acid	0.848	0.193	12.7	3.4	7.5	1.7
Glutamine	0.689	0.174	10.3	3.1	12.2	3.0
Valine	0.413	0.246	6.2	4.3	3.7	2.1
α-Amino butyric acid	0.287	0.075	4.3	1.3	2.5	0.6
Glycine	0.285	0.380	4.3	6.6	2.5	3.3
Alanine	0.212	0.360	3.2	6.3	1.9	3.1
Serine	0.175	0.375	2.6	6.6	1.6	3.2
Threonine	0.074	0.091	1.1	1.6	0.6	0.8
Lysine	0.046	0.098	0.7	1.7	0.8	1.4
Ornithine	0.043	0.174	0.6	3.0	0.6	3.0
Histidine	0.029	0.059	0.4	1.0	0.8	1.7
Phenylalanine	0.021	0.043	0.3	0.8	0.2	0.4
Leucine	0	0.048	0	0.8	0	0.4
Isoleucine	0	0.041	0	0.7	0	0.3
Tyrosine	0	0.019	0	0.3	0	0.1
Total	6.683	5.723	100.0	100.0	100.0	100.0

* per g fresh nodule wt.
** percentage of the total N in amino acids.

D. cannabina (Fig. 5). The infectivity of this inoculum was, however, lower than that of crushed *A. nitida* nodules (Table 1).

Free-amino acids in root nodules
Significant differences were found in the composition of soluble amino acids in root nodules of *D. cannabina* and *A. nitida* (Table 4). In the latter species citrulline was the predominant amino acid, while arginine and glutamic acid were also present in relatively large quantities. In *D. cannabina* nodules the predominant amino acid was arginine followed by glutamic acid, aspartic acid and glutamine. No citrulline was found in *D. cannabina*.

DAP content of root nodules
Root nodules of *D. cannabina* contained rather large quantities of DAP, *i.e.* 5.8 μmol DAP. g⁻¹ nodule dry weight (or 1.4 μmol. g⁻¹ nodule fresh weight). All DAP was particle bound. Roots contained very low quantities of DAP, *i.e.* 18 nmol. g⁻¹ root dry weight, probably associated with contaminants at the root surface.

Table 5. Nitrogenase (C_2H_2 reduction) activity of field-grown *D. cannabina, A. nitida, E. umbellata* and *H. rhamnoides* in Swat

Species	Plant height (cm)	Nodule diameter (cm)	μmol C_2H_4 g nodule dry wt^{-1}.h^{-1}
D. cannabina	10−15	0.1−0.2	38.0*
	25	0.2−0.8	15.2
A. nitida	30	0.2−0.5	36.0
	200	2.0−4.0	6.0
E. umbellata	150	0.2−0.4	23.2
	150	1.0−2.0	6.2
H. rhamnoides	100	0.1−0.5	14.4
	100	1.0−2.0	4.2

* Average of 15 plants. Intact plants were incubated. In other experiments detached nodules derived from one plant were incubated.

Table 6. Uptake of H_2 and O_2 by detached nodules of *D. cannabina* and *A. nitida*

Species	Date	H_2 uptake μmol g^{-1}.h^{-1}	O_2 uptake μmol g^{-1}.h^{-1}
D. cannabina	6 June 1981	0.81	17.00
	9 November 1981	0.94	16.27
A. nitida	9 November 1981	1.29	14.08

Each value denotes the average of four samples of nodules. Nodules (1.5 g fresh wt.) were incubated for one hour in 13.5 ml vials, containing air with 1% H_2. Uptake of H_2 and O_2 are expressed per g nodule fresh wt.

Nitrogenase activity of field-grown nodules

Field-grown *D. cannabina* nodules showed high acetylene reduction activity. The activity decreased with the age of the nodules, as has also been observed with other actinorhizal plants from the same area (Table 5).

Hydrogen and oxygen uptake by root nodules

Uptake of H_2 and O_2 by nodules from field-grown plants of *D. cannabina* and *A. nitida* was measured at different times of the year. Both types of nodules showed significant uptake of H_2 (Table 6). Nodules from *D. cannabina* plants which were collected in summer in Swat and transported to Islamabad usually lost most of the hydrogenase activity when the plants were cultivated at ambient temperatures (maximum day temperature 35°C). Under these conditions nodules evolved H_2 produced by nitrogenase.

Homogenates of *D. cannabina* nodules utilized H_2 when incubated anaerobically in the presence of MB, although the *in vitro* hydrogenase activity was lower than of intact nodules (Fig. 6). About one third of the *in vitro* activity was found in the 20 μm residue, *i.e.*, the fraction containing intact vesicle clusters (Fig. 7). Hydrogenase activity was

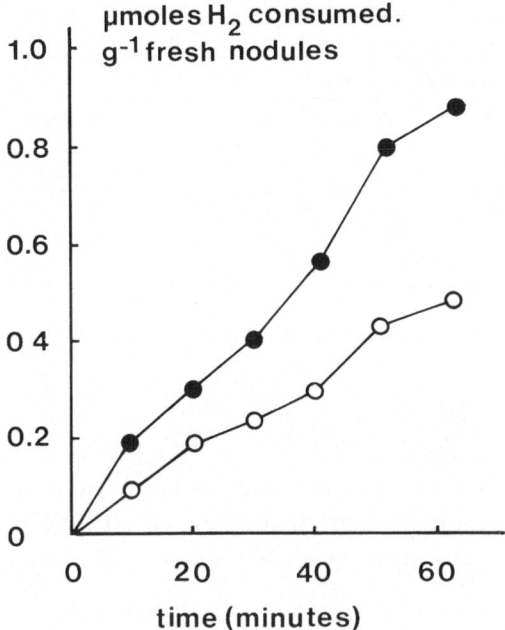

Fig. 6. H$_2$ uptake of root nodules of field-grown *D. cannabina* (●) and nodule homogenates with MB as electron acceptor (○).

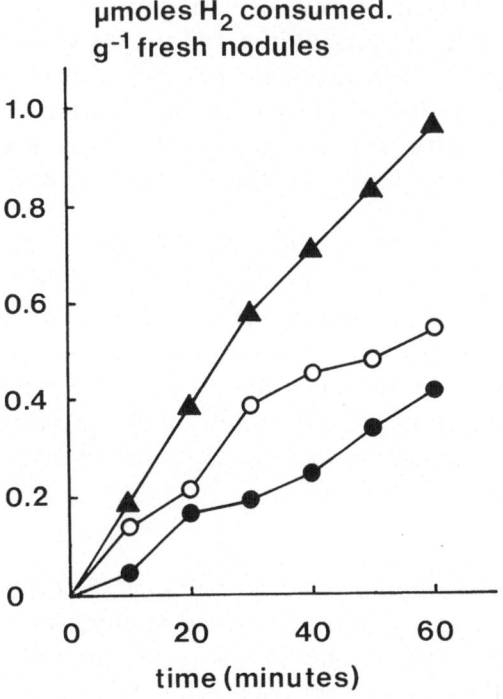

Fig. 7. H$_2$ uptake by 20 μm residue (●) and 20 μm filtrate (○) of a nodule homogenate (▲) of *D cannabina* with MB as electron acceptor. Summer harvested nodules.

Table 7. Oxygen uptake by $20\,\mu$m residue of root nodule homogenates of *D. cannabina* and *A. nitida* with different substrates added

Species	Date	nanomoles O_2 consumed. g^{-1} . min^{-1} *			
		NADH	Succinate	Succinate + ADP	Glutamate + NAD + malate
D. cannabina	May 1981	50.0	11.0	20.0	ND
	Sept. 1982**	21.2	18.8	25.5	2.8
A. nitida	Sept. 1982***	14.1	1.4	1.4	3.3

  * Activity expressed per g fresh nodule weight.
    Final concentrations of NADH ($50\,mM$), succinate ($10\,mM$), malate ($5\,mM$), glutamate ($5\,mM$), NAD ($1\,mM$), ADP ($1\,mM$).
    ND: not determined
 ** Nitrogenase activity of intact nodules was $2.0\,\mu$mol C_2H_4 .g fresh wt^{-1} .h^{-1} .
*** Nitrogenase activity of intact nodules was $4.6\,\mu$mol C_2H_4 .g fresh wt^{-1} .h^{-1} .

particle bound and no activity was found in the soluble nodule fraction. Activities somewhat lower than those presented in Figs. 6 and 7 were obtained when PMS or DCPIP were used as electron acceptors (data not shown here).

Tetrazolium reduction and O_2 uptake by root nodule homogenates

When the $20\,\mu$m residue of nodule homogenates of *D. cannabina* was incubated with TTC and NADH, significant formation of formazan crystals occurred inside the vesicle clusters of the endophyte within 30 min. Formazan was also formed when succinate was added instead of NADH, although the activity was lower. The formation of formazan confirmed that dehydrogenase activity was localized within the vesicle clusters. No reduction of TTC was found in the controls without added NADH or succinate.

The respiratory activity was also determined by measuring the uptake of O_2 by suspensions of vesicle clusters. As shown in Table 7, the $20\,\mu$m residue of both *D. cannabina* and *A. nitida* nodule homogenates showed high NADH-dependent O_2 uptake. The former one also showed significant O_2 uptake in the presence of succinate. This activity could be enhanced by addition of ADP. Addition of a mixture of malate, glutamate and NAD resulted only in a low O_2 consumption.

Discussion

Datisca species have little taxonomic relationships with other plant taxa and have therefore been classified into one family, Datiscaceae with one genus. The only two species of the genus are nodulated[13,14,36] although both of them have a restricted but widely separated distribution, *viz., D. glomerata* in California and *D. cannabina* in the Himalaya region of Pakistan. In the past several studies had been made on the morphology of these plants[1, 15, 16, 21, 24, 25, 27, 32, 33, 34] and in particular *D. cannabina* has received special attention of phytochemists because of the occurrence of the intense yellow dye datiscin and derivates[19]. In the Himalaya region this natural dye has traditionally been used. Since this dye has now mainly been replaced by synthetic dyes and since *Datisca* is not eaten by the cattle, it has only restricted value to the local population as an additional source of fuel. However, the abilities to colonize and stabilize steep hills and add nitrogen to the ecosystem are important enough to upgrade the usefulness of this plant.

D. cannabina plants (local names in Swat 'Karati' and in Kaghan 'Kalbir') were profusely nodulated. The number and dry weight of nodules increased rapidly with the age of the plants, while the relative weight of nodule tissue with respect to the dry weight of the plants rapidly decreased within three years from 8.5–0.8%. Since the nitrogenase activity, based on nodule weight, also decreased with age, the contribution of nitrogen fixation to the total N-supply of the plant decreased with age of the plants. This phenomenon has also been described for *A. glutinosa* and *Hippophaë rhamnoides*[2, 9] and may be a general feature of perennial nodulated plants.

The nodulation studies with seedlings of *D. cannabina* and *A. nitida* showed that crushed nodule suspensions are highly infective when added to their original hosts. Although exact determinations of the infective potential of the nodule homogenates still have to be made, it is evident that the values for *D. cannabina* and *A. nitida* approximate the values reported for *A. glutinosa* with spore-positive nodules[3, 4] and for *H. rhamnoides* with spore-negative nodules[29, 30]. Nodules of *D. cannabina* and *A. nitida* showed marked differences in the composition of the free amino acids. In *A. nitida* citrulline was the predominant amino acid, as has also been reported for other *Alnus* spp. (*e.g. A. inokumai*[35] and *A. glutinosa*[7, 8, 23, 26]). In *D. cannabina* no citrulline was detectable and the chief amino acids were arginine, followed by glutamic acid and glutamine. A similar pattern has also been observed in nodules of *Coriaria myrtifolia*[35]. Since *Datisca* and *Coriaria* spp. are the only examples in which the orientation of the endophyte inside the host cell is reversed, a further study to the relationships between both types of symbioses will be worthwhile.

The results presented in this paper indicate that both *D. cannabina* and *A. nitida* nodules have an active H_2 uptake system. Hydrogenase activity of both summer and winter harvested nodules of *D. cannabina* was almost constant, although in summer the results were often disturbed by the temperature shock during transportation of the plants from the mountains (20–25°C) to Islamabad (30–40°C day temperature). The rate of H_2 consumption in winter harvested nodules of *A. nitida* was identical to that of summer harvested *A. glutinosa* in The Netherlands[31], indicating the importance of a system that can recycle H_2 which is formed by nitrogenase activity. The hydrogenase is localized in the endophyte and can be used as a marker enzyme of *Frankia*. Since only one third of the *in vitro* activity was found in the 20 µm residue it can be concluded that a significant part of the endophyte tissue was disintegrated during homogenization as has also been observed in *A. glutinosa*[31].

Vesicle clusters from nodules of both species showed NADH-supported O_2 uptake. In *D. cannabina* the 20 µm residue also showed succinate enhanced O_2 consumption which could be stimulated by addition of ADP. This indicates that the membranes are still coupled. These effects have not yet been found in vesicle clusters of *A. glutinosa*, *H. rhamnoides*[5,6] and *A. nitida* (this paper) and in part can be explained by the aberrant ultrastructure of the endophyte of *Datisca* and the complex mixture of endophyte and plant mitochondria within the clusters[6,18].

Acknowledgements Thanks are due to M. Sajjad Mirza, Nazeer Ahmed (Islamabad) and Wim Roelofsen (Wageningen) for the technical assistance. This research was supported by United States Department of Agriculture under PL-480 programme grants No. FG-Pa-231 and 347, by UNESCO/United Nationals Development Programme (Project PAK/77/010) and by the Agricultural University of Wageningen (fellowship to the first author).

References

1   Abrams L 1952 An illustrated Flora of the Pacific States Washington, Oregon and California. 3, 143 Family 101, Stanford Univ. Press.

2   Akkermans A D L 1971 Nitrogen fixation and nodulation of *Alnus* and *Hippophaë* under natural conditions. Thesis University Leiden, The Netherlands.

3   Akkermans A D L and Dijk C van 1981 Non-leguminous root-nodule symbioses with actinomycetes and *Rhizobium*. *In* Nitrogen Fixation Vol 1, Ecology, Ed. W J Broughton, Oxford University Press, Oxford, pp 57–103.

4   Akkermans A D L and Houwers A 1979 Symbiotic nitrogen fixers available in temperate forestry. *In* Symbiotic Nitrogen Fixation in the Management of Temperate Forests. Eds. J C Gordon, C T Wheeler and D A Perry, Corvallis, Oregon USA, pp 23–35.

5   Akkermans A D L, Huss-Dannell K and Roelofsen W 1981 Enzymes of the citric acid cycle and the malate-aspartate shuttle in the N_2-fixing endophyte of *Alnus glutinosa*. Physiol. Plant. 53, 289–294.

6   Akkermans A D L, Roelofsen W, Blom J, Huss-Danell K and Harkink R 1983 Utilization of carbon and nitrogen compounds by *Frankia* in synthetic media and in root nodules of *Alnus glutinosa*, *Hippophaë rhamnoides* and *Datisca cannabina*. Can. J. Bot. 61, 2793–2800.

7   Aseeva K B, Evstigneeva Z G and Kretovich V L 1966 Dokl. Akad. Nauk. SSR. 169, 463.
8   Blom J, Roelofsen W and Akkermans A D L 1981 Assimilation of nitrogen in root nodules of alder (*Alnus glutinosa*). New Phytol. 89, 321–326.
9   Bond G 1958 Symbiotic nitrogen fixation by non-legumes. *In* Nutrition of the legumes. Ed. E G Hallsworth, Butterworth and Co., London, pp 216–231.
10   Calvert H E, Chaudhary A H and Lalonde M 1979 Structure of an unusual root nodule symbiosis in a non-leguminous herbaceous dicotyledon. *In* Symbiotic Nitrogen fixation in the Management of Temperate Forests. Eds. J C Gordon, C T Wheeler and D A Perry, Corvallis, Oregon, USA, pp 474–475.
11   Chaudhary A H 1975 Annual Second Research Report. PL-480 Research Project, Pk-FS-32 (Grant No. FG-Pa-231), Department of Biological Sciences, Quaid-i-Azam University, Islamabad, Pakistan, p 55.
12   Chaudhary A H 1978 The discovery of root nodules on new species of non-leguminous angiosperms from Pakistan and their significance. *In* Limitations and Potentials for Biological Nitrogen Fixation in the Tropics. Eds. J Dobereiner, R H Burris and A Hollaender, Plenum Press, New York and London, p 359.
13   Chaudhary A H 1979 Nitrogen-fixing root nodules in *Datisca cannabina* L. Plant and Soil 51, 163–165.
14   Chaudhary A H, Khokhar S N, Zafar Y and Hafeez F 1981 Actinomycetous root nodules in angiosperms of Pakistan. Plant and Soil 60, 341–348.
15   Davidson C 1973 An anatomical and morphological study of Datiscaceae. Aliso 8 (no 1), 49–110.
16   Davidson C 1976 Anatomy of xylem and phloem of the Datiscaceae. *In* Contributions in Science no 280, Natural History Museum of Los Angeles County.
17   Dijk C van and Merkus E 1976 A microscopic study of the development of a spore-like stage in the life cycle of the root-nodule endophyte of *Alnus glutinosa* (L.) Gaertn. New Phytol. 77, 73–91.
18   Hafeez F, Akkermans A D L and Chaudhary A H 1984 Observations on the ultrastructure of *Frankia* sp. in root nodules of *Datisca cannabina* L. Plant and Soil *In press*.
19   Hafeez F, Akkermans A D L and Chaudhary A H 1982 Some studies on the respiratory metabolism of the endophyte from the root nodules of *Datisca cannabina* and *Alnus nitida*. Pak. J. Bot. 14, 35.
20   Hegnauer R 1966 Chemotaxonomie der Pflanzen. 4, 11–12. Birkhauser Verlag Basel.
21   Hooker J D 1897 The Flora of British India. 2, 656–657. L. Reeve and Co Covent Garden London.
22   Humphries E C 1956 Determination of total nitrogen. *In* Modern Methods of Plant Analysis. Eds. K Peach and M V Tracey, Springer Verlag Berlin. Göttingen-Heidelberg, 1, pp 479–481.
23   Leaf G, Gardner I C and Bond G 1958 Observations on the composition and metabolism of the nitrogen-fixing root nodules of *Alnus*. J. Exp. Bot. 9, 320–331.
24   Mason H L 1957 A Flora of the Marshes of California. Univ. Calif. Press Berkeley and Los Angeles, p 593.
25   Metcalfe C R and Chalk L 1957 Anatomy of Dicotyledons 1, 695–699. Oxford at the Clarendon Press.
26   Miettienen J K and Virtanen A I 1952 The free amino acids in the leaves, roots and root nodules of the alder (*Alnus*). Physiol. Plant. 5, 540–557.
27   Munz P A and Keck D D 1959 A California Flora. Univ. Calif. Press, p 177.
28   Noel A R A 1964 A staining and mounting combination of plant tissues. Stain Technol. 39, 324.
29   Oremus P A I 1980 Occurrence and infective potential of the endophyte of *Hippophaë rhamnoides* L. ssp. *rhamnoides* in coastal sand-dune areas. Plant and Soil 56, 123–139.
30   Oremus P A I, Akkermans A D L and Nijholt L 1981 The effect of disintegration on the infective potential of the root-nodule endophyte of *Hippophaë rhamnoides* L. ssp. *rhamnoides*. Verhandelingen der Koninklijke Akademie van Wetenschappen, afd. Natuurkunde, 77, 45–50.

31 Roelofsen W and Akkermans A D L 1979 Uptake and evolution of H_2 and reduction
    of C_2H_2 by root nodules and nodule homogenates of *Alnus glutinosa*. Plant and Soil 52,
    571–578.

32 Siddiqi M A 1973 Datiscaceae. *In* Flora of West Pakistan Eds. E Nasir and S I Ali, 37,
    pp 1–3.

33 Severinį G 1922 Sui tubercoli radicali di *Datisca cannabina*. Annali di Bot., Roma 15,
    29–52.

34 Thomas 1961 Flora of the Santa Cruz Mountains of California. Stanford Univ. Press,
    p 243.

35 Wheeler C T and Bond G 1969 The amino acids of non-legume root nodules. Phyto-
    chemistry 9, 705–709.

36 Winship L J and Chaudhary A H 1979 Nitrogen fixation by *Datisca glomerata:* a new
    addition to the list of actinorhizal diazotrophic plants. *In* Symbiotic Nitrogen Fixation in
    the Management of Temperate Forests. Eds. J C Gordon, C T Wheeler and D A Perry,
    Corvallis, Oregon, USA, p 485.

Plant and Soil 78, 147–158 (1984).

Ms. Fr 16

Growth, nitrogen fixation and relative efficiency of nitrogenase in *Alnus incana* grown in different cultivation systems

ANITA SELLSTEDT and KERSTIN HUSS-DANELL
Department of Plant Physiology, University of Umeå, S-901 87 Umeå, Sweden

Key words *Alnus incana* C_2H_2-reduction Cultivation systems *Frankia* H_2-evolution in air Hydrogenase Nitrogen content Relative efficiency

Summary Three cultivation systems were compared. In one system the alders were grown hydroponically. In the two other systems the alders were planted in gravel and either given water and nutrients at intervals or the nutrient solution was continuously supplied. Alders continuously supplied with nutrients and water showed a significantly more rapid growth, higher biomass production and higher nitrogen content than did alders given nutrients and water at intervals or alders hydroponically grown. Alders continuously supplied with water and nutrients had a constant RE (relative efficiency of nitrogenase) of about 0.80 throughout the experimental period while alders supplied with water and nutrients at intervals showed a slight decrease in RE at the end of the experimental period. No strict relationship was found between RE and nitrogen content or between RE and plant productivity.

Introduction

For studies on physiological aspects of nitrogen fixation in higher plants, *e.g.* energy relationships and photosynthetic supply, it is important to have a cultivation system where neither water nor nutrients are limiting or fluctuating. Such cultivation systems are already described[2,10,13,17]. But, none of these systems was designed to permit repeated measurements of nitrogen fixation in intact plants, which is greatly needed in many kinds of physiological studies on nitrogen fixation. Growing the plants in a hydroponic system as done by Imsande and Ralston[17], fulfilled all the requirements stated, but had the disadvantage of disturbing the plants when the nitrogenase activity was measured.

Nitrogen fixation is always accompanied with a reduction of H^+ to H_2. The uptake of H_2 by a hydrogenase occurs frequently occurring in nitrogen-fixing organisms, but is lacking in some strains of *Rhizobium*[9].

The two hydrogen reactions occur also in actinomycete nodulated plants, but the H_2-evolution in air is reported to be lower in these symbioses than in legumes[22,23,25]. Recent work on leguminous plants revealed that plant ontogeny[4,11] altered the relative efficiency of nitrogenase (RE; electrons used for nitrogen reduction as part of total electron flux available for nitrogenase activity). Data on *Alnus glutinosa* indicate that there are fluctuations in H_2-evolution in air during the year[22].

The aim of this work on the *Alnus incana — Frankia* symbiosis was (1) to find a cultivation system suitable for studies on energy demand of nitrogen fixation and (2) to study how growth conditions and plant development affected the relative efficiency of nitrogenase. Three cultivation systems were evaluated for growth, nitrogen fixation and nitrogen content of the alders. Two of these cultivation systems were also evaluated for relative efficiency of nitrogenase.

Materials and methods

Plant material and growth chamber conditions

Green cuttings of one clone of *Alnus incana* (L.) Moench were rooted for 21 days in an aerated, diluted nutrient solution[14] complemented with $0.358 \, mM \, NH_4NO_3$.[15] The rooting solution was not renewed during the rooting period. The rooted cuttings were inoculated with a water suspension of crushed nodules from alders of the same clone. The inoculum consisted of 0.1 g (fresh weight nodules) per ml distilled water, and 1 ml was given to each cutting immediately after transferring them to the different cultivation systems. During rooting as well as during growth the alders were in a controlled environment growth chamber with 17 h light, a thermoperiod of 17/7 h of 25/15°C and a relative air humidity of 75%. The light source was Osram HQI 400 W-70 halogen lamps giving a photon flux density of 200 μmol m^{-2} s^{-1} (Lambda Quantummeter LI-185 A).

Cultivation systems

In all cultivation systems the above-mentioned nutrient solution (pH *ca.* 7) was used. It was complemented with $0.358 \, mM \, NH_4NO_3$ until root nodules were visible. Three cultivation systems were compared. In one system the plants were grown hydroponically in black 1-litre polystyrene pots with one plant per pot and the nutrient solution diluted to 1/10 of full strength. The nutrient solution was renewed three times a week when also the pH was measured. In the two other systems the rooted cuttings were planted singly in pots with gravel. The bottom of the pots was covered with a fine mesh net to prevent the roots from growing through the drainage holes. One group of plants, hereafter called the traditional system, received water twice a day and full strength nutrient solution once a day[14]. The other group of plants was continuously supplied with aerated, circulating, diluted nutrient solution. For this purpose an air-lift technique was used. The potted plants were kept above a tank with nutrient solution. For every pot there were two glass tubes, which had the lower end down in the nutrient solution and the upper end emerging into the pot. A smaller tube was inserted into the lower end of the glass tube. A stream of air was pumped (RENA 301 aquarium pump) through the smaller tube, and the air pressed the nutrient solution up through the glass tube and into the pot. A flow of 0.6 l h^{-1} was delivered to each pot and the drainage was collected in the tank, which contained 10 l of nutrient solution. Five plants were kept over each tank. The nutrient solution was renewed three times a week, when also the pH of the solution was measured. Two strengths of the nutrient solution were tested, viz. 1/4 and 1/10 of full strength.

The system with circulating nutrient solution was later on modified. In the modified continuous flow system the pots were kept over a black catchment tank. The nutrient solution was aerated with the aquarium pump and lifted (Iwasaki magnet pump) to the reservoir, which was placed higher than the plants. The solution was pumped to the upper reservoir faster than it could flow out through the distribution system and thereby a constant level of nutrient solution was kept in the upper reservoir. The outlet of each distribution tube was fitted with a removable syringe needle[13]. The needle served as a restricted outlet and gave back pressure so that all outlets delivered equal amounts of nutrient solution (0.6 l h^{-1} to each pot)[12]. The solution was complemented with only $0.072 \, mM \, NH_4NO_3$ until root nodules were visible.

Measurements

C_2H_2 *reduction* Nitrogenase activity was measured on intact plants[14] as C_2H_2-dependent C_2H_4-production. Ten % (v/v) of the air in the incubation chamber was withdrawn and replaced with C_2H_2. During the incubation period (1.5 h) gas samples of 0.5 ml were taken at intervals and immediately determined by gas chromatography[14]. The plants were kept in the growth chamber during the C_2H_2-incubations which were always made the same time of the day. The measurements were made on each plant once a week, except for the hydroponically grown plants, whose C_2H_2-reduction was measured only at the end of the experimental period.

Relative efficiency of nitrogenase Measurements were made repeatedly and always at the same time of the day on intact plants during the experimental period. The plants were kept in the climate chamber during the incubations. H_2-evolution in air was measured in gas samples (0.2 ml) taken at intervals during the incubation period (1.5 h). H_2 was determined in a Varian 3700 gas chromatograph with a thermal conductivity detector and a 2 m stainless steel column (inner diameter 3.2 mm) containing Molecular Sieve 5 A (80–100 mesh). The carrier gas was N_2 at a flow rate of 0.33 ml s^{-1}, the column temperature was 80°C, the detector temperature 130°C and the filament temperature 250°C. The amount of H_2 was calculated by comparison with a standard mixture of H_2 in N_2 (AGA Specialgas, Lidingö, Sweden). H_2-evolution was measured on plants grown in the traditional and in the modified continuously circulating system. The H_2-evolution had a constant rate for at least 1 h. After measurements of H_2-evolution in air C_2H_2-dependent C_2H_4-production was measured with 10% (v/v) C_2H_2 in air as described above. Relative efficiency of nitrogenase (RE) was calculated as $1 - \left(\dfrac{H_2\text{-evolution in air}}{C_2H_2\text{-reduction}} \right)$ according to Schubert and Evans[23].

H_2 *uptake* H_2 uptake was measured on intact plants as well as in nodule homogenates[22]. Intact plants were incubated in 10% (v/v) C_2H_2[5] and either 0.1, 0.7[22], 1.7, 2.0[8] or 2.4% (v/v) H_2 in air. For nodule homogenates methylene blue as well as phenazine methosulphate were tried as electron acceptors[22] and the H_2-concentration was 1.3% (v/v). Gas chromatography was as described above but Ar served as carrier gas and the filament temperature was 280°C.

Growth and biomass production In general only one shoot developed on each plant and growth was therefore estimated as shoot length. Leaf areas were measured at harvest on detached leaves in a leaf area meter (Lambda LI-COR 3000) with a transparent belt conveyor accessory. The plants were harvested 45–47 days after planting and biomass was measured as dry weight (70°C, 24 h) separately for leaves, stem, roots and nodules.

Nitrogen content The dried plant organs were ground in a ball mill (Retsch Schwingmühle MM) and redried (70°C, 1 h). Samples of 150 mg were analysed for content of Kjeldahl-nitrogen using Cu and Se as catalysts in the digestion and colorimetric determination[6,19] of the ammonia.

Statistical treatments
Significant differences were evaluated by using the Mann-Whitney U-test according to Siegel[26], with $P < 0.05$ as significance level.

Results

Nutrient solution in the continuously circulating system

Two strengths of the nutrient solution were tested, *viz.* 1/4 and 1/10 of full strength. Of these concentrations the less diluted nutrient solution resulted in significantly higher shoot lengths and leaf areas (Table 1). The biomass production (Table 1) was also greater in alders

Table 1. Growth, nitrogenase activity and nitrogen content of *A. incana* grown in a continuously circulating system with two different strengths of nutrient solutions. The values are from the end of the growth period (45 days after planting). $\bar{x} \pm SE$; n = 4

Plant characteristic	Strength of nutrient solution	
	diluted to 1/4	diluted to 1/10
Shoot length (cm)	54.6 ± 3.1	41.6 ± 2.3
Leaf area (cm^2)	662.6 ± 65.4	391.8 ± 49.9
Dry weight (g): root system	0.951 ± 0.120	0.651 ± 0.057
shoot	4.300 ± 0.668	2.558 ± 0.331
whole plant	5.251 ± 0.783	3.209 ± 0.381
Nitrogenase activity (μmol C$_2$H$_4$ plant^{-1} h^{-1})	64.72 ± 4.73	41.96 ± 7.72
Nitrogenase activity (μmol C$_2$H$_4$ g (dry wt nodule)$^{-1}$ h^{-1})	311.9 ± 21.9	317.6 ± 27.7
Nitrogen content of total plant (mg)	125.8 ± 17.1	74.1 ± 10.1
Nitrogen content of total plant (% of dry weight)	2.41 ± 0.08	2.30 ± 0.10

supplied with 1/4 of the full strength nutrient solution, though the difference was not statistically significant. The nitrogenase activity measured as C$_2$H$_4$-production per plant and hour showed a 54% greater value for plants in the less diluted solution but when the nitrogenase activity was related to the dry weight of the nodules there was no difference (Table 1). The nitrogen content was significantly higher in leaves, stem and roots as well as in the whole plant in the nutrient solution diluted to 1/4. However, there was no significant difference when the nitrogen content was related to the dry weight (Table 1). The nutrient solution diluted to 1/4 of full strength was used hereafter.

Comparison of cultivation systems

Alders grown in the continuously circulating systems were visibly nodulated after 14 days, *i.e.* about five days earlier than alders in the other cultivation systems. Alders grown in the hydroponical and in the continuously circulating systems showed no lag phase in growth, as traditionally grown alders did (Fig. 1). The hydroponically grown alders developed deficiency symptoms. The leaves became all yellow except for the outer edges which became brownish. At the end of the growth period the plants recovered somewhat and developed new green leaves. The root nodules had plenty of callus growth through the lenticels.

The alders in the other cultivation systems held green leaves during the whole experimental period. Alders continuously supplied with nutrient solution grew much faster than hydroponically or traditionally grown alders. At the end of the experimental period the alders in the continuously circulating system had developed stem branches and had

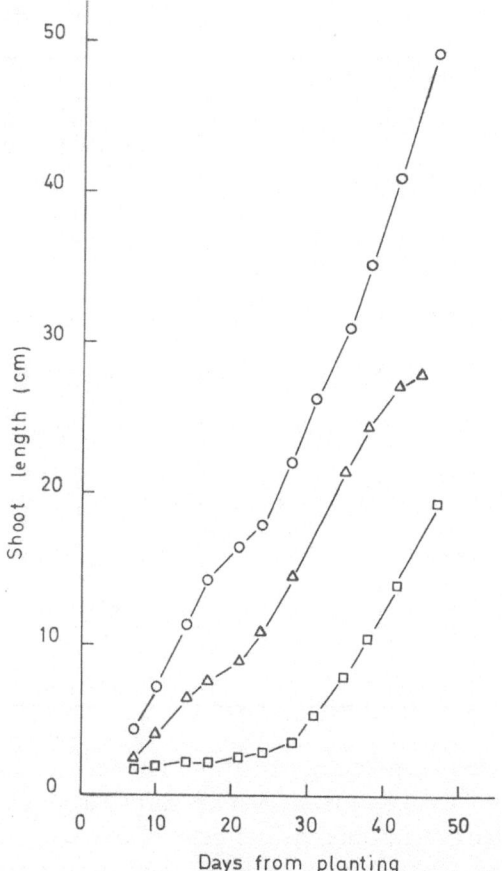

Fig. 1. Development of shoot length in *A. incana* grown in three different cultivation systems. x̄, n = 5, SE was 5 to 26% of x̄, with the highest SE values on day 7.
∘ continuously circulating system
□ traditional system
△ hydroponic system

shoot lengths (main shoot) of 49.2 cm, *i.e.* about twice as long shoots as the other alders (Fig. 1). Leaf areas showed values of 796.4 cm² for alders in the continuously circulating system compared to only 209.5 cm² and 152.6 cm² for alders in the hydroponics and the traditional growth system, respectively (Table 2).

Also, the biomass production was significantly higher in all plant organs in alders grown in the continuously circulating system (Table 2). Although all rooted cuttings were inoculated with the same amount of *Frankia* (root nodules) the dry weight of the nodules was about five times higher in the continuously circulating system than in the traditional system (Table 2). The plants grown in the continuously

Table 2. Dry weight, leaf area, nitrogenase activity and nitrogen content of *A. incana* grown in three different cultivation systems. The values are from the end of the growth period (45–47 days after planting). $\bar{x} \pm SE$; $n = 5$

Plant characteristic	Cultivation system		
	Hydroponics	Traditional	Continuously circulating
Dry weight (g): shoot	1.473 ± 0.196	1.082 ± 0.279	5.351 ± 0.928
root	0.425 ± 0.040	0.283 ± 0.040	0.801 ± 0.109
nodules	0.047 ± 0.007	0.056 ± 0.020	0.313 ± 0.030
plant	1.946 ± 0.234	1.422 ± 0.338	0.685 ± 0.120
Root/shoot ratio	0.34 ± 0.03	0.35 ± 0.03	0.22 ± 0.02
Leaf area (cm²)	209.5 ± 27.0	152.6 ± 41.7	796.4 ± 110.1
Nitrogenase activity			
(μmol C_2H_4 plant^{-1} h^{-1})	0.20 ± 0.06	16.92 ± 7.14	75.35 ± 10.18
Nitrogenase activity			
(μmol C_2H_4 g(dry wt nodule)$^{-1}$ h^{-1}	6.0 ± 2.6	281.4 ± 19.5	254.6 ± 12.0
Nitrogen content (mg):			
Total plant	27.00 ± 2.68	26.21 ± 8.98	131.4 ± 22.6
Leaves	11.41 ± 2.06	14.48 ± 5.57	82.18 ± 15.94
Roots	7.36 ± 0.51	4.47 ± 0.71	13.19 ± 1.57
Nodules	1.63 ± 0.33	2.09 ± 0.84	11.38 ± 1.65
Nitrogen content (% of dry weight):			
Total plant	1.44 ± 0.16	1.67 ± 0.18	1.95 ± 0.16
Leaves	1.33 ± 0.28	1.84 ± 0.27	3.29 ± 0.53
Roots	1.75 ± 0.07	1.57 ± 0.04	1.66 ± 0.03
Nodules	3.34 ± 0.21	3.51 ± 0.15	3.87 ± 0.07

circulating system developed a lower root/shoot ratio than the plants grown in the traditional and the hydroponical system did (Table 2). The nodule percentage (dry weight of nodules as per cent of dry weight of the total plant) of the alders continuously supplied with nutrients was 4.5 ± 0.4 compared to 3.5 ± 0.6 for the traditionally grown and 2.4 ± 0.2 for the hydroponics (n = 5 in all cases). The values were significantly higher for alders continuously supplied with nutrients compared to the hydroponics, otherwise the differences were not significant.

The nitrogenase activity measured per plant and hour was significantly higher in alders continuously supplied with nutrients than in those grown hydroponically or traditionally (Fig. 2). When nitrogenase activity was expressed on a nodule dry weight basis, plants from the continuously circulating and the traditional system showed similar values, indicating a close relation between nodule dry weight and nitrogenase activity (Table 2). The hydroponically grown alders differed much with their very low nitrogenase activity, only 6.0 μmol C_2H_4 g^{-1} (dry wt nodule) $\cdot$ h^{-1}. They were apparently disturbed by the movement from the growth conditions to the measurement conditions in air.

The amount of nitrogen in plants from the continuously circulating

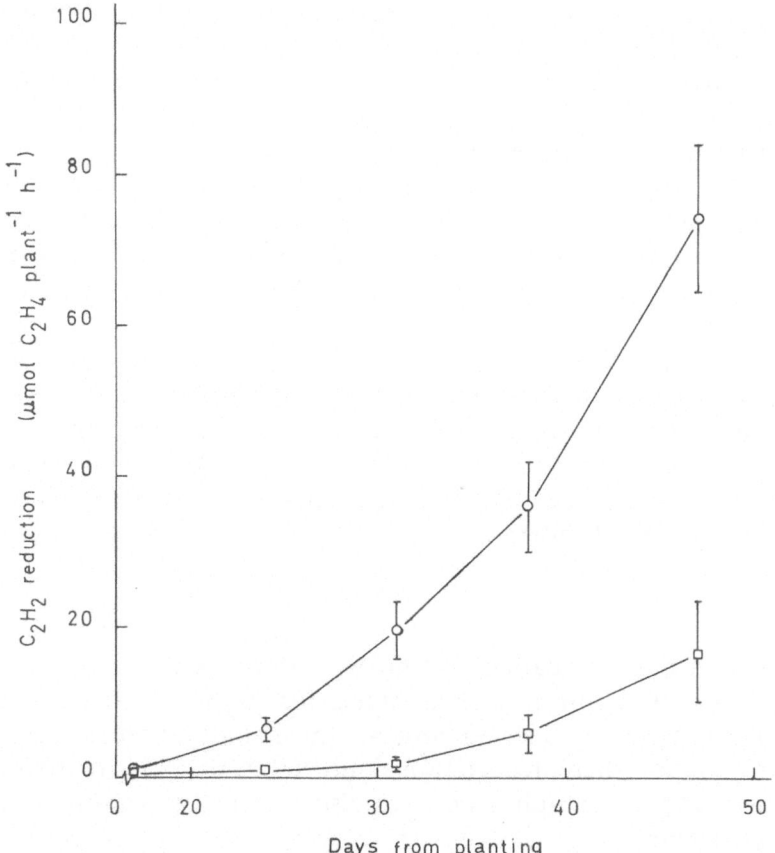

Fig. 2. Development of nitrogenase activity (C_2H_2-reduction) in *A. incana* grown in two different cultivation systems. $\bar{x} \pm SE; n = 5$.
○ continuously circulating system
□ traditional system

system exceeded that of plants from the two other cultivation systems in all organs (Table 2). For example, the nitrogen content in the leaves was 82.18 mg compared to only 14.48 and 11.41 mg for the continuously circulating system, the traditional and the hydroponics system, respectively (Table 2). There was significantly more nitrogen per dry weight in the nodules of the alders continuously supplied with nutrients than in the nodules of the traditionally and hydroponically grown alders (Table 2). There was also more nitrogen per dry weight in the leaves of the continuously supplied alders (Table 2), although the difference was not statistically significant.

The modified continuously circulating system was more easily operated than the continuously circulating system. Alders grown in the modified system still showed significantly higher values of biomass,

shoot length, nitrogen content and nitrogenase activity than traditionally grown alders, but not as high values as alders in the continuously
circulating system (Table 3).

Relative efficiency of nitrogenase

When C_2H_2-reduction increased also H_2-evolution in air increased
(Fig. 3B). In alders grown in the modified continuously circulating
system RE was fairly constant and ranged only from 0.75 to 0.83
(Fig. 3A). In the traditionally grown alders RE decreased slightly from
0.88 to 0.67 (Fig. 3A).

Neither in intact plants nor in nodule homogenates could any biological H_2-uptake (hydrogenase activity) be demonstrated. The decrease
in H_2 in the gas phase was of the same rate in incubation tubes with
only buffer and electron acceptor as in tubes with nodule homogenate
and electron acceptor. The nodule homogenates were capable to reduce
C_2H_2 *in vitro* (data not shown).

Discussion

The continuously circulating systems were developed in an attempt
to obtain unstressed nitrogen-fixing alders for studies on the energy
demand of the nitrogen fixation process. Such studies also requires a
cultivation system where repeated measurements of C_2H_2-reduction
can be made and this requirement was also met in the continuously
circulating systems.

Alders continuously supplied with nutrients and water were visibly
nodulated 5 days earlier than alders in the two other cultivation systems. This earlier nodule formation means an earlier start of nitrogen
fixation and can also explain the absent lag phase in shoot growth

Table 3. Shoot length, dry weight, leaf area, nitrogenase activity and nitrogen content in two
different cultivation systems. The values are from the end of the growth period (47 days from
planting). $\bar{x} \pm$ SE, n = 4–6

Plant characteristics	Cultivation systems	
	Traditional	Modified continuously circulating system
Shoot length	21.7 ± 4.9	39.0 ± 2.6
Leaf area (cm^2)	184.3 ± 58.3	383.8 ± 30.1
Dry weight (g): total plant	1.552 ± 0.486	3.686 ± 0.328
Root/shoot ratio	0.36 ± 0.02	0.28 ± 0.01
Nitrogenase activity		
(μmol C_2H_4 plant^{-1}h^{-1})	19.61 ± 7.21	42.03 ± 4.02
Nitrogen content (mg):		
total plant	37.8 ± 12.9	75.3 ± 8.0

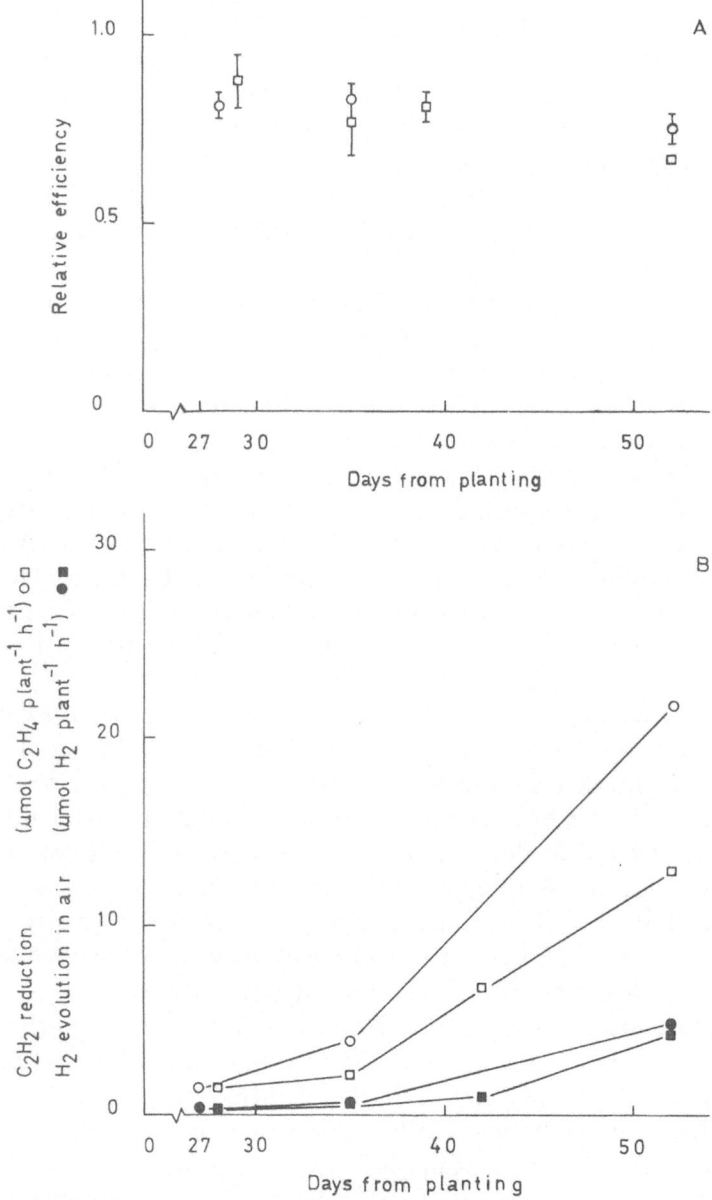

Fig. 3A. Development of relative efficiency of nitrogenase in *A. incana* grown in two different cultivation systems. $\bar{x} \pm SE$, n = 4. Bars indicate SE, unless SE is smaller than the dimension of the symbol.

○ continuously circulating system

□ traditional system

B. Development of C_2H_2-reduction (○,□) and H_2 evolution (●,■) in air in *A. incana* grown in two different cultivation systems. $\bar{x}$, n = 4, SE was 5 to 31% of $\bar{x}$, with the highest SE values in the traditional system.

○ continuously circulating system

□ traditional system

(Fig. 1), as nitrogen content and growth are strongly correlated[18]. Alders grown in the continuously circulating system developed stem branches at the end of the experimental period. Therefore, the biomass or the leaf area gives a better description of the plant development than the shoot length does at this stage. The root/shoot ratio of the alders in the continuously circulating system indicates easily available nutrients (Table 2). Branched shoots is also a response to good nutrient status[7].

Thus, the continuous supply of water and nutrients permitted these plants to grow faster[13] and to have more nitrogen than plants in the two other cultivation systems studied. Plants in the traditional system received a limited supply of nutrients, like *e.g.* phosphorus, a nutrient of great important for nitrogen fixation. Together with the later onset of nitrogen fixation this led to a lower content of nitrogen, and a lower growth rate. In spite of a continuous supply of water and nutrients, the hydroponic system was less successful. The continuously circulating system provided a well aerated environment for the root system. The aeration of the solution in the hydroponic system was not sufficient, as indicated by the large amount of callus growth through the nodule lenticels.

The relative efficiency of nitrogenase in the *A. incana-Frankia* symbiosis ranged from 0.67 to 0.88. RE did not change during the part of ontogeny studied in the alders continuously supplied with nutrients and water. A slight decrease was observed in the traditionally grown alders. However, this decrease was smaller than the variation in RE reported for pea by Bethlenfalvay[4]. Such differences may well occur when symbioses with *Frankia* and *Rhizobium* are compared. The period studied is a limited part of the ontogeny in *Alnus,* while the studies made on legumes covered several ontogenic stages.

This whole plant study on the *Alnus-Frankia* symbiosis showed that irrespective of C_2H_2-reduction rate RE held a constant value during the growth period studied in the modified continuously circulating system (Fig. 3B). The results on traditionally grown alders were more like data on cowpea for which Rainbird *et al.*[21] reported maximum relative efficiency at low rates of N_2-fixation. It was not possible in the study by Rainbird *et al.*[21] to separate the extent to which changes in electron allocation by nitrogenase from N_2-reduction to H^+-reduction or changes in hydrogenase activity were contributing to the change in RE. In this study no hydrogenase activity was detectable and the changes in RE are therefore considered due to changes in electron allocations from N_2-reduction to H^+-reduction.

The relative efficiency differed between the two cultivation systems studied, showing a slight decrease in RE in the traditionally grown alders while RE was constant in alders in the continuously circulating system. This indicates that the growth conditions affect the efficiency of nitrogen fixation. No strict relationship was found between RE and nitrogen content in this study, which is in contrast to value presented in a study of soybean[1,24] and cowpea[24]. If RE would have been the same in alders in the continuously circulating system and in the traditional system, the nitrogen content of the traditionally grown alders would have been about 6% greater. The value is estimated from the total C_2H_2-reduction integrated over the whole growth period. This nitrogen content is calculated from C_2H_2 reduction and H_2 evolution in air by the formula $(C_2H_2$ reduced/1.5) × RE. RE is easily overestimated when an uptake hydrogenase is present. But, in studies like the present one, where no hydrogenase was detected, RE seems to be a good tool for estimation of nitrogen fixation from C_2H_2-reduction measurements. Thus, the almost half as high nitrogen content (given by the analyses of Kjeldahl-N) of the traditionally grown alders compared to those in the continuously circulating system cannot be explained by the difference in RE. The amount of *Frankia* and nitrogenase is a more likely explanation to the difference in nitrogen content.

Growth, biomass production and nitrogen content are generally correlated. RE and nitrogen content were not closely related in this study. We conclude that the difference in RE between alders in the traditional and the modified continuously circulating system was too small to explain the obtained difference in biomass in these plants. Therefore, it is not expected that RE and plant productivity should be closely related to each other. Such results are in accordance with studies on pea[3,20]. RE might also give a better description of electron allocation to N_2 or H^+ than it describes the capacity of the symbiosis to fix nitrogen.

Acknowledgements This study was financially supported by Magn. Bergvalls Stiftelse, the Swedish Council for Forestry and Agricultural Research and the Swedish Natural Science Research Council.

References

1 Albrecht S L, Maier R J, Hanus J F, Russel S A, Emerich D W and Evans H J 1979 Hydrogenase in *Rhizobium japonicum* increases nitrogen fixation by nodulated soybeans. Science 203, 1255–1257.
2 Asher C J, Ozanne P G and Loneragan J F 1965 A method for controlling the ionic environment of plant roots. Soil Science 100, 149–156.
3 Bethlenfalvay G J and Phillips D A 1977 Effect of light intensity on efficiency of carbon dioxide and nitrogen reduction in *Pisum sativum* L. Plant Physiol. 60, 868–871.
4 Bethlenfalvay G J and Phillips D A 1977 Ontogenetic interactions between photosynthesis and symbiotic nitrogen fixation in legumes. Plant Physiol. 60, 419–421.

5   Bethlenfalvay G J and Phillips D A 1979 Variation in nitrogenase and hydrogenase activity
    of Alaska pea root nodules. Plant Physiol. 63, 816–820.
6   Chaney A L and Marbach E P 1962 Modified reagents for determinations of urea and
    ammonia. Clin. Chem. 8, 130–132.
7   Chapin 1980 The mineral nutrition of wild plants. Annu. Rev. Ecol. Syst. 11, 233–60.
8   Dixon R O D 1967 Hydrogen uptake and exchange by pea root nodules. Ann. Bot. 31,
    179–188.
9   Dixon R O D 1972 Hydrogenase in legume root nodule bacteroids: occurrence and proper-
    ties. Arch. Microbiol. 85, 193–201.
10 Eaton F M and Bernadin J E 1962 Soxhlet-type automatic sand cultures. Plant Physiol.
    37, 357.
11 Edie S A 1983 Acetylene reduction and hydrogen evolution by nitrogenase in a *Rhizo-
    bium*- legume symbiosis. Can. J. Bot. 62, 780–785.
12 Edwards D G and Asher C J 1974 The significance of solution flow rate in flowing culture
    experiments. Plant and Soil 41, 161–175.
13 Hubick T, Drakeford D R and Reid D M 1982 A comparison of two techniques for
    growing minimally water-stressed plants. Can. J. Bot. 60, 219–223.
14 Huss-Danell K 1978 Nitrogenase activity measurements in intact plants of *Alnus incana*.
    Physiol. Plant. 43, 372–376.
15 Huss-Danell K 1981 Clonal differences in rooting of *Alnus incana* leafy cuttings. Plant and
    Soil 59, 193–199.
16 Imsande J and Ralston E J 1981 Hydroponic growth and nondestructive assay for dini-
    trogen fixation. Plant Physiol. 68, 1380–1384.
17 Ingestad T and Lund A-B 1979 Nitrogen stress in birch seedlings I. Growth technique and
    growth. Physiol. Plant. 45, 137–148.
18 Ingestad T 1979 Nitrogen stress in birch seedlings II. N, K, P, Ca and Mg nutrition. Physiol.
    Plant. 45, 149–157.
19 Koroleff F 1970 Direct determination of ammonia in natural waters as indophenol blue.
    Conseil international pour l'exploration de la mer. Interlaboratory Report 3, 19–22.
20 Nelson L 1983 Hydrogen recycling by *Rhizobium leguminosarum* isolates and growth and
    nitrogen contents of pea plants (*Pisum sativum* L.). Appl. Environ. Microbiol. 45, 856–
    861.
21 Rainbird R M, Atkins C A, Pate J S and Sanford P 1983 Significance of hydrogen evo-
    lution in the carbon and nitrogen economy of nodulated cowpea. Plant Physiol. 71, 122–
    127.
22 Roelofsen W and Akkermanns A D L 1979 Uptake and evolution of H_2 and reduction of
    C_2H_2 by root nodule and nodule homogenates of *Alnus glutinosa*. Plant and Soil 52,
    571–578.
23 Schubert K R and Evans H J 1976 Hydrogen evolution: A major factor affecting the
    efficiency of nitrogen fixation in nodulated symbionts. Proc. Nat. Acad. Sci. USA 73,
    1207–1211.
24 Schubert K R Jennings N T and Evans H J 1978 Hydrogen reactions of nodulated legu-
    minous plants. II. Effects on dry matter accumulations and nitrogen fixation. Plant
    Physiol. 61, 398–401.
25 Schubert K R and Evans H J 1977 The relation of hydrogen reactions to nitrogen fixation
    in nodulated symbionts. *In* Recent Developments in Nitrogen Fixation. Eds. N Newton,
    J R Postgate and C Rodriguez-Barrueco. Academic Press London. pp 469–485.
26 Siegel S 1956 Nonparametric statistics for the behavioral sciences. McGraw-Hill Kugakusha,
    Tokyo, 312 p.

Plant and Soil 78, 159–170 (1984).
© 1984 *Martinus Nijhoff/Dr W. Junk Publishers, The Hague.*

Ms. Fr 17

Nitrogenase activity in root nodule homogenates of *Alnus incana*

KERSTIN HUSS-DANELL and ANN-SOFI AHLQVIST
Department of Plant Physiology, University of Umeå, S-901 87 UMEÅ, Sweden

Key words *Alnus incana Alnus glutinosa Frankia* Intact plants Nitrogenase activity Nodule homogenates Root nodules

Summary Root nodule homogenates of actinorhizal plants may represent *Frankia* in a symbiotic stage but released from environmental influence of the host plant. Anaerobic homogenization with a blender in buffer supplied with sucrose, polyvinylpyrrolidone and reducing substances gave three times higher yields of nitrogenase activity (C_2H_2-reduction) than crushing the nodules in liquid nitrogen. The activity in the homogenates was very reproducible and was, on average, nearly twice as high as the activity in excised nodules and c. 10% of the activity in intact plants. The difference in activity between excised nodules and intact plants was, roughly by halves, due to removal of the root system from the pot and to excision of the nodules. The nitrogenase activity in the homogenates was slightly higher when nodule excision was done in Ar or under water as well as after treatment of the homogenate with toluene or Triton X-100 or osmotic shock. These gains in activity were considered too small to outweigh the increased complications of preparing homogenates for routine use. Due to the reproducible recovery of nitrogenase in the homogenates the technique seems useful for physiological studies on nitrogen fixation in *Alnus incana*.

Introduction

In the symbiosis of *Alnus incana* with *Frankia* sp. the nitrogenase activity measured on intact plants is easily disturbed by *e.g.* ammonium additions to the root system[11]. To understand the reason for such lowered nitrogenase activities it would be desirable to measure the potential nitrogenase activity of *Frankia, i.e.* to measure nitrogenase activity in a stage where the host plant is no longer regulating the environmental conditions of the microsymbiont. This approach has already been used for Rhizobium in symbiosis with Pisum[7] and was made possible by the technique for obtaining bacteroid suspensions described by Straten and Roelofsen[15].

The attempts to measure nitrogenase activity in root nodule homogenates of Frankia-symbioses gave initially very low activities[14]. Later work[1,3,4,5,16] showed that nitrogenase activity can be measured if the whole procedure is done under strictly anaerobic conditions. The nodules were either homogenized with a blender in buffer[1,16] or crushed in liquid nitrogen in a mortar[3,4,5]. Each of the used techniques was applied to several actinorhizal species but did not include *A. incana*. For the use of nodule homogenates in physiological studies it is also necessary to have a procedure which gives high and reproducible recoveries of nitrogenase activity in the nodule homogenates. We report here on our attempts to meet these requirements for nodule homogenates of *A. incana.*

Materials and methods

Plant material

One clone of grey alder (*Alnus incana* (L.) Moench) from southern Sweden (59° 37′N, 12° 58′E) was used. One-leaf internode cuttings were rooted, inoculated with a suspension of crushed root nodules and grown in a climate chamber as described previously[9]. The plants were used when 7 to 11 weeks old. Black alder (*A. glutinosa* (L.) Gaertn.) from southern Sweden (56° 38′N, 16° 28′E) was propagated vegetatively. The rooted green cuttings were inoculated with the same nodule suspension as was used for *A. incana* and were grown as described above. The plants were used when 8 weeks old.

Excision of root nodules

The root system and gravel was carefully poured out of the pot. With a fine pair of tweezers the nodules were excised rapidly and quantitatively, free from root pieces. This operation was usually done in normal laboratory air, but in some experiments in Ar or humid air (25°C, 75% RH) or under water. The root nodules were kept on wet filter papers until all nodules were collected and then weighed. Nodules from one or two, sometimes up to four plants were pooled. A representative sample was used for measurement of nitrogenase activity in excised nodules and the rest of the nodules homogenized.

Preparation of root nodule homogenates

The homogenization technique was essentially as described by Akkermans et al.[1] and van Straten et al.[16]. The nodules (0.4 to 1.0 g) were placed in a gas-tight, custom-made chamber of Plexiglass surrounding the homogenizing rod of an Ultra-Turrax (TP 18–10, Janke & Kunkel KG) homogenizer. The chamber was continuously flushed with Ar through an inlet in the bottom and with excess of gas vented out at the top of the chamber. The flushing of the chamber and homogenizer with Ar started before the nodules were placed in the chamber. Five minutes after the nodules were placed in the chamber about 20 ml ice-cold buffer was added with a syringe through a rubber membrane stopper in the wall of the chamber. The buffer was Ar-flushed 0.05 M Tris-HCl (pH 8.0–8.1) which contained 0.6 M sucrose, 4% (w/v) PVP (polyvinylpyrrolidone K25; Roth), 100 mM $Na_2S_2O_4$ and 5 mM DTT (DL-dithiothreitol; Sigma). After 5 minutes in the buffer the nodules were homogenized for 4 s. The foam was allowed to settle during 30–60 s and the homogenization was repeated for 4 s. The suspension was transferred with a syringe to an Ar-flushed Sartorius SM 16 510 filter apparatus with a 100 μm nylon filter. The filtrate was centrifuged anaerobically for 10 minutes at 5°C and 27 000 × g. The pellet was suspended in the buffer to a concentration of about 0.05 g (fresh weight nodule) per ml. Unless otherwise given this fraction (no. 1 in Fig. 1A) was used in the experiments.

A simplified form of the technique of Benson and Eveleigh[3] was used in a comparative study. The excised, weighed nodules were plunged into liquid N_2 in a glass homogenizer. The nodules were crushed and while liquid N_2 was still present the piston was removed and the homogenizer capped with a membrane stopper. Continuous Ar-flushing was started before all liquid N_2 was vented off. Buffer (composition as above) was added to give a suspension of about 0.05 g (fresh weight nodules) per ml. Compared to the original description[3] we thus avoided transfers of crushed nodules from a mortar to a tube and thereby we increased the possibilities to work quantitatively.

Treatments of nodule homogenates

Treatments with toluene and with toluene + EDTA were as proposed for pea bacteroids[15]. One ml homogenate was shaken with 0.02 ml toluene or 0.02 ml toluene + 1 ml 1 mM EDTA for 1 minute and the phases were allowed to separate for 1 minute before the aqueous phase was used. Osmotic shock was done by suspending the pellet in a sucrose-free buffer instead of the normal concentration of $0.6\,M$ sucrose (fraction 1, Fig. 1A). Lysozyme treatment was done hypotonically by suspending the pellet in sucrose-free buffer supplemented with lysozyme (Sigma L6876; $1\,mg\cdot ml^{-1}$) and $1\,mM$ $MgCl_2$. In the experiment with DMSO (dimethyl sulfoxide) treatment the incubation mixture for C_2H_2-reduction measurement was supplemented with 7.5% (v/v) DMSO (Fisher Scientific Co.). The treatments with Triton X-100 were done similarly by including Triton X-100 (Sigma T6878) in the incubation mixture for C_2H_2-reduction measurement.

Measurements

The nitrogenase activity was measured as C_2H_2-dependent C_2H_4-production with C_2H_4 quantified as described previously[8]. Intact potted plants and nodulated roots were incubated in Plexiglass cuvettes[8] of 600 ml gas volume with 10% (v/v) C_2H_2 in air and were kept in the growth chamber during the incubations. Excised nodules were incubated in test tubes (30 ml gas volume) with a few drops of water and with 10% (v/v) C_2H_2 in air at room temperature.

For the nodule homogenates the complete reaction mixture (final volume 2 ml) contained, unless otherwise indicated, $20\,mM$ Na_2-ATP (pH 8.0; Sigma A5394 or USB 24915), $10\,mM$ $MgCl_2$ (pH 8.0), $100\,mM$ $Na_2S_2O_4$ and 1.0 ml homogenate in tubes with 7 ml gas phase of 10% (v/v) C_2H_2 in Ar. The reaction was started with the addition of nodule homogenate and the tubes were kept on a rotary shaker at room temperature. In experiments with both excised nodules and nodule homogenates from the same plants the C_2H_2-reduction was started simultaneously for nodules and nodule homogenates. The C_2H_2-reduction was followed for 1 to 1.5 h and at least four gas samples were taken at intervals during this period for calculation of μmol $C_2H_4\cdot g^{-1}$ (fresh weight nodule) $\cdot h^{-1}$. The gas samples from the incubated nodule homogenates were taken very cautiously so that air was not introduced into the tubes. Three incubation tubes were run for each experiment. They gave always very similar results, and the mean of the three have been used in the calculations. At the end of the C_2H_4-measurements the pH of the incubation mixture was checked. A pH-value lower than 7.0 was regarded as abnormal (probably due to leakage or insufficient Ar-flushing) and such occasional tubes were excluded from calculation of nitrogenase activity.

All values expressed on a nodule fresh weight basis can easily be converted to a nodule dry weight basis by using the ratio fresh weight: dry weight $= 5:1$. Statistical treatments were according to Daniel[6] with $P = 0.05$ as significance level.

Results and discussion

Activity in nodule homogenates

The nitrogenase activity was measured in nodule homogenates of *A. incana* prepared from plants having nitrogenase activities of 10 to $40\,\mu$mol C_2H_4 plant^{-1}.h^{-1} as intact plants. After a short lag period of 5 to 10 minutes the activity was constant over a period of at least 1.5 hours. In an initial series of 10 experiments the homogenates had on average an activity of $4.09\,\mu$mol $C_2H_4\cdot g^{-1}\cdot h^{-1}$ (range 3.07 to $5.14\,\mu$mol C_2H_4 .g^{-1}h^{-1}). Higher values were occasionally recorded later on but the given range of activities can be regarded as typical.

The homogenization procedure (Fig. 1A) was effective in releasing the active tissue from the root nodules. Fig. 1 B–C shows that most of the activity was measured in the pelleted 100 μm filtrate fraction (fraction 1, Fig 1A), while little activity was left in the normally discarded

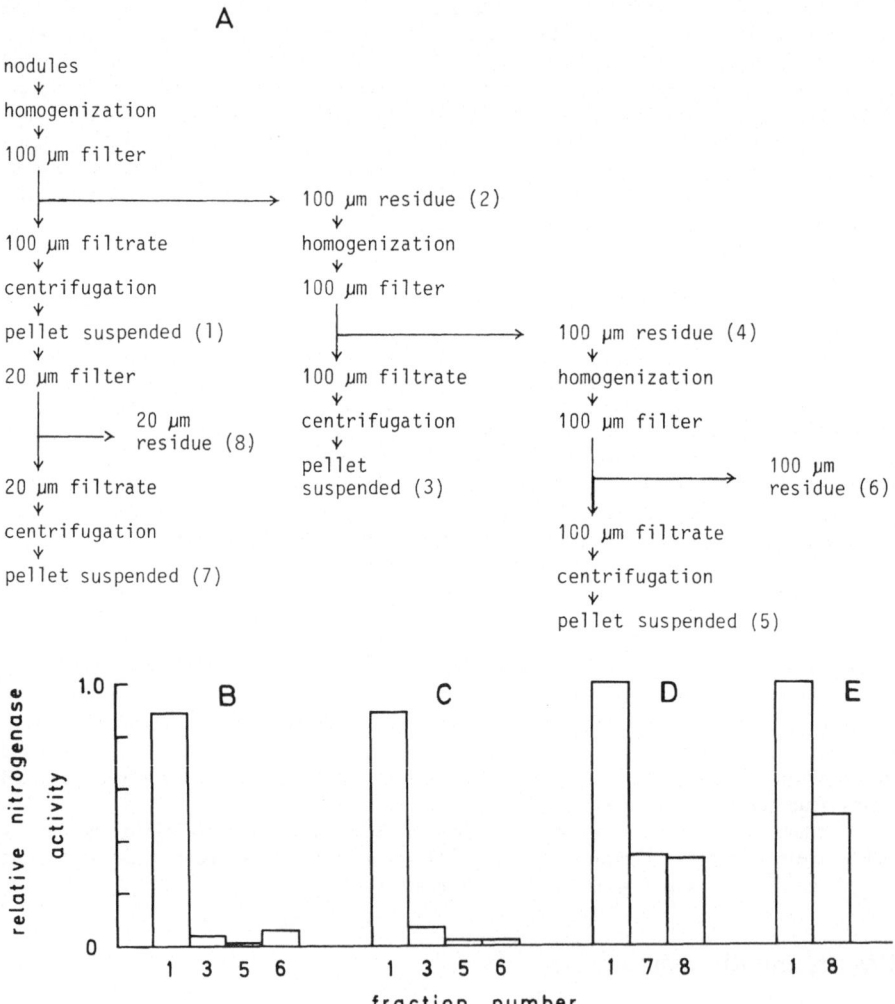

Fig. 1. A, Fractionation of root nodule homogenates of *A. incana*. B–E, Relative nitrogenase activity in different fractions (numbers refer to A) of the root nodule homogenates. Each histogram represents one experiment.

$100\,\mu m$ residue (fraction 2, Fig. 1A). No activity was found in the supernatants obtained at $27\,000 \times g$. When a further filtration step was used (Fig. 1A) nearly half the activity in the $100\,\mu m$ filtrate (fraction 1) was found to be located in particles having a size between 20 and $100\,\mu m$ (Fig. 1 D–E). In the light microscope this fraction consisted almost entirely of vesicle clusters of *Frankia* and had only small amounts of plant cell material. This information and earlier studies[1, 2, 11, 16] mean that there are great similarities between the nodule organization in *A. incana* and in *A. glutinosa*.

Activity in nodule homogentates compared to nodules and intact plants

The nitrogenase activity in the root nodule homogenates was usually about twice as high as the activity in excised nodules from the same plant ($\bar{x} \pm SE = 217 \pm 70\%$ of activity in excised nodules; range 56 to 470%; n = 6). This means that the conditions for nitrogenase activity were improved in the homogenates compared to the conditions in excised nodules. Also, the recovery of activity (per cent of activity in intact plant) was generally more reproducible in nodule homogenates than in excised nodules (Table 2). In other studies on Frankia-symbioses (Table 1) the activity in the homogenates was always much lower than the activity in excised nodules. Our data on *A. glutinosa* (Table 1) also show higher activity in the homogenates than in the excised nodules. This suggests that our recovery of activity in nodule homogenates of *A. incana* is not solely due to this particular species but rather to the procedure used. The lower activities per g nodule in the homogenates of *A. glutinosa* than in *A. incana* may be due to the use of the same *Frankia* inoculum for the two alder species and that the inoculum gives better results on its normal host species than on *A. glutinosa*.

In actinorhizas[17] as well as in nodulated legumes[12] most of the nitrogenase activity measured in intact plants is lost when the measurement is done on excised nodules. On average the excised nodules of *A. incana* showed 7.67% (range 3.28 to 14.6%) of the activity in intact plants (Table 2). This means that although the activity was higher in homogenates than in excised nodules the activity in homogenates did never reach 100% of the activity in intact plants, but stayed reproducibly at a level of 10.4% (range 7.46 to 15.4%; $SE/\bar{x} = 0.080$) in 10

Table 1. Nitrogenase activity and recovery of nitrogenase activity in root nodule homogenates of *Frankia*-symbioses. The maximal values of activity per g fresh weight of nodules are not always correlated to the maximal value of recovery reported in or calculated from each study. [a] nodule excision in Ar [b] homogenate treated with osmotic shock

	Maximal activity (μmol C_2H_4 · $g^{-1} \cdot h^{-1}$)	Maximal recovery; % of activity measured in		
		excised nodules	intact plants	References
Alnus incana	10.5[a]	470	20.5[b]	this study
A. glutinosa	1.65	193	6.5	this study
	1.353	12.8	–	1
	5.19	26.6	–	16
A. rubra	0.93	14.5	–	5
Hippophaë rhamnoides	3.23	26.2	–	16
Myrica gale	0.72	87.8	–	16
M. pennsylvanica	1.6	–	–	3
Shepherdia canadensis	1.18	41.1	–	16

Table 2. Nitrogenase activity in intact plants of *A. incana* compared to activity after handling the root system, excision of root nodules and homogenization of nodules. The first incubation (= before treatment) was always done in the morning on intact potted plants and the second incubation was done in the afternoon on the same plants after the treatment indicated. Each value is the mean (± SE) of n experiments. (a) = gas volume of cuvette not measured by replacement with water after first incubation; in all other cases the gas volume was measured after first incubation. The activity in intact plants was $10-40 \mu mol\, C_2H_4 \cdot plant^{-1} \cdot h^{-1}$

Treatment	Activity after treatment as % of activity before treatment	n
Intact plant		
none (a)	99.3 ± 6.2	8
none	94.2 ± 6.0	8
plant removed from pot and immediately replanted without touching the nodules	73.7 ± 5.0	8
plant removed from pot and nodules touched as if they were excised before replanting	52.6 ± 6.2	12
Nodulated roots		
roots cut off from plant and all gravel shaken off the roots; nodules not touched	82.4 ± 5.9	7
Excised nodules		
excision in normal air	7.67 ± 1.79	6
excision in humid air	13.4	1
excision under water	4.1	3
excision in Ar	2.00	2
Nodule homogenates		
excision in normal air	10.4 ± 0.8	10
excision in humid air	10.2	1
excision under water	13.8	3
excision in Ar	12.9	2

experiments (Table 2). This reproducible recovery of activity in the homogenates may be due to the standardized growth of the plants. For *Pisum* the recovery of nitrogenase activity in bacteroid suspensions was about 2.5 times higher for plants grown in gravel than for plants grown in liquid culture[7].

We looked for significant correlations (Spearman's rank correlation coefficient) between recovery of activity in the homogenates (% of activity in intact plant) and (1) nitrogenase activity in intact plant; (2) age of the plants (range 7 to 11 weeks); (3) recovery of activity in excised nodules, and (4) time from start of nodule excision to start of homogenization of nodules (range 10 to 45 minutes). Only the last mentioned parameter was significantly correlated to the recovery of activity in the homogenates ($r_s = -0.80$; $0.001 < P < 0.005$; $n = 10$). However, this correlation was not linear ($r = 0.155$ for the line $y = 0.029 x + 11.647$) and therefore difficult to interpret. Although the nodule size was not actually measured we did not observe any relation between recovery of activity in the homogenates and size of the nodules that were homogenized.

Table 2 shows that there was almost no effect of repeated measurements of nitrogenase activity on the same intact plants. But, when the plant was removed from the pot and its roots and nodules were touched (as if the nodules were excised) and the intact plant was replanted before measurement of nitrogenase activity only about 50% of the activity was left (Table 2). Touching nodules appeared to be a critical step. If plants were removed from the pot but the nodules were not touched before replanting about 75% of the activity was left (Table 2). Likewise, nodulated roots that were prepared without touching the nodules showed about 80% of the activity in the intact plant (Table 2). Some loss of activity after excision of the shoot from an otherwise intact alder plant was also recorded earlier[8]. We conclude that the low recovery of activity in excised nodules (about 7%) is, roughly by halves, due to touching the nodules prior to and during excision and to the excision process in itself.

Attempts to improve conditions during nodule excision

Moisture was expected to be a critical factor during nodule excision. Therefore we tried working in humid air or under water when the plant was removed from the pot and the nodules were excised. None of these conditions gave any substantial gain in activity in excised nodules or in nodule homogenates (Table 2). Since anaerobic conditions have been proven to be important when preparing nodule homogenates[1, 3, 16] we tried working in Ar also when the plant was removed from the pot and the nodules were excised. The increased recovery of activity in nodule homogenates was insignificant (Wilcoxon matched-pairs signed-ranks test; $P = 0.10$) and for excised nodules the treatment was negative (Table 2). The negative effect of Ar for the excised nodules is most likely due to depletion of available oxygen in the nodules and delayed diffusion of air into the nodules during the subsequent incubation with C_2H_2 in air.

Thus, the losses of activity caused by handling the nodules and detaching the nodules from the roots are not easily overcome. At present, we have no good explanation for the decrease in nitrogenase activity. In soybeans, where lenticels are in close proximity to vascular strands in the nodules, it was shown that excision changed the pressure inside the nodules and thereby the gas diffusion decreased[13]. Better conditions during excision of nodules will not help entirely as long as 50% of the activity in *A. incana* is lost already by removing the plant from the pot and touching the nodules.

Attempts to improve conditions during nodule homogenization

One reason for a recovery much lower than 100% in the homogenates might be that the vesicles of *Frankia* were damaged during the homogenization procedure. We tried to avoid this in an experiment where the homogenization was performed in the filter apparatus. The lid of the upper part of the apparatus was modified so that the penetration by the homogenizing rod was gas tight. The width and volume of the upper part of the apparatus was decreased by an inserted Plexiglass cylinder.The homogenate was then rapidly passed down through the $100 \mu m$ filter and thus the released vesicles were protected from mechanical disruption when a new volume of buffer was added to the upper container, and the nodule pieces were rehomogenized and filtered. The homogenization time was 2 times 2 s in each of the buffer volumes and the $100 \mu m$ filtrates were pooled. However, no gain in recovery of activity was obtained (10% of activity in intact plant).

A totally different homogenization procedure was also tried, *viz.* crushing the nodules in liquid nitrogen. This technique yielded activities of only about $1 \mu mol\ C_2H_4 \cdot g^{-1} \cdot h^{-1}$ or, on average, 3.5% of the activity in intact plants (Table 3). Although this technique is somewhat more simple to perform we found the low recovery of nitrogenase activity disadvantageous.

Attempts to improve the incubation conditions for nodule homogenates

To optimize the incubation conditions for nodule homogenates we changed the molar ratio of Mg^{2+}/ATP and the concentration of ATP in

Table 3. Nitrogenase activity in root nodule homogenates of *A. incana* prepared by grinding the nodules in liquid nitrogen

Experiment no	Nitrogenase activity		
	$\mu mol\ C_2H_4$ $\cdot g^{-1} \cdot h^{-1}$	% of activity in	
		excised nodules	intact plant
1	1.12	5.1	2.46
2	1.22	530	5.32
3	0.94	38.3	2.90

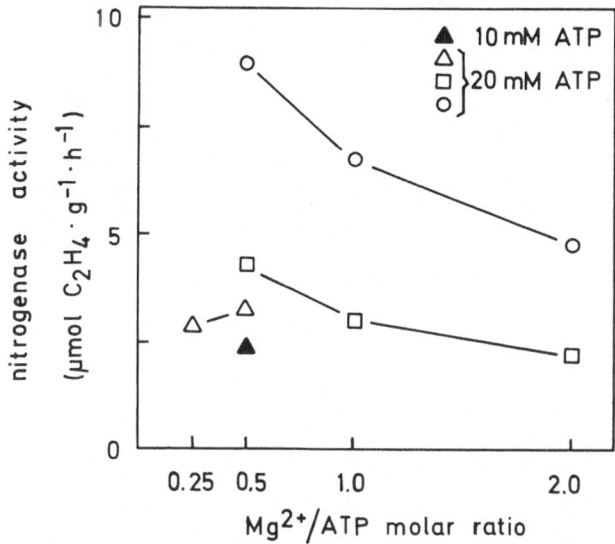

Fig. 2. Nitrogenase activity in root nodule homogenates of *A. incana* measured at different Mg^{2+}/ATP molar ratios. The three experiments are represented by circles, squares and triangles, respectively.

the incubation mixture. When the molar ratio was changed from 0.5 to 1.0 and to 2.0 the nitrogenase activity was lower (Fig. 2). A change in molar ratio from 0.5 to 0.25 did not give higher nitrogenase activity (Fig. 2). When the molar ratio was kept at 0.5 the nitrogenase activity was slightly higher at 20 than at 10 mM ATP (Fig. 2). Thus, we saw no need to change the incubation mixture in these respects.

Low activities in nodule homogenates can be due to permeability limitations, possibly of ATP, through the cell membrane and/or cell wall of *Frankia*. Treatments of bacteroid suspensions with toluene and EDTA raised the recovery of nitrogenase activity 10 times in *Pisum* and made it possible to detect nitrogenase activity in preparations from soybean[15]. Nodule homogenates of *A. incana* showed only slightly higher nitrogenase activity after toluene treatment but not after treatment with toluene + EDTA (Table 4). This is similar to data on nodule homogenates of *A. glutinosa*[1] and *Myrica pennsylvanica*[3]. The different response by *Rhizobium* and *Frankia* can be due to the differences in cell wall composition. Rhizobia are Gram-negative bacteria and have a high content of lipids in their cell wall while *Frankia* are Gram-positive bacteria with very low content of lipids in their cell wall. The toluene treatment is therefore expected to increase the cell wall permeability much more in *Rhizobium* than in *Frankia*.

Osmotic shock of root nodule homogenates of *A. incana* increased the activity from about 7 to 9 μmol $C_2H_4 \cdot g^{-1} \cdot h^{-1}$, that is by 29% (Table 4). The addition of lysozyme to the sugar-free buffer had a

Table 4. Effects of various treatments on nitrogenase activity in root nodule homogenates of *A. incana*. The values are from single experiments. n.d. = not determined. [a]incubation period 15−35 min, [b]35−55 min, [c]55−75 min

Treatment	Nitrogenase activity		
	μmol C_2H_4 $\cdot g^{-1} \cdot h^{-1}$	% of activity in intact plant	increase (% of control)
None (control)	7.80	12.7	
Toluene	9.11	14.8	17
Toluene + EDTA	7.13	11.6	−
None	7.07	15.9	
Osmotic shock	9.11	20.5	29
None	2.88	n.d.	
Lysozyme + osmotic shock	2.50[a]	n.d.	−
	1.40[b]	n.d.	−
	0.59[c]	n.d.	−
None	7.07	15.9	
DMSO	6.87	15.5	−
None	2.87	8.98	
Triton X-100 (0.1%)	3.45	10.8	20
None	4.05	15.2	
Triton X-100 (0.1%)	4.57	17.1	13
Triton X-100 (1.0%)	4.00	15.0	−

negative effect on the nitrogenase activity of the homogenate. While the control, as usual, showed a constant activity for more than 1 h the lysozyme treated homogenate lost activity during the whole incubation period (Table 4). Treatment with DMSO gave no effect but Triton X-100 increased the nitrogenase activity of nodule homogenates with about 20% (Table 4).

Use of nodule homogenates in physiological studies

For *A. incana* we conclude that reproducible, controlled growth conditions and careful handling of the nodulated root system during excision of the root nodules were important to obtain reproducible recoveries of nitrogenase activity in the nodule homogenates. The slightly higher activity obtained after excision in Ar or under water and after treatment of the homogenates with toluene, Triton X-100 or osmotic shock appear too small to outweigh the increased complications of the technique for routine use. So far, the carbon metabolism of symbiotic *Frankia* has been studied on fractionated nodule homogenates[2, 10]. Based on the experience presented here it seems that nodule homogenates can also be used for physiological studies on nitrogen fixation in *A. incana*. Nitrogenase activity measurements may also be combined with studies of other Frankia-activities like hydrogenase activity. In many cases an intact nodulated plant is a far too complex system to work with. On the other hand, pure cultures of free-living *Frankia* do not always give a clear picture of the conditions in a root nodule since biochemical changes are known to occur when *Frankia* develops from a free-living to a symbiotic stage[10]. Therefore, properly prepared root nodule homogenates represent *Frankia* in a symbiotic stage but released from the influence of the host plant.

Acknowledgements The study was financially supported by the Swedish Natural Science Research Council. We thank A. Akkermans, K. Gezelius and A. Sellstedt for valuable discussions and A. Flower-Ellis for technical help.

References

1 Akkermans A D L, Straten J van and Roelofsen W 1977 Nitrogenase activity in nodule homogenates of *Alnus glutinosa:* A comparison with the *Rhizobium*-pea system. *In* Recent Developments in Nitrogen Fixation. Eds: W Newton, J R Postgate and C Rodriguez-Barrueco, Academic Press, London, p. 591–603.

2 Akkermans A D L, Huss-Danell K and Roelofsen W 1981 Enzymes of the tricarboxylic acid cycle and the malate-aspartate shuttle in the N_2-fixing endophyte of *Alnus glutinosa*. Physiol. Plant. 53, 289–294.

3 Benson D R and Eveleigh D E 1979 Nitrogen-fixing homogenates of *Myrica pennsylvanica* (bayberry) non-legume root nodules. Soil Biol. Biochem. 11, 331–334.

4 Benson D R, Arp D J and Burris R H 1979 Cell-free nitrogenase and hydrogenase from actinorhizal root nodules. Science 205, 688–689.

5 Benson D R, Arp D J and Burris R H 1980 Hydrogenase in actinorhizal root nodules and root nodule homogenates. J. Bacteriol. 142, 138–144.

6 Daniel W W 1978 Applied nonparametric Statistics. Houghton Mifflin Co., Boston, 510 p.

7 Houwaard F 1978 Influence of ammonium chloride on the nitrogenase activity of nodulated pea plants (*Pisum sativum*). Appl. Environ. Microbiol. 35, 1061–1065.

8 Huss-Danell K 1978 Nitrogenase activity measurements in intact plants of *Alnus incana*. Physiol. Plant. 43, 372–376.

9 Huss-Danell K and Sellstedt A 1983 Nitrogenase activity in response to restricted shoot growth in *Alnus incana*. Can. J. Bot. 61, 2949–2955.

10 Huss-Danell K, Roelofsen W, Akkermans A D L and Meijer P 1982 Carbon metabolism of *Frankia* spp. in root nodules of *Alnus glutinosa* and *Hippophaë rhamnoides*. Physiol. Plant. 54, 461–466.

11 Huss-Danell K, Sellstedt A, Flower-Ellis A and Sjöström M 1982 Ammonium effects on function and structure of nitrogen-fixing root nodules of *Alnus incana* (L.) Moench. Planta 156, 332–340.

12 Mague T H and Burris R H 1972 Reduction of acetylene and nitrogen by field-grown soybeans. New Phytol. 71, 275–286.

13 Ralston E J and Imsande J 1982 Entry of oxygen and nitrogen into intact soybean nodules. J. Exp. Bot. 33, 208–214.

14 Sloger C and Silver W S 1965 Note on nitrogen fixation by excised root nodules and nodular homogenates of *Myrica cerifera* L. *In* Non-heme iron proteins: Role in energy conversion. Ed: A San Pietro. Antioch Press, Yellow Springs, Ohio, p. 299–313.

15 Straten J van and Roelofsen W 1976 Improved method for preparing anaerobic bacteroid suspensions of *Rhizobium leguminosarum* for the acetylene reduction assay. Appl. Environ. Microbiol. 31, 859–863.

16   Straten J van, Akkermans A D L and Roelofsen W 1977 Nitrogenase activity of endophyte
     suspensions derived from root nodules of *Alnus, Hippophaë, Shepherdia* and *Myrica* spp.
     Nature 266, 257–258.
17   Wheeler C T, Cameron E M and Gordon J C 1978 Effects of handling and surgical treat-
     ments on nitrogenase activity in root nodules of *Alnus glutinosa,* with special reference to
     the application of indole-acetic acid. New Phytol. 80, 175–178.

Plant and Soil 78, 171–179 (1984).   Ms. Fr 18
© 1984 Martinus Nijhoff/Dr W. Junk Publishers, The Hague.

Selection and micropropagation of nodulating and non-nodulating clones of *Alnus crispa* (Ait.) Pursh

F. M. TREMBLAY, X. NESME and M. LALONDE
Faculté de Foresterie, Université Laval, Ste-Foy, Québec, Canada G1K 7P4

Key words Actinorhizae *Alnus crispa Alnus viridis* spp. *crispa* Clones *Frankia* Nodulation *In vitro* culture.

Summary 600,000 seedlings of *Alnus crispa* were inoculated with a 1:1:1 mixture of the *Frankia* strains ACN1^{AG}, AGN1$^{AG}_{exo}$ and MGP10i. After 3 successive inoculations and screenings, one individual, AC-4, was selected as non-nodulating (Nod$^-$) with Frankiae. This selected individual AC-4 (Nod$^-$) and two other clones of *A. crispa*, AC-2 and AC-5, known for their ability to nodulate (Nod$^+$) with *Frankia* were *in vitro* propagated. The different clones of *A. crispa* in culture required different kinds and concentrations of sugar during the in vitro multiplication and rooting stages. Nodulation tests using 7 *Frankia* strains indicated that the clone AC-4 (Nod$^-$) was non-nodulating with 6 of the 7 *Frankia* strains tested. One strain, *Frankia* ANNI, isolated from one unique nodule produced on the mother-plant AC-4, induced 38% of the AC-4 plantlets to nodulate but with a number of nodules 10 to 20 times less than the clones AC-2 (Nod$^+$) and AC-5 (Nod$^+$). Morphological observations of the roots of AC-4 (Nod$^-$) indicated that this clone had few and abnormally short root hairs.

Introduction

The nitrogen-fixing symbiosis with *Frankia* has been particularly studied with regard to the actinomycete while less work has been done on the genetics of the host-plant[5]. Little is known about the genetic mechanisms, the nodulation genes, which assure to the host-plants the ability to associate with *Frankia*. More than 15 genera can associate with *Frankia*[1]. One can assume that the 'nodulation genes' are a common part of the genome of all these different host-plants assuring them the ability for symbiotic association. A way to study these nodulation genes should be to select a non-nodulating individual of a normally nodulating species. This paper deals with the selection of a non-nodulating individual of the normally nodulating species *Alnus crispa*. This species lives in the North of Canada and is characterized by a high cold hardiness and drought tolerance, which are interesting characteristics for breeding programs with southern species. The selection work resulted in one non-nodulating individual, AC-4 (Nod$^-$), which was then submitted to micropropagation. The *in vitro* propagated plantlets from this clone were then used in nodulation tests with pure *Frankia* strains and compared with known nodulating clones of *A. crispa* AC-2 (Nod$^+$) and AC-5 (Nod$^+$). The selection method used and the different stages of *in vitro* propagation are presented, and the results of the nodulation tests are discussed.

Methods

Selection

Seeds of *A. crispa* (Ait.) Pursh from different provenances, supplied by Société d'Energie de la Baie James (Montréal, Québec) or harvested in the field in Quebec (1981–82) were directly germinated in the greenhouse in trays (54 cm × 28 cm × 6 cm) containing Turface (International Minerals and Chemicals Corp., Des Plaines, IL) saturated with the nitrogen-free Crone's solution[10]. Plantlets bearing 1–2 leaves (3–4 week-old) were each inoculated with 0.5 ml of a 1:1:1 mixture of cultures of the *Frankia* strains ACN1^{AG} (ref.[9]) AGN1$^{AG}_{exo}$ (ref.[13]) at an O.D. 0.03 at 620 nm[11] in N-free Crone's solution. One to two months after inoculation of more than 600,000 seedlings, plantlets with green leaves and nodules, indicating nitrogen fixation by *Frankia*, were eliminated while those showing symptoms of nitrogen deficiency were inoculated for a second time. After a second screening nine plantlets indicating no or little N_2 fixation were selected and transferred to individual pots. They were reinoculated and checked after 3 weeks for the absence of nodules. Between each three successive nodulation tests, plantlets were fertilized with Crone's solution[10] containing nitrogen (KNO_3, 0.5 g/l) for 4 weeks and then N-starved again for 4 weeks before reinoculation.

Micropropagation

Three clones of *A. crispa* were used for micropropagation and nodulation experiments, AC-2 and AC-5 able to nodulate (Nod$^+$) with *Frankia*, and clone AC-4 selected as a presumably non-nodulating clone (Nod$^-$).

The two Nod$^+$ clones AC-2 and AC-5 were initiated from 1–8 month-old seedlings while the clone AC-4 Nod$^-$) was initiated from the pre-selected 8 month-old mother-plant kept in a growth chamber at 24°C under 10 Klux. The mother-plant AC-4 was regularly fertilized with the nitrogen-containing Crone's solution. Actively growing shoot tips were surface sterilized according to Jones *et al.*[6] and initiated on the Murashige and Skoog (MS) medium[14] containing 1 μM of BAP (benzyl adenine purine), 0.5 μM of IBA (indole butyric acid), 87.5 mM of sucrose and 0.7% Bacto Difco agar. After 8–10 weeks on this medium, the cultures were transferred to the multiplication medium. The multiplication medium consisted of the MS medium including 5 μM of BAP. The medium was supplemented with 175 mM of either sucrose or glucose depending of the clone (Table 1). The rooting medium containing the MS salts at half strength including 1 μM of IBA and different concentrations of glucose depending on the clone (Table 1). The pH of all media was adjusted with NaOH to 5.5 before autoclaving. The initiation and rooting media (15 ml) was kept in 20 × 150 mm culture tubes and the multiplication medium (250 ml) was in 1 liter Mason jars. The media were autoclaved at 121°C at 1.05 kg · cm^{-2} for 12 min for tubes and 15 min for jars.

The cultures were grown in a growth chamber at a day/night thermoperiod of 26/22°C (± 1°C). The light intensity for the initiation and multiplication steps was 2.5 Klux given by Vita-Lite$^{®}$ fluorescent tubes (Duro-Test Electric Ltd, Ont.), while the rooting treatment was conducted under 7.5 Klux given by a mixture of Cool-White and Gro-WS fluorescent tubes (Sylvania, Drummondville, Qué.) in the ratio 2:1.

Soil transfers

After 21 days on the rooting media, the rooted shoots were washed in tap water and transferred into PlantCon$^{®}$ containers (9 × 9 × 9 cm, Flow Laboratory Inc., Virginia) including a mixture of Turface:vermiculite:peat moss (4:1:1). The substrate was saturated with N-free Crone's solution before autoclaving at 121°C, 1.05 kg · cm^{-2} for 1 h at least 2 weeks before utilization.

Inoculation

The *in vitro*-propagated plantlets were inoculated 21 days after their transfer to soil. Some plantlets of each 3 clones were transferred alone into the containers to verify their nodulation ability but the majority of the plantlets were planted two per container, one non-nodulating (AC-4) with one nodulation (AC-2 or AC-5) plantlet, as a positive control of the inoculum.

Finally, plantlets of the nodulating clones AC-2 and AC-5 were not inoculated as negative controls of the system used for any *Frankia* contamination. Twenty-one or 28 days after inoculation, the root system of each plantlet was washed under tap water and the root system observed under a stereo-microscope (Leitz, Wild) for the presence of nodules.

Three inoculation tests were performed using different *Frankia* strains (Table 2) in order to verify a possible specificity in the non-nodulation of AC-4 (Nod⁻). All *Frankia* strains were grown in the Qmod B nutrient medium[11] containing 5 mg/l of L-α-lecithin but without any calcium carbonate. The inoculum was prepared to an O.D. of 0.03 at 620 nm[11] and 3 ml/ PlantCon® was injected around the plantlets into the soil. The nodulation was evaluated 21 or 28 days after inoculation following the appearance of green leaves. The nitrogen fixation was verified in the second inoculation test as acetylene reduction measured by gas chromatography[7].

Results

Following the screening of more than 600,000 seedlings of *A. crispa*, nine seedlings were selected at the second of three successive screenings. From them, one seedling was visually identified as a non-nodulating (Nod⁻) individual in presence of a mixture of the *Frankia* strains $ACN1^{AG}$, $AGN1^{AG}_{exo}$, $MGP10_i$, in its rhizosphere for more than 8 months (Fig. 1). This individual, named AC-4, was submitted to *in vitro* propagation to test its ability to nodulate with *Frankia*.

Micropropagation

On the initiation medium, the explants developed as multiple shoot clumps ready to divide after 8–10 weeks. After this time, a subculture on the same medium lead to the production of basal callus and important necrosis. The leaves developed abnormally with large, warped and crisped laminae. These symptoms were corrected by raising the sucrose concentration to 175 mM for AC-2, and by replacing sucrose by glucose at 175 mM for AC-4 and AC-5. The multiplication rate of each clone on its optimal medium was 5–7 every 3 weeks (Figs 2 and 3).

The shoots produced were excised and placed in tubes containing the roots medium. At 1 μM of IBA, no basal callus was produced and rooting occurred in less than 3 weeks (Fig. 4). The highest percentage of rooting, *i.e.* 80–100%, was obtained with glucose at 87.5 mM for AC-2 and at 175 mM for AC-5, and with glucose or sucrose at 87.5 mM for AC-4 (Table 1).

Soil transfers

We obtained 100% survival of the rooted plantlets transferred to PlantCon® containers (Fig. 5 and 6) with a nitrogen-free Turface substrate. These containers assured 100% relative humidity around the plantlets, which was necessary for survival when plantlets were transferred to soil. This closed system was used during the whole nodulation tests, thereby avoiding watering during the tests.

Fig. 1. Screening stage for a nodulating *A. crispa* seedling. On the left a nodulated seedling showing dark (green) leaves and on the right, the presumptive non-nodulating individual AC-4, showing light (yellow) leaves. The *Frankia* strain ANNI was isolated from this seedling. Bar scale = 10 cm.

Fig. 2. Multiplication stage of *A. crispa* in a 1 liter Mason jar. Bar scale = 10 cm.

Fig. 3. Multiple shoots formation of clone AC-4 (Nod⁻) on multiplication medium with 5 μM of BAP and 175 mM of glucose after 3 weeks of culture. (The shoots are on a Petri dish cover for illustration purpose). Bar scale = 1 cm.

Fig. 4. Rooted plantlets of the *A. crispa* clone AC-4, 3 weeks after transfer into rooting medium containing 1 μM of IBA and 87.5 mM of glucose. Bar scale = 1 cm.

Inoculation

Seven *Frankia* strains were tested on the nodulating clones AC-2 and AC-5 and on the presumptive non-nodulating clone AC-4. One hundred percent of the plantlets of the AC-2 (Nod⁺) and AC-5 (Nod⁺) clones became nodulated with the seven *Frankia* strains tested, while no nodules appeared on the plantlets AC-4 (Nod⁻) for the 6 *Frankia* strains ACN14a, AGN1$_{exo}^{AG}$, TN18^{AC}, ARbn4b, ARI3 and ArgN22d (Table 3). However the *Frankia* strain ANNI (*A*lnus *N*on-*N*odulating *I*solate) induced the nodulation of 38% of the plantlets of the clone AC-4. This *Frankia* endophyte was isolated using the technique of Lalonde et al.[12] from one nodule produced on the mother-plant AC-4 after one year in growth chamber. This nodule was the only one ever produced on the mother-plant AC-4. With this *Frankia* strain ANNI, the mean number of

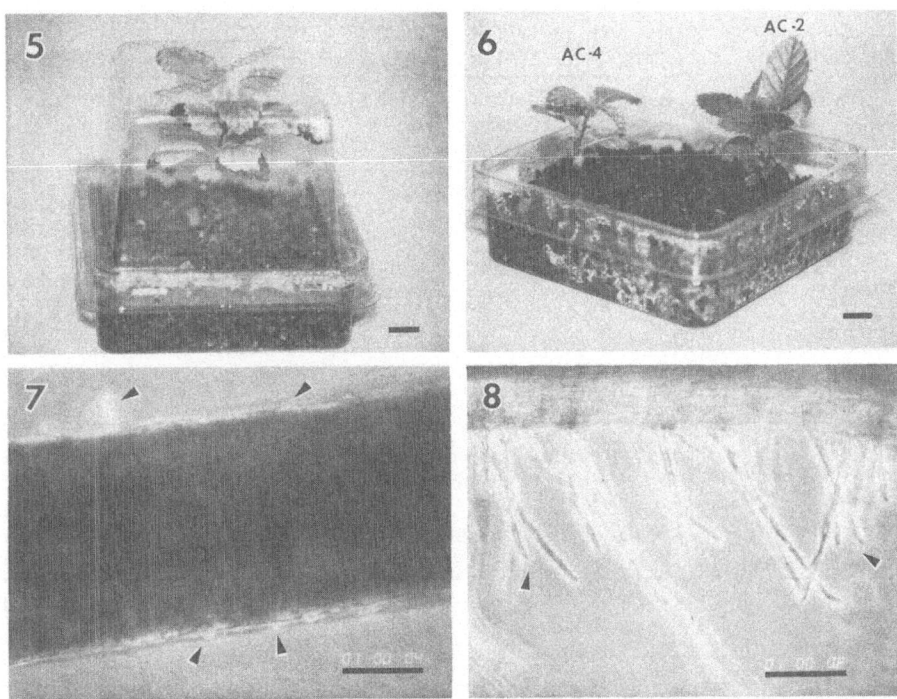

Fig. 5. A non-inoculated control plantlet of th clone AC-2 (Nod⁺) growing in N-free Turface substrate. Symptoms of leaf chlorosis are visibl ?ar = 1 cm.
Fig. 6. A PlantCon without its lid, containing on the left, a non-nodulating plantlet AC-4 (Nod⁻) and on the right a plantlet AC-2 (Nod⁺) 28 days after inoculation with the *Frankia* strain ACN14a. Bar scale = 1 cm.
Fig. 7. Root surface of the clone AC-4 (Nod⁻) showing scarce and short root hairs (arrows). Bar scale = 1 mm.
Fig. 8. Root surface of the clone AC-2 (Nod⁻) showing numerous and typical long root hairs (arrows). Bar scale = 1 mm.

Table 1. Culture media used for *in vitro* propagation of the three clones of *A. crispa* AC-2 (Nod⁺), AC-5 (Nod⁺) and AC-4 (Nod⁻).

Stage	Nodulating clones		Non-nodulating clone
	AC-2	AC-5	AC-4
Initiation	MS + BAP (1 μ*M*) + IBA (0.5 μ*M*) + sucrose (87.5 m*M*)	as for AC-2	as for AC-2
Multiplication	MS + BAP (5 μ*M*) + sucrose (175 m*M*)	MS + BAP (5 μ*M*) + glucose (175 m*M*)	MS + BAP (5 μ*M*) + glucose (175 m*M*)
Rooting	½ MS + IBA (1 μ*M*) + glucose (87.5 m*M*)	½ MS + IBA (1 μ*M*) + glucose (175 m*M*)	½ MS + IBA (1 μ*M*) + glucose or sucrose (87.5 m*M*)

Table 2. Strains of *Frankia* used and number of plantlets of the nodulating clones AC-2 and AC-5 and of the non-nodulating clone AC-4 in the 3 inoculation tests

Inoculation test	*Frankia* strain	Source of *Frankia* (ref. no.)	Nodulating clones		Non-nodulating clone
			AC-2	AC-5	AC-4
1	ACN14$_a$	16	11	19	10
2	ANNI	*	11	21	42
3	AGN1$^{AG}_{exo}$	9	2	3	5
	TN18$^{AC}_a$	16	0	5	5
	ARbN4$_b$	12	0	5	5
	ARI3	2	2	3	5
	ARgN22$_d$	16	2	3	5
total	7		28	59	77
	none (controls)		3	9	

*strain isolated by the OsO$_4$ technique[10] from the one nodule ever produced on the mother-plant AC-4 (Nod⁻) after 1 year in growth chamber

Table 3. Percentage of nodulation of two nodulating clones of *A. crispa* (AC-2 Nod⁺ and AC-5 Nod⁺) and one non-nodulating clone (AC-4 Nod⁻) with different *Frankia* strains

Frankia strains	Nodulating clones					Non-nodulating clone	
	AC-2†		AC-5†			AC-4†	
	NNP*/ NIP	Nodulation %	NNP/ NIP	Nodulation %	Control not inoculated %	NNP/ NIP	%
ACN14$^°_a$ *Frankia*	11/11	100	19/19	100	100	0/10	0
ANNI°	11/11	100	21/21	100	100	16/42	38
ARI3	2/2	100	3/3	100		0/5	0
TN18$^{AC}_a$	–	–	5/5	100		0/5	0
AGN1$^{AG}_{exo}$	2/2	100	2/3	100		0/5	0
ARbN4$_b$	–	–	5/5	100		0/5	0
ARgN22$_d$	2/2	100	3/3	100		0/5	0
Total	28/28	100	59/59	100		16/77	20

†Nitrogen fixation was expressed as visible young green leaves °Nitrogen fixation verified by the acetylene reduction test; in other cases *NNP/NIP = Number of nodulated plantlets/number of inoculated plantlets

nodules formed per plant on 10 nodulated plantlets of each clone was 1.9 nodules on AC-4, 20.6 on AC-5 and 40.0 on AC-2 (Table 4). The acetylene reduction assay performed on nodules from each 3 clones AC-2, AC-5 and AC-4, showed an average rate of $0.5-15\,\mu$mol ethylene $\cdot$ h^{-1} $\cdot$ g (dry weight of nodules)$^{-1}$.

Microscopic observations of the root systems of the plantlets showed the presence of very few root hairs on AC-4 (Nod⁻) plantlets compared with the nodulating clones AC-2 and AC-5. Furthermore, when present the root hairs were abnormally short (Fig. 7 and 8).

Table 4. Mean numbers of nodules (evaluated on 10 nodulated root systems) on nodulated AC-2 (Nod+), AC-5 (Nod+) and AC-4 (Nod−) inoculated with *Frankia* strain ANNI.

| | Number of nodules per plantlet | | |
| | Nodulating clones | | Non-nodulating clone |
Frankia strain	AC-2	AC-5	AC-4
ANNI	40.0	20.6	1.9

Discussion

Following a selection pressure on an actinorhizal host-plant in order to identify a non-nodulating *Alnus* in presence of *Frankia*, the present work on *A. crispa* indicates that some non-nodulating individuals can appear in a population but at a very low occurrence, *i.e.* one in 600,000. The *in vitro* propagation of *A. crispa* has previously been reported[18] but only for the multiplication stage, the rooting being done as cuttings under mist conditions. We now report all *in vitro* stages, including *in vitro* rooting, of *A. crispa* micropropagation. According to our data, different clones of *A. crispa* required different culture media for the multiplication stage as well as for the rooting stage. Consequently, *A. crispa* might be more variable than other species of *Alnus* in culture[4, 17]. The type and concentration of sugar used were essential to define for each clone under study to avoid abnormalities in development, to permit a good multiplication rate and to optimize rooting.

The nodulation tests performed to verify the non-nodulation character of the clone AC-4 (Nod−) compared to the two nodulating clones AC-2 (Nod+) and AC-5 (Nod+) confirmed the abnormality of AC-4 in relation with Frankiae. Six of the seven *Frankia* tested and known for the infectivity and effectivity on *A. crispa*[2, 9, 12, 16] were unable to nodulate with AC-4 (Nod−) while they formed actionorhizae on 100% of the plantlets of the nodulating clones AC-2 (Nod+) and AC-5 (Nod+). The use of positive controls of the inoculum by transferring 2 plantlets (one Nod+, one Nod−) to each container permitted us to conclude that the non-nodulation of the clone AC-4 (Nod−) was not due to the experimental conditions. The *Frankia* strain ANNI isolated from the mother-plant AC-4 (Nod−) was able to nodulate 38% of the AC-4 (Nod−) plantlets, but induced a lower number of nodules, *i.e.* 10 to 20 times less than the number of nodules produced on AC-5 (Nod+) and AC-2 (Nod+) with the same *Frankia* strain ANNI. The differences in the number of nodules on AC-2 (Nod+) and AC-5 (Nod+) confirmed the observations of Dawson *et al.*[3] on the clonal variation between host-plants of the same species in relation with the nodulation. The few nodules produced on AC-4 (Nod−) plantlets by the *Frankia* strain ANNI indicated that the AC-4 (Nod−) clone had not lost its ability to be receptive to *Frankia*. The positive acetylene reduction assay indicated that the AC-4 (Nod⁻) clone, when infected, was still an active host in respect to N_2 fixation by *Frankia*. The *Frankia* strain ANNI was able to nodulate the AC-4 (Nod⁻) clone while the 6 other Frankiae known to be infective on *A. crispa* were not. This *Frankia* strain ANNI might be assumed to be much more aggressive than the 6 other strains tested or was presumably more adapted to the AC-4 (Nod⁻) clone as it was isolated from the initial mother-plant of the AC-4 clone.

Abnormalities in the root hairs formation of AC-4 (Nod⁻) plantlets, *i.e.* few and short root hairs, might explain the difficulty of *Frankia* to infect AC-4 (Nod⁻). With the living root hairs being the site of infection of actinorhizal host-plant by *Frankia*[8, 19], their scarcity and abnormality probably explain the lack of nodulation with 6 strains of *Frankia* tested. Nambiar *et al.*[15] reported for *Rhizobium* a case of non-nodulation on the legume *Arachis hypogeae* lacking root hairs. They suggested an association between the non-nodulation and the absence of root hairs on the host-plant. The low number of nodules produced on the nodulated plantlets AC-4 (Nod⁻) by the *Frankia* strain ANNI might indicate that the AC-4 plantlets still have limited infection sites and, that it can be colonized by a more aggressive *Frankia* strain. This *Alnus* clone AC-4 (Nod⁻) should constitute an important tool to study the root hairs implications in the nodulation process of an actinorhizal host while the *Frankia* strain 'ANNI', which is probably a variant of the original mixture of strains (ACN1^{AG}, AGN1$^{AG}_{exo}$ and MGP10$_i$) used in the initial screenings tests, needs further characterisation to explain its increased infectivity compared to the 6 other strains tested.

Acknowledgements We greatly appreciate the assistance of the Louis St-Laurent with the isolation of the *Frankia* strain ANNI, Luc Simon with the gas chromatography and Danielle Lamontagne with the tissue culture of the *Alnus* clones.

This paper was made possible by an ENFOR project (P-232) from Environment Canada To M.L.

References

1   Akkermans A D L and Houwers A 1979 Symbiotic nitrogen-fixers for use in temperate forestry. *In* Symbiotic Nitrogen Fixation in the Management of temperature Forests. Eds. J C Gordon, C T Wheeler and D A Perry. Oregon State Univ., Corvallis, OR, pp 23–37.

2   Berry A and Torrey J G 1979 Isolation and characterization *in vivo* and *in vitro* of an actinomycetous endophyte from *Alnus rubra* Bong. *In* Symbiotic Nitrogen Fixation in the Management of temperate Forests. Eds. J C Gordon, C T Wheeler and D A Perry. Oregon State University, Corvallis, OR, pp 69–83.

3   Dawson J O and Sun S H 1981 The effect of *Frankia* isolates from *Comptonia peregrina* and *Alnus crispa* on the growth of *Alnus glutinosa, A. cordata* and *A. incana* clones. Can. J. For. Res. 11, 758–762.

4   Garton S, Hosier N A, Read P E, and Farnham R S 1981 *In vitro* propagation on *Alnus glutinosa* Gaertn. Hortic. Sci. 16, 758–759.

5   Hall R B and Maynard C A 1979 Considerations in the genetic improvement of alder. *In* Symbiotic Nitrogen Fixation in the Management of temperate Forests. Eds. J C Gordon, C T Wheeler and D A Perry. Oregon State University, Corvallis, OR. pp 322–344.

6   Jones O P, Hopgood M E and O'Farrell D 1977 Propagation *in vitro* of M26 apple rootstocks. J. Hortic. Sci. 52, 235–238.

7   Koch B and Evans H J 1966 Reduction of acetylene to ethylene by soybean root nodules. Plant Physiol. 41, 1748–1750.

8   Lalonde M 1977 Infection process of the *Alnus* root nodule symbiosis. *In* Recent Developments in Nitrogen Fixation. Eds. W E Newton, J R Postgate and C Rodriguez-Barrueco. Ac. Press, London. pp 569–589.

9   Lalonde M 1979 A simple and rapid method for the isolation, cultivation *in vitro* and characterization of *Frankia* strains from *Alnus* root nodules. *In* Symbiotic Nitrogen Fixation in the Management of temperate Forests. Eds. J C Gordon, C T Wheeler and D A Perry. Oregon State University, Corvallis, OR., p 480.

10 Lalonde M 1979 Techniques and observations of the nitrogen-fixing *Alnus* root nodules symbiosis. *In* Recent Advances in biological Nitrogen Fixation. Ed. N.S. Subba Rao, Oxford & IBH Publ., New Delhi. pp 421–434.

11 Lalonde M and Calvert H E 1979 Production of *Frankia* hyphae and spores as an infective inoculant for *Alnus* species. *In* Symbiotic Nitrogen Fixation in the Management of temperate Forests. Eds. J C Gordon, C T Wheeler and D A Perry. Oregon State University, Corvallis, OR, pp 95–110.

12 Lalonde M, Calvert H E and Pine S 1981 Isolation and use of *Frankia* strains in actinorhizae formation. *In* Current Perspectives in Nitrogen Fixation. Eds. A H Gibson and W E Newton. Australian Academy of Science, Canberra. pp 296–299.

13 Mort A, Normand P and Lalonde M 1983 2-O-Methyl-D-mannose, a key sugar in the taxonomy of *Frankia*. Can. J. Microbiol. 29, 00–00.

14 Murashige T and Skoog F 1962 A revised medium for rapid growth and bioassays with tobacco tissue culture. Physiol. Plant 15, 473–497.

15 Nambiar P T C, Nigam S N, Dart P J and Gibbons R W 1983 Absence of root hairs in non-nodulating groundnut, *Arachis hypogeae* L. J. Exp. Bot. 34, 484–488.

16 Normand P, and Lalonde M 1982 Evaluation of *Frankia* strains isolated from provenances of two *Alnus* species. Can. J. Microbiol. 28, 1133–1142.

17 Perinet P and Lalonde M 1983 *In vitro* propagation and nodulation of the actinorhizal host plant *Alnus glutinosa* (L.) Gaertn. Plant Sci. Lett. 29, 9–17.

18 Read P E, Garton S, Louis K and Zimmerman E S 1982 *In vitro* propagation of species for bioenergy plantations. *In* Proc. 5th Intern. Cong. Plant Tissue and Cell Culture. Ed. Akio Fujiwara. pp 757–758.

19 Torrey J G and Callaham D 1979 Early nodule development in *Myrica gale*. Bot. Gaz. 140, (S), S10–S14.

Plant and Soil 78, 181–188 (1984).
© 1984 *Martinus Nijhoff/Dr W. Junk Publishers, The Hague.*

Ms. Fr 20

Seasonal variations of the sexual reproductive growth and nitrogenase activity (C₂ H₂) in mature *Alnus glutinosa*

G. PIZELLE
Physiologie végétale, Faculté des Sciences, B.P. n° 239, F. 54506 Vandoeuvre les Nancy Cédex, France

Key words *Alnus glutinosa* Carbon-nitrogen Competition *Frankia* symbioses Nitrogenase activity Sexual reproductive growth

Summary The seasonal variations of the growth of sexual reproductive organs and of the nitrogenase activity (acetylene reduction) of root nodules are surveyed in mature field alders (*Alnus glutinosa*). The growth of female catkins – pollinated in February-early March – takes place chiefly from June to August and the growth of immature male catkins from July to September. The nitrogenase activity steadily shows two periods of high rate – the first from late April to early June, the second in September-October – and a summer period of low rate when the female catkins and the seeds achieve the most part of their growth.

The seasonal fluctuations of the *in vitro/in vivo* nitrogenase activity ratio showing the supply of metabolic factors in the root nodules as a likely cause of the variations of the *in vivo* nitrogenase activity, the possible competition for photosynthate allocation between the production of sexual organs and the nitrogen-fixing capacity in mature field alders is discussed.

Introduction

Several studies conducted with glasshouse-grown plants show that the nitrogenase activity in root nodules or actinorhizae of immature *Alnus glutinosa* are in close relation with other physiological processes of the host-plant, especially with the photosynthesis; indeed, every factor altering the photosynthetic capacity (light intensity, foliar area) or the translocation of photosynthate to nodules (girdling) alters the nitrogenase activity of young alders in the same way [4,6,12].

In field mature *Alnus glutinosa,* the seasonal variations of the nitrogenase activity have been studied on whole nodules or nodular lobes[1,3,9,10], but the relations of this enzymatic activity with the other physiological processes, particularly with the sexual reproductive cycle, are not yet documented. Now, the annual curves of nitrogenase activity fluctuate greatly. In winter the enzymatic activity is absent; during the growing season, the curves show two peaks, one in spring between late April and mid-June the other in September and October, separated by a low level in July and August. The summer depression of the enzymatic activity, which does not appear in data of other authors[1,3] is not accidental in mature alders since it has been observed during several years[9,10].

The present study examines possible causes of these important fluctuations of the nitrogenase activity. At first, the *in vivo* and *in vitro* nitrogenase activities are compared during the year in order to know whether the seasonal enzymatic variations are due to a variation of the amount of active enzyme or to the availability of metabolic factors necessary for the enzymatic activity. After which, the possible relation between growth of the sexual organs and nitrogenase activity is investigated.

Materials and methods

Plant material

Nodules and reproductive organs were harvested on twenty to thirty-year-old *Alnus glutinosa* naturally growing near Nancy in the Moselle valley. Nodular lobes were dissociated from nodules of four or five trees and pooled before measurement of nitrogenase activity. The samples of reproductive organs were composed of five clusters of male or female catkins taken from different trees. The dry matter of plant material was obtained by desiccation in a ventilated oven at 80°C. The nitrogenase activities were estimated by the acetylene reduction method.

In vivo *nitrogenase activity*

Ethylene production by 0.1 to 1.0 g of nodular lobes was measured after 20 min of incubation in experimental conditions previously described[10].

In vitro *nitrogenase activity*

1500 mg of fresh nodular lobes were homogenized in 20 ml of tris-HC1 buffer $0.05 M$ (pH 8) containing soluble polyvinylpyrrolidones (360, 40, 10) 1% of each, sucrose $0.15 M$, dithioerythritol 5 mM, sodium dithionite 100 mM, normal octyl alcohol (anti-foam) 4 drops.

The nodular lobes were crushed with an Ultra-Turrax T 35 in a tight vial flushed with argon and dipped in ice water; homogenization time was 45 sec divided in three periods of 15 sec at 10,000 rev/min. A beige colored suspension was obtained in which the most part of *Frankia* vesicles were released from host nodular cells.

The incubation vials (screw cap septum vials of 16.5 ml) contained 1.9 ml of a suspension including 0.9 ml of homogenate (about 67 mg of nodular freshweight) and 1 ml of a solution of such composition that the final concentration of incubation medium was: tris-HC1 buffer $0.05 M$ (pH 8), PVPs (360, 40, 10) 1% of each, sucrose $0.075 M$, dithioerythritol 2.5 mM, sodium dithionite 100 mM. ATP (neutralized by KOH) 10 mM, MgC1$_2$ 10 mM. The gas phase was 10% acetylene in argon.

Ethylene produced at 20−22°C was measured after 20 min of incubation with agitation (120 osc./min).

The *in vivo* and *in vitro* nitrogenase activities are expressed in micromoles $C_2H_4 \cdot g^{-1}$ nodular dry weight. h^{-1}.

Total carbon and nitrogen

In reproductive organs, the total carbon and nitrogen contents were estimated by the Anne method (bichromate-sulfuric acid oxidation) and the Kjeldahl method respectively.

Results and discussion

Seasonal variations of the in vivo *and* in vitro *nitrogenase activities in mature* Alnus glutinosa

The *in vitro* nitrogenase activity measured on nodular homogenates in definite and constant medium evaluates the amount of active enzyme present in the nodules; whereas the *in vivo* nitrogenase activity measured on nodular lobes evaluates both the amount of active enzyme and the nodular metabolic supply needed for the enzymatic activity as

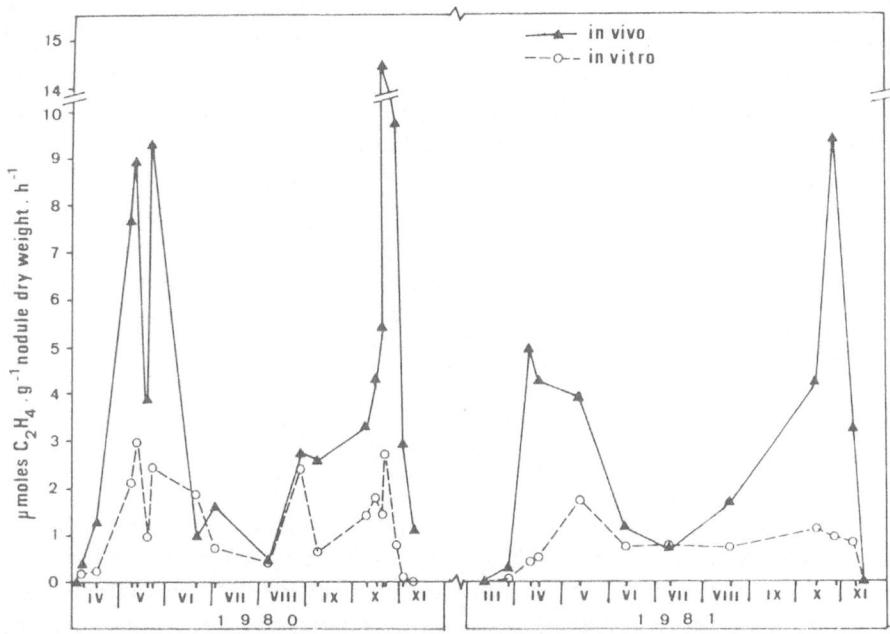

Fig. 1. Seasonal variations of the *in vivo* and *in vitro* nitrogenase activity of root nodules of mature *Alnus glutinosa*. Each point represents the mean of two measurements on pooled nodular lobes of five trees.

a whole. If the seasonal variations of the *in vivo* nitrogenase activity are only due to a variation of the amount of enzyme, the *in vitro/in vivo* nitrogenase activity ratio should be approximately constant during the year; but if they are, partly at least, due to a variation of the availability of necessary metabolites, the ratio should vary according to this availability.

In practice, the *in vivo* and *in vitro* nitrogenase activities have been followed during two years (Fig. 1). It appeared that no active nitrogenase existed in nodules in winter, since both activities, *in vivo* and *in vitro,* were absent. During the growing season, the curve of the *in vivo* nitrogenase activity showed the two typical previously observed[9,10]. Moreover it can be noted that the *in vitro* nitrogenase activity fluctuated less than the *in vivo* one did. In spring and autumn the curve of the *in vivo* enzymatic activity went far above the other curve; while in summer, when *in vivo* nitrogenase activity is low, both curves came close. In the data already published the *in vitro* nitrogenase activity of nodular homogenates of *Alnus glutinosa* was about three to four times lower than the *in vivo* activity of the unbroken nodules[2,11]. In fact, the data of Fig. 1 show that the *in vitro/in vivo* nitrogenase activity ratio greatly fluctuated during the growing season with minimal values —

usually lower than 0.5 — in spring and autumn and maximal values — possibly higher than 1 — in summer. These fluctuations indicate that the seasonal variations of the *in vivo* nitrogenase activity may be caused by metabolic limiting factor(s), the intranodular availability of which might be greater in spring and autumn than in summer.

Seeing that photosynthate is accepted as a "major limiting factor" of the nitrogenase activity in *Alnus* and other symbiotic diazotrophes[7,12], it seems attractive to investigate the possible relations between the nitrogenase activity and a physiological process likely competitive for the photosynthate supply to the different parts of mature *Alnus:* the development of the sexual organs.

Seasonal variations of sexual reproductive growth and in vivo *nitrogenase activity in mature* Alnus glutinosa

As may be seen in Table 1 and Fig. 2, the life cycle of sexual reproductive organs — from initiation of male and female catkins to release of mature seeds — lasts about fifteen months in *Alnus glutinosa.* The male catkins decayed about nine months after their initiation, one generation of these organs may be found on the trees between July and the early spring. On the other hand, the shedding of dead female catkins occurring several months after complete release of seeds, the trees usually bear three successive generations of female organs — empty of seeds, maturing, and unfertilized catkins respectively — in summer and early autumn.

Except during the periods of winter frost, the male catkins grow continuously from the initiation to the pollination with most increase in dry weight from late August to early October. They are chlorophyllous until December, then become brownish-red before pollination. The significant growth of male organs in winter indicates that reserves of the tree can be mobilized for sexual reproductive growth long before the return of vegetative growth.

Unlike the male catkins, the female catkins show no growth in winter; their notable enlargement begins after pollination, and the greatest increase in size and dry weight takes place from June to August. They remain chlorophyllous until the release of mature seeds which begins in October.

The growth of seeds — usually 60 to 70 seeds per catkin — is rapid in July and August; the mass of seeds is about a quarter of a mass of whole immature catkin. From early September the seeds inside the catkins are already mature and able to germinate without cold treatment (Table 1); the browning of their pericarp coincides with maturation which occurs almost two months before the opening of catkins.

Table 1. Phenology of reproductive development in mature *Alnus glutinosa* growing near Nancy (France)

Date	Male catkins	Female catkins	Seeds = Achenes
Early Jul.	Initiation	Initiation	–
From Jul. to Feb.	Growth ± continuous	Dormant	–
Feb.-early Mar.	Anthesis, pollination	Anthesis, stigma expansion	–
Apr.	Abscission complete	? Fertilization	
May	–	Beginning of growth	Pericarp white
From Jun. to Aug.	–	Most increase in dry weight	Most increase in dry weight
Late Aug.-early Sep.	–	Growth reduced	Browning of the pericarp maturation complete
Oct.	–	Growth complete	Beginning of release

An assay with seeds extracted from catkins in early September 1982 gave 10%, 13%, 17%, and 34% of germination after 4, 6, 7, and 10 days respectively.

Table 2. Dry weight(g) of the different organs borne on twigs of mature *Alnus glutinosa* (17 Aug. 1983)

Sample	Twig	Leaves	Female catkins	Male catkins
1	8.7	4.7	3.6 (20)	0.044 (3)
2	14.3	7.2	4.7 (27)	0.131 (17)
3	14.7	11.2	10.9 (38)	0.322 (16)
4	31.1	30.0	7.0 (37)	0.429 (35)

Female catkins initiated in July 1982. Male catkins initiated in July 1983; the female catkins of same age were too small to be harvested. The number of catkins is given in brackets.

The part of sexual organs in the total dry matter of the twigs is noteworthy at some periods of the year: in summer and early autumn the mass of female catkins may be equal to the foliar mass (Table 2).

The total carbon content of sexual organs is generally high, namely near or higher than 50% except for the dead female catkins after complete seed release; it varies less than the total nitrogen content during catkin development (Table 3). The total nitrogen content of the male catkins clearly decreases after scattering of the pollen which is likely richer in nitrogen than the rest of the organ. The rapid growth of female catkins after fertilization coincides with a drop of nitrogen percentage affecting the catkin parts (scales and axis) other than the seeds. It may be noted here that the dead catkins have an higher C/N ratio and are poorer in total nitrogen than the leaves in the litter of *Alnus glutinosa*.

From these quantitative data it appears that the deciduous catkins of mature alders, especially the female catkins, accumulate an important amount of organic carbon. Though these organs are chlorophyllous till maturation, their respiratory activity likely prevails far over their

Table 3. Total carbon and nitrogen content (%) of male and female catkins at different developmental stages in mature *Alnus glutinosa*

Organ	Date	Stage	%C	%N	C/N
Male catkins	July 20, 82	Immature°	55.8	2.5	22.3
	Sept. 13, 82	Immature°	56.4	2.3	24.5
	Nov. 3, 82	Immature°	56.4	2.0	28.2
	Jan. 9, 83	Before pollination°	56.7	2.3	24.7
	April 5, 82	After pollination°°	53.4	1.7	31.4
Female catkins	Jan. 9, 83	Before fertilization°	57.3	3.1	18.5
	May 5, 82	Just fertilized°°	54.6	2.7	20.2
	July 6, 82	Rapid growth°°	52.8	1.4	37.7
	Sept. 13, 82	Reduced growth°°	49.2(55.5)	1.4(2.4)	35.1(23.1)
	Oct. 4, 82	Seeds mature in catkins°°	51.3(54.6)	1.3(2.5)	39.5(21.8)
	July 6, 82	Empty dead catkins°°°	44.1	1.6	27.6

°Catkins initiated in July 1982. °°Catkins initiated in July 1981. °°°Catkins initiated in July 1980. Data in brackets are given by the seeds extracted from catkins.

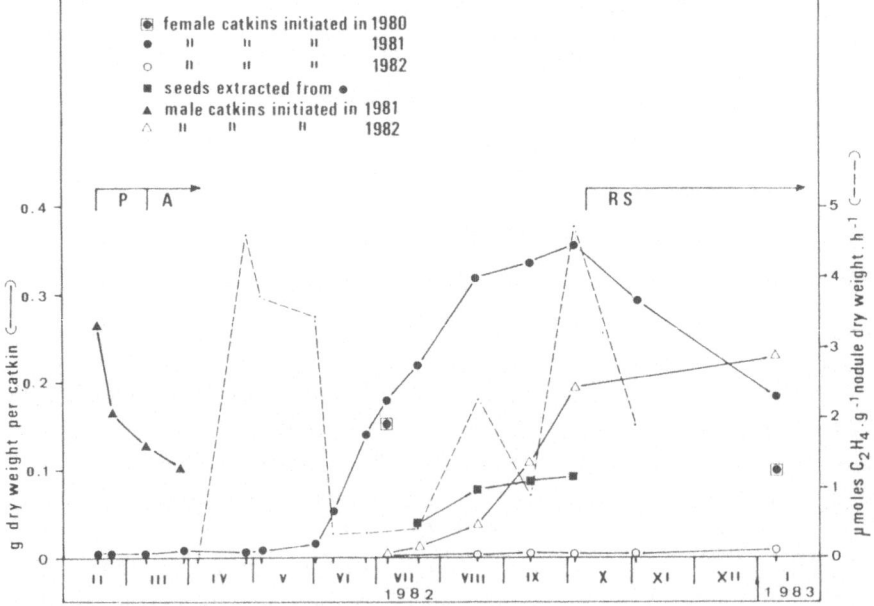

Fig. 2. Seasonal variations of the sexual reproductive growth and *in vivo* nodular nitrogenase activity in mature *Alnus glutinosa*. The dry weight of female catkins includes the mass of seeds kept inside the catkins. P: pollination, A: abscission of male catkins initiated in July 1981, RS: release of seeds of female catkins initiated in July 1981.

photosynthetic activity owing to the low surface to volume ratio of cylindrical male catkins and of ovoid female catkins. So, the growing reproductive tissues probably mobilize an important amount of photoassimilates.

The juxtaposition of the annual curves of sexual reproductive growth and of *in vivo* nitrogenase activity (Fig. 2) brings to light several typical sequences during the growing season.

In spring, the peak of nitrogenase activity occurs between the abscission of male catkins and the beginning of enlargement of the female catkins, while the foliar expansion is rapid and the mass of living sexual organs borne by the tree is minimal.

In early summer, the nitrogenase activity drops while the enlargement of female catkins begins; the enzymatic activity remains low as long as the increase in dry weight of these organs is rapid.

Afterwards, in late summer, the nitrogenase activity rises again and reaches a second peak between late September and mid-October when the maturation of seeds is complete and the growth of immature male catkins reduced.

These alternate periods of high nitrogenase activity and of rapid growth of catkins suggest that the sexual reproductive development could control the nitrogenase activity through competition for photosynthetic products. The mobilization of photosynthate respectively by the catkins, shoots, trunk, roots and nodules of mature *Alnus glutinosa* should be measured during the year to confirm this suggestion; but the data showing the growing reproductive tissues as major carbohydrate sinks in forest trees[5,8] support such a conclusion.

Whatever the causes, the practical fact remains that the nitrogen-fixing capacity of mature field *Alnus glutinosa* is probably lower in summer than in spring and in autumn.

The present work finally underlines that, in the study of *Frankia*-woody plant symbioses, the results given by the immature glasshouse-grown plant are not directly applicable to the mature field tree because major biological and physiological differences separate the two developmental stages, especially the amount of reserves, the distances of metabolite translocations, and the sexual reproductive growth.

Acknowledgements The study was partly supported by aid n° 80 54 229 from COMES. The technical assistance of J. Banvoy in C and N analyses is gratefully acknowledged.

References

1   Akkermans A D L 1971 Nitrogen fixation and nodulation of *Alnus* and *Hippophaë* under natural conditions. Thesis, Leiden, 85p.
2   Akkermans A D L, Roelofsen W and Blom J 1979 Dinitrogen fixation and ammonia assimilation in actinomycetous root nodules of *Alnus glutinosa*. *In* Symbiotic Nitrogen Fixation in the Management of temperate Forests. Eds. J C Gordon, C T Wheeler and D A Perry. Forest Res. Lab., Oregon State Univ., Corvallis, pp 160–174.
3   Akkermans A D L and Van Dijk C 1976 The formation and nitrogen-fixing activity of the root nodules of *Alnus glutinosa* under field conditions. *In* Symbiotic Nitrogen Fixation in Plants. Ed. P S Nutman, Cambridge Univ. Press, Cambridge, pp 511–520.
4   Dawson J O and Gordon J C 1979 Nitrogen fixation in relation to photosynthesis in *Alnus glutinosa*. Bot. Gaz., 140 suppl., 70–75.
5   Dickmann D I and Kozlowsky T T 1968 Mobilization by *Pinus resinosa* cones and shoots of C^{14} photosynthate from needles of different ages. Am. J. Bot. 55, 900–906.

6   Gordon J C and Wheeler C T 1978 Whole plant studies on photosynthesis and acetylene reduction in *Alnus glutinosa*. New Phytol. 80, 179–186.
7   Hardy R W F and Havelka U D 1976 Photosynthate as a major factor limiting nitrogen fixation by field-grown legumes with emphasis on soybeans. *In* Symbiotic Nitrogen Fixation in Plants. Ed P S Nutman. Cambridge Univ. Press, Cambridge, pp 421–439.
8   Kramer P J and Kozlowsky T T 1979 Physiology of Woody Plants. Academic Press, New York, pp 400.
9   Pizelle G 1975 Variations saisonniéres de l'activité nitrogénasique des nodules d'*Alnus glutinosa* (L.) Gaertn., d'*Alnus incana* (L.) Moench et d'*Alnus cordata* (Lois.) Desf.. C.R. Acad. Sc., Paris, D, 281, 1829–1832.
10 Pizelle G and Thiery G A 1977 Variations saisonniéres des activités nitrogénase et nitrate réductase chez l'Aune glutineux (*Alnus glutinosa* L. Gaertn.). Physiol. Veg., 15, 333–342.
11 Straten J van, Akkermans A D L and Roelofsen W 1977 Nitrogenase activity of endophyte suspensions derived from root nodules of *Alnus, Hippophaë, Shepherdia* and *Myrica* spp. Nature London 266, 257–258.
12 Wheeler C T and Lawrie A C 1976 Nitrogen fixation in root nodules of alder and pea in relation to the supply of photosynthetic assimilates. *In* Symbiotic Nitrogen Fixation in Plants. Ed. P S Nutman, Cambridge Univ. Press, Cambridge, pp 497–510.

Plant and Soil 78, 189–199 (1984).
© 1984 *Martinus Nijhoff/Dr W. Junk Publishers, The Hague.*

Ms. Fr 2

Mycorrhizal improvement in non-leguminous nitrogen fixing associations with particular reference to *Hippophaë rhamnoides* L.

I. C. GARDNER, D. M. CLELLAND and A. SCOTT
Biology Division, Department of Bioscience and Biotechnology, University of Strathclyde, Glasgow G1 1XW, UK

Key words *Alnus* *Hippophaë* Mycorrhiza *Myrica* Nitrogenase Phosphate Triple symbiosis Ultrastructure

Summary The roots of *Hippophaë rhamnoides* which regularly bear actinomycete induced nodules when growing on Scottish sand dunes have also been found to support an endomycorrhizal association with *Glomus fasciculatus*. Ultrastructural and cytochemical studies carried out on the indigenous infections of established *Hippophaë* mycorrhizal roots would support the postulate that transport is indeed occurring between the fungal symbiont and the host plant and vice versa in respect of phosphate and carbohydrate. Experiments using various inoculation regimes, demonstrated the significant improvement in the mycorrhizal/nodulated plants compared to the nodulated-only and the mycorrhizal-only plants with respect to plant growth, uptake of phosphate and nitrogenase activity, when grown in a medium poor in combined nitrogen and soluble phosphate. Preliminary work on *Alnus* and *Myrica* species growing in Central Scotland indicates that the mycorrhizae associated with these nodulated root systems exhibit a different interaction pattern which may be dependent on habitat type and associated angiosperm species.

Introduction

An adequate supply of phosphate is an essential requirement for biological nitrogen fixation. That a mycorrhizal association can improve the uptake of phosphate by a plant is now well established[15] and it has been shown by many workers that the presence of a vesicular-arbuscular mycorrhiza in the roots of leguminous plants growing in phosphate poor media increases phosphate uptake and stimulates nitrogenase activity[2,4,5]. Non-leguminous plants possessing actinorhizal nodules typically colonise nutrient-deficient habitats where the presence of a mycorrhizal association to enhance mineral uptake would be highly beneficial. Comparatively little is known of the mycorrhizal status of such plants. Mycorrhizal associations have been reported for some actinorhizal plants in U.S.A.[11,21,23,24]. It has further been reported that mycorrhizal-nodulated plants of *Ceanothus velutinus* showed increased growth, higher levels of N, Ca^{2+} and P and an increased nitrogenase activity[22]. Recently the present authors[7] have shown that *Hippophaë rhamnoides,* growing on the early colonisation stages of sand dunes in Scotland, supports a vesicular-arbuscular mycorrhiza and the fungal endophyte has been identified as *Glomus fasciculatus*

Table 1. Type of mycorrhiza associated with actinorhizal plants native to Scotland

Host plant	Location (National Grid Ref)	Habitat	Type
Hippophaë rhamnoides	Tentsmuir (NO/505267)	Sand dune complex	VA
	Gullane (NT/435768)	Sand dune complex	VA
	Prestwick (NS/339284)	Sand dune complex	VA
Alnus glutinosa	Kilsyth (NS/776714)	Coal tip	Ecto-
	Stockiemuir (NS/848494)	Stream side	Ecto-
	Stirling (NS/923799)	Urban parkland	VA and Ecto-
Myrica gale	Stockiemuir (NS/811508)	*Calluna* bog	Endo-, not VA

The present work extends these studies on *Hippophaë*. The role of the arbuscule in metabolite exchange has been examined using ultra-structural and cytochemical techniques and the effect of the my-corrhizal infection on the growth, phosphate uptake and nitrogenase activity of the plants has been investigated. In addition the mycorrhizal status of the other actinorhizal species native to Britain, *Alnus glutinosa* and *Myrica gale,* has been examined.

Materials and methods

Examination of roots for mycorrhizal infection

Roots of *Hippophaë rhamnoides, Alnus glutinosa* and *Myrica gale* were collected from various sites in central Scotland (Table 1). The roots were first cleared and stained with Azure blue in lactic acid[20]. A root slide technique was then employed to assess the mycorrhizal status of the roots[10].

Ultrastructural studies

For SEM (scanning electron microscopy), segments of mycorrhizal roots were first fixed as for TEM (transmission electron microscopy). These were then cut along the longitudinal axis and the ruptured host cells cleared of cytoplasm[17]. The tissue was subsequently critical-point dried[1], coated with a 40 nm layer of gold/palladium and examined in a Jeol JSM 35 microscope operating at 10 kV.

For TEM the tissue was fixed in 2.5% glutaraldehyde in 0.1 M cacodylate buffer pH 7.4 for a minimum of 4 hours at 4°C, rinsed in several changes of fresh cacodylate buffer and post-fixed in aqueous osmium tetroxide at room temperature for 1 hour. After dehydration, the tissue was unfiltrated with and embedded in Taab resin formulated to give hard block [25 parts resin: 18 parts DDSA (dodecenylsuccinic anhydride): 7 parts MNA (methyl nadic anhydride): 1 part DMP (2,4,6 tri (dimethylamino-methyl) phenol)]. Sections, on copper grids, were stained routinely with uranyl and lead and viewed in an AEI EM6B microscope operating at 50 kV.

Localisation of glycogen was carried out using Thiery's method as modified for use in our laboratory[18]. Alkaline phosphatase was localised using a modified Gomori technique[14].

Growth studies

Vesicular-arbuscular mycorrhizal fungi are obligate biotrophs but they can be "cultured" and maintained in the roots of *Zea mays* plants[8]. The endophyte from field *Hippophaë* was cultured thus and an inoculum, prepared by crushing the *Zea* roots, was added to the sand in 5 inch pots together with the *Hippophaë* seed.

Four growth regimes were employed – mycorrhizal/nodulated, nodulated only, mycorrhizal only and non-mycorrhizal/non-nodulated. Nodulation was achieved using a crushed nodule suspension as inoculum. This inoculum was also added to the sand with the *Hippophaë* seeds. Since it is known that high levels of soluble phosphate inhibit the establishment of a vesicular-arbuscular mycorrhizal association and that high levels of combined nitrogen inhibit nodulation and nitrogenase activity, all plants were grown in washed sterilised sand watered twice weekly with half strength Long Ashton nutrient solution minus both phosphate and nitrate. A source of "insoluble" or less available phosphate, in the form of Ca_3 $(PO_4)_2$, was added to all pots at the rate of 61.8 mg per kg sand at the beginning of the experiment[9]. The plants were grown in a cool greenhouse and harvested after twelve weeks.

At harvest the plants were carefully removed from the sand and placed in 500 ml conical flasks fitted with rubber membrane stoppers. Ten per cent (v/v) of the gas phase in these flasks was replaced by acetylene and the nitrogenase activity of the plants assayed by means of acetylene reduction[3].

Plant fresh and dry weights were then recorded and total phosphate estimated colorimetrically[12].

Results

Ultrastructural studies

Arbuscules were found predominantly in the inner cortical cells of the root in close proximity to the vascular tissue. SEM of these host cells, ruptured and cleared of cytoplasm, revealed that the arbuscules occupied a substantial proportion of the host cell volume. Fig. 1 clearly shows the complex three dimensional structure of the arbuscule. The repeated dichotomous branching of the intracellular hypha results in a succession of progressively narrower hyphae being produced, those of smallest diameter occurring around the periphery of the arbuscule. This extensive dichotomy of the hypha substantially increases the surface area of the host/fungal interface.

TEM showed that these cells containing arbuscules had a markedly increased volume of host cytoplasm (Fig. 2) compared to the adjacent uninfected cells which were highly vacuolate. The cytoplasm of the infected cells contained numerous mitochondria, plastids, endoplasmic reticulum and lipid droplets (Fig. 3). A membrane of host origin separated the arbuscule from the host cytoplasm. The main trunk hyphae of the arbuscules contained abundant polyphosphate granules and rosettes of glycogen whereas the fine terminal branches were depleted of polyphosphate and had less obvious accumulations of glycogen (Figs. 2, 3 and 5). The electron dense polyphosphate granules

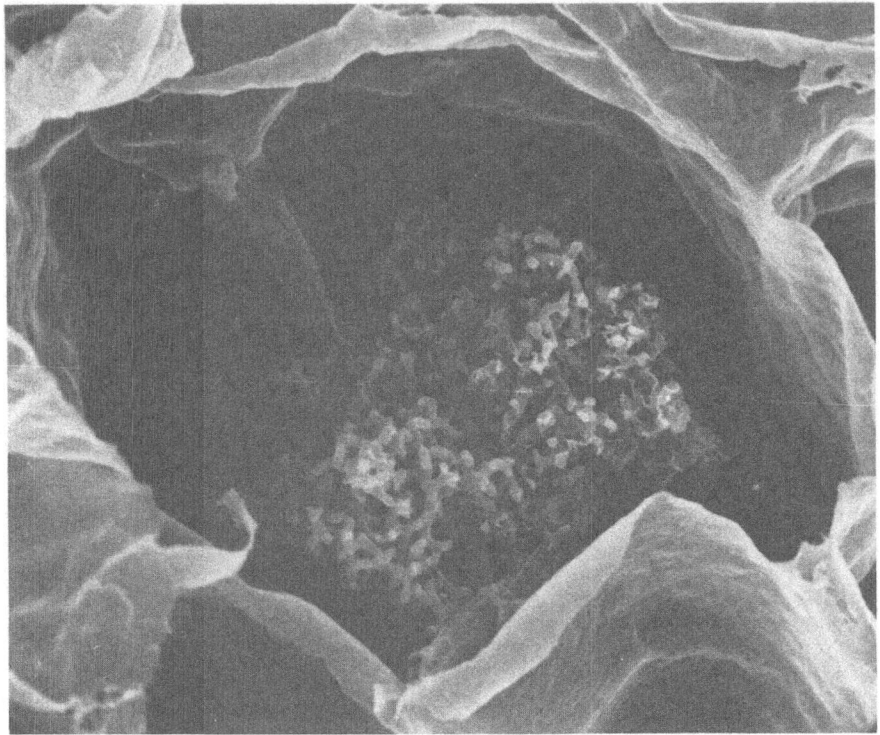

Fig. 1. Scanning micrograph showing a cell from the inner cortical region of the root which has been ruptured and cleared of cytoplasm. The main trunk hypha of the fungus can be seen entering the cell at the top left and the extensive dichotomous branching of the arbuscule is clearly evident. × 3,300.

occurred within the extensive vacuolar system of the arbuscular hyphae (Fig. 3). Alkaline phosphatase was also localised within these vacuoles, occurring most prominently in those near the hyphal tips in mature arbuscules (Fig. 4). Lipid droplets were seen throughout the hyphal cytoplasm but occurred most abundantly in the main trunk hyphae and in the intercellular hyphae where rosettes of glycogen were in close approximation to the lipid (Fig. 5). Complex membranous structures were present in the cytoplasm of the fine hyphae. These arose from invaginations of the fungal plasma membrane.

Growth studies

At harvest the root systems of all the plants were examined for the occurrence of mycorrhizal infection and nodulation. None of the uninoculated plants had in any way developed any infection and there was no increase in nodule number in those nodulated plants which had acquired a mycorrhizal association. Those mycorrhizal/nodulated plants, however, sustained a 34% mycorrhizal infection as compared to a 17% infection for the mycorrhizal only plants.

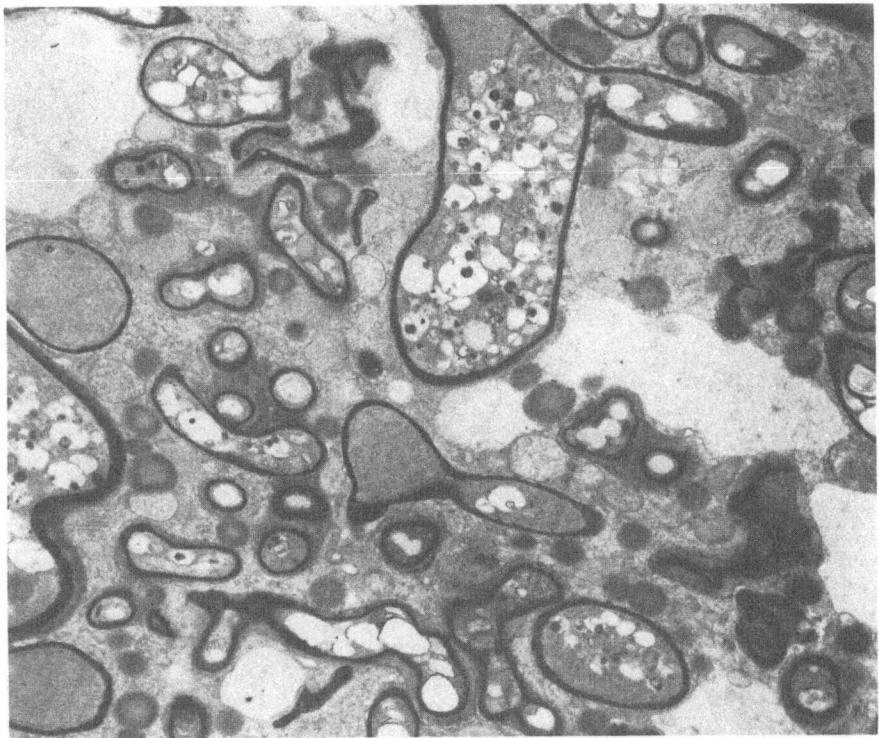

Fig. 2. Low power transmission electron micrograph of part of a host cell containing a fungal arbuscule. A section through the main trunk hypha can be seen at the top middle and sections of the finer arbuscule branches are scattered through the host cytoplasm. × 8,100.

Fig. 6 shows the plant growth, uptake of phosphate and nitrogenase activity for the different plant regimes. Results in each case were calculated on the basis that the mycorrhizal/nodulated group represents 100% for each parameter considered. On a plant fresh weight basis, there was a significant difference between the two nodulated groups and the two non-nodulated groups ($P = 0.01$) but the difference between the mycorrhizal/nodulated and the nodulated only plants was significant only at $P = 0.05$. There was, however, a significant ($P = 0.01$) improvement in the mycorrhizal/nodulated plants compared to those of all the other regimes with respect to plant dry weight and uptake of phosphate. Additionally there was a very greatly enhanced nitrogenase activity in the mycorrhizal/nodulated plants compared to plants nodulated only.

Mycorrhizal status

In every case of roots sampled from the three actinorhizal species native to Britain, a mycorrhizal infection was found (Table 1). In

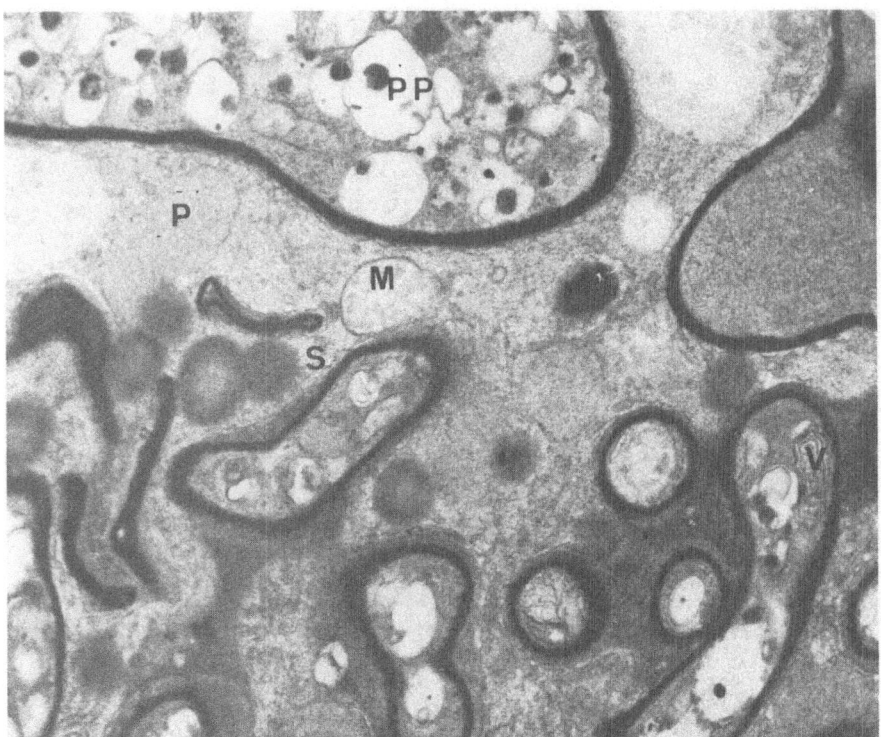

Fig. 3. Part of Fig. 2 at higher magnification to show details of host and fungal cytoplasms, mitochondria **(M)**, plastids **(P)**, multi-vesicular bodies **(V)**, polyphosphate granules **(PP)** and the membrane separating host and fungus **(S)**. × 18,100.

Hippophaë rhamnoides where the habitat was the sand dune complex and the samples were all taken from the early colonisation stages the mycorrhizal type found was vesicular-arbuscular and in each case the fungal species was *Glomus fasciculatus.* In *Alnus glutinosa* ecto-mycorrhizal infections were found in samples from three different habitats, namely, coaltip, stream side and urban parkland. Only in *Alnus* from this last site was a vesicular-arbuscular infection also present. In *Myrica gale* samples from *Calluna* bog the infection was endo-mycorrhizal but not of the vesicular-arbuscular type. No attempts have been made to date to identify the fungal species in *M. gale* and *A. glutinosa.*

Discussion

The mycorrhizal status of the actinorhizal plants investigated in Scotland is rather complex. Only in *Hippophaë* is the infection exclusively vesicular-arbuscular and caused by a single fungal species, *G. fasciculatus.* In the case of *A. glutinosa,* plants growing in different

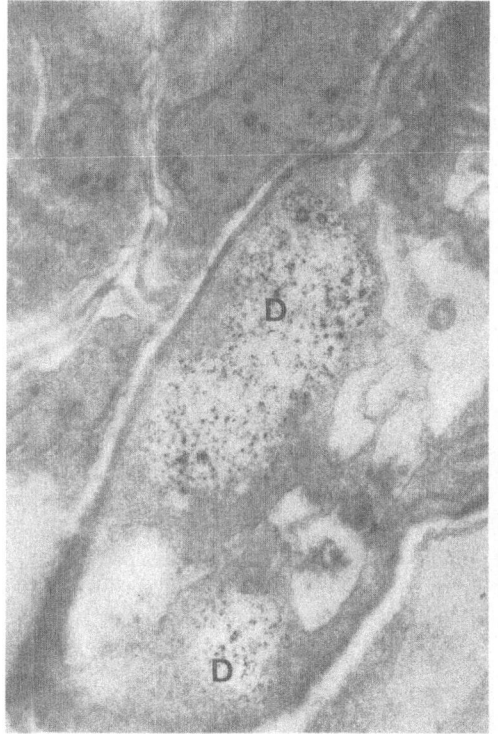

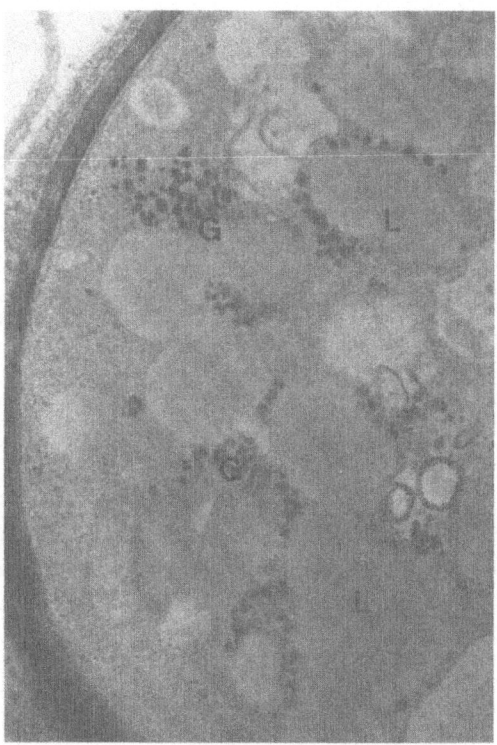

Fig. 4. Material incubated for the presence of alkaline phosphatase. Section through the hyphal tip of a mature arbuscule shows the electron dense reaction product of the alkaline phosphatase localised within the hyphal vacuoles **(D)**. × 30,300.

Fig. 5. Material treated with Thiery stain to localise carbohydrate. Section through the main trunk hypha of an arbuscule shows the rosettes of glycogen **(G)** in close proximity to the lipid droplets **(L)**. × 32,000.

habitats supported different types of mycorrhiza while in *M.gale* the association was endomycorrhizal although not of the vesicular-arbuscular type. Reports of other workers reveal a similar variation. In coastal sand dune areas in the Netherlands nodulated *H. rhamnoides* showed well developed ectomycorrhiza[19] and *A. glutinosa* growing on sandy aluvium in England is reported as having a vesicular-arbuscular mycorrhiza while *A. rubra* on coastal dunes in Oregon and *A. incana* on sandy river aluvium in Oregon have both ecto and vesicular-arbuscular systems on their roots[21]. *A. glutinosa* on bituminous coal wastes in Pennsylvania again had only a vesicular-arbuscular association[11] as did *M. gale* growing in peat and *Sphagnum* bogs in Massachusetts[21]. It would thus appear that the mycorrhizas associated with these nodulated root systems exhibit a varied interaction pattern which may be dependent on habitat type and on the associated

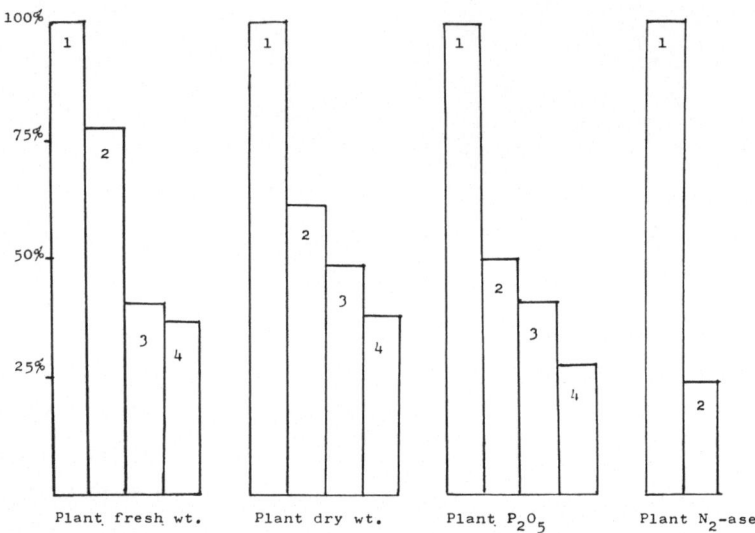

Fig. 6. Effect of various culture regimes on plant growth, uptake of phosphate and nitrogenase activity. 1, mycorrhizal/nodulated; 2, nodulated only; 3, mycorrhizal only; 4, non-mycorrhizal/non-nodulated.

angiosperm species. Indeed it is of interest that the *Ammophila* and *Agropyron* species growing with *Hippophaë* on the pioneer dunes in Scotland were both found to possess *G. fasciculatus* associations[6]. This could have important implications with respect to nutrient transfer if the root systems of these three plants were to be connected by a network of *Glomus* mycelium.

The marked increase in the volume of cytoplasm and in the number of organelles in the root cells containing arbuscules together with the increased surface area of the host/endophyte interface afforded by arbuscular formation, would substantiate the proposal that the arbuscule is the main site of metabolite exchange for the *Hippophaë* association. The fact that the main trunk hyphae of the arbuscules contain abundant polyphosphate granules and rosettes of glycogen whereas the fine terminal branches are depleted of polyphosphate and have less obvious accumulations of glycogen is again consistent with such a role for the arbuscule and suggests that metabolite exchange is most active at the hyphal tips.

Little is known of the actual mechanisms of movement of either phosphate or carbohydrate between the symbionts in mycorrhizal associations in general. The occurrence of polyphosphate in the vacuolar system of the *Hippophaë* endophyte is consistent with its location in other fungi which store polyphosphate. The location of alkaline phosphatase also within these vacuoles in the mature arbuscules is in agreement with other work on vesicular arbuscular associations[13]. Alkaline phosphatase activity specific to the mycorrhizal condition has been reported in onions. This enzyme activity was closely linked to the arbuscular phase of the infection and the enzyme was thought to be fungal in origin. More recently a non-specific alkaline phosphatase was located at the ultrastructural level within the vacuoles of mature arbuscules and the intercellular hyphae in the onion roots[14]. Although no definite conclusion was reached as to the precise relationship between the two it is considered that they may be the same enzyme and it was concluded that, in onions, vacuolar alkaline phosphatase is involved in the active mechanism of phosphate transport within the hyphae of the mycorrhizal fungus. A similar situation may well exist in the *Hippophaë* association.

Rosettes of glycogen are found scattered randomly in the cytoplasm of the arbuscular hyphae in the *Hippophaë* roots but in the main trunk hyphae of the arbuscules and in the intercellular hyphae these glycogen rosettes are found predominantly in close approximation to the lipid droplets. It is thought that this glycogen represents a transition storage compound between the photosynthate which crosses the host/fungal interface and lipid which is the predominant storage form of the microsymbiont[6].

Growth experiments with the lettuce/*Glomus mosseae* association have shown that at low concentrations of phosphate the infection levels in the lettuce roots appeared to be related to the nitrogen content of the host tissue. Mycorrhizal infections in low-nitrogen treatments were poorly developed with few arbuscules[16]. This could explain the low infection rate of mycorrhizal-only *Hippophaë* plants compared to mycorrhizal/nodulated plants in a medium deficient in both soluble phosphate and combined nitrogen.

The stimulation of nitrogen fixation found in the mycorrhizal *Hippophaë* plants can be attributed to the increased phosphate uptake by these roots, a situation analagous to that in nodulated legumes colonised by vesicular-arbuscular fungi[2,4,5]. The greatly enhanced nitrogenase activity in the mycorrhizal/nodulated plants compared to plants nodulated only is not, however, completely reflected in the overall growth of these two groups of plants. It could be that assimilation of the fixed nitrogen in the double inoculated plants is limited by the availability of carbon compounds with excess fixed nitrogen being excreted from the roots. In a sand dune system such excreted nitrogen could benefit the associated grass species.

The enhancement of growth and fixation as seen in the experimentally grown *Hippophaë* could be expected to be found in the dune habitat where nitrogen and phosphorus levels are also low. This, combined with the propensity of the mycorrhizal extra-radical mycelium for sand aggregation and the capacity of the fungus to resist desiccation, makes the triple symbiosis a feature of prime importance in the development and stabilisation of dune systems.

Acknowledgements The authors thank Mr Dennis McLoughlin for technical assistance and are grateful to the Natural Environment Research Council for a research studentship to D.M.C.

References

1   Anderson T F 1951 Techniques for the preservation of three dimensional structure in preparing specimens for the electron microscope. Trans. N. Y. Acad. Sci. 13, 130–136.
2   Asimi S, Gianinazzi-Pearson V and Gianinazzi S 1980 Influence of increasing soil phosphorus levels on interactions between vesicular-arbuscular mycorrhizal and *Rhizobium* in soybeans. Can. J. Bot. 58, 2200–2206.
3   Bergersen F J 1980 Methods for evaluating nitrogen fixation. Wiley Interscience, London, 563 p.
4   Bethlenfalvay G J and Yoder J F 1981 The *Glycine-Glomus-Rhizobium* symbiosis. 1. Phosphorus effect on nitrogen fixation and mycorrhiza infection. Physiol. Plant. 52, 141–145.
5   Carling D E, Riehle W G, Brown M F and Johnson D R 1978 Effects of a vesicular arbuscular mycorrhizal fungus on nitrate reductase and nitrogenase activity in nodulating and non-nodulating soybeans. Phytopathology 68, 1590–1596.
6   Clelland D M 1983 Ultrastructural, cytological and physiological studies on the mycorrhizal association in *Hippophaë rhamnoides* L. Ph.D. Thesis, University of Strathclyde, Glasgow, U.K.
7   Clelland D C, Gardner I C and Scott A 1983 The occurrence of *Glomus fasciculatus*, a mycorrhizal endophyte, in the nitrogen fixing non-legume *Hippophaë rhamnoides*. Microbios Letters. 24, 107–113.
8   Daft M J and Nicolson T H 1966 Effect of *Endogone* mycorrhiza on plant growth. New Phytol. 65, 343–351.
9   Daft M J and Nicolson T H 1969 Effect of *Endogone* mycorrhiza on plant growth. II. Influence of soluble phosphate on endophyte and host in maize. New Phytol. 68, 945–952.
10 Daft M J and Nicolson T H 1972 Effect of *Endogone* mycorrhiza on plant growth. IV. Quantitative relationships between the growth of the host and the development of the endophyte in tomato and maize. New Phytol. 71, 287–295.
11 Daft M J and Hacskaylo E 1976 Arbuscular mycorrhizas in the anthracite and bituminous coal wastes of Pennsylvania. J. Appl. Ecol. 13, 523–531.
12 Fogg D N and Wilkinson N T 1958 The colorimetric determination of phosphorus. Analyst 83, 406–408.
13 Gianinazzi-Pearson V and Gianinazzi S 1978 Enzymatic studies on the metabolism of vesicular-arbuscular mycorrhiza. II. Soluble alkaline phosphatase specific to mycorrhizal infection in onion roots. Physiol. Plant Pathol. 12, 45–51.
14 Gianinazzi S, Gianinazzi-Pearson V and Dexheimer J 1979 Enzymic studies on the metabolism of vesicular-arbuscular mycorrhiza. III. Ultrastructural localisation of acid and alkaline phosphatase in onion roots infected by *Glomus mosseae* (Nicol. and Gerd.). New Phytol. 82, 127–130.
15 Hayman D S 1980 Mycorrhiza and crop production. Nature London 287, 687–688.
16 Hepper C M 1983 The effect of nitrate and phosphate on the vesicular-arbuscular mycorrhizal infection of lettuce. New Phytol. 92, 389–399.

17 Kinden D A and Brown M F 1975 Technique for scanning electron microscopy of fungal structures within plant cells. Phytopathology 65, 74–76.

18 Miller I M, Scott A and Gardner I C 1983 The development, structure and function of dendroid colleters in *Psychotria kirkii* Hiern (Rubiaceae). Ann. Bot. 41, 621–630.

19 Oremus P A I 1980 Occurrence and infective potential of the endophyte of *Hippophaë rhamnoides* L. sap *rhamnoides* in coastal sand-dune areas. Plant and Soil 56, 123–139.

20 Phillips J M and Hayman D S 1970 Improved procedures for clearing roots and staining parasitic and vesicular-arbuscular mycorrhizal fungi for rapid assessment of infection. Trans. Br. Mycol. Soc. 55, 158–163.

21 Rose S L 1980 Mycorrhizal associations of some actinomycete nodulated nitrogen fixing plants. Can. J. Bot. 58, 1449–1454.

22 Rose S L and Youngberg C T 1981 Tripartite associations in snowbrush (*Ceanothus velutinus*): effect of vesicular-arbuscular mycorrhizae on growth, nodulation and nitrogen fixation. Can. J. Bot. 59, 34–39.

23 Williams S E and Aldon E F 1976 Endomycorrhizal (VA) associations of some arid zone shrubs. Southwest. Nat. 20, 437–444.

24 Williams S E 1979 Vesicular-arbuscular mycorrhizae associated with actinomycete-nodulated shrubs, *Cercocarpus montanus* Raf. and *Purshia tridentata* (Pursh) DC. Bot. Gaz. 140 (Suppl.), S115–S119.

Plant and Soil 78, 201–208 (1984).
© 1984 *Martinus Nijhoff/Dr W. Junk Publishers, The Hague.*

Seasonal fluctuations of the mineral concentration of alder (*Alnus glutinosa* (L.) Gaertn.) from the field

C. RODRÍGUEZ-BARRUECO, C. MIGUEL and P. SUBRAMANIAM
Unit of Nitrogen Fixation, Centro de Edafología y Biología Aplicada, C.S.I.C, Salamanca, Spain

Key words *Alnus glutinosa* *Frankia* Nutrient cycling Seasonal fluctuation

Summary The seasonal fluctuation of N, P, K, Ca, Mg, Fe, Mn, Mo, and Co, in leaves, roots and nodules of 40–50 year old *Alnus glutinosa* trees growing at four different locations along the banks of the Tormes river, in the province of Salamanca, was studied. Also, the evolution of the soil organic matter under the trees sampled was evaluated. The data obtained for the various nutrient elements in the three plant parts are statistically treated at the significance levels of 99–95 per cent, and some remarks as to the nutritional status of the European alder in respect to the nutrients and its contribution to soil nutrient-cycling are provided. A positive correlation was found between N-P, N-K, N-Mg, and N-Mo, in leaves, and between N-P, N-K, N-Fe, N-Mn, and N-Mo in root nodules. In roots only, no significance at any level was obtained between N and any of the elements analyzed.

Introduction

The nitrogen fixed in the root-nodules of actinomycete-nodulated plants is in part translocated eventually to the rest of the plant. The nitrogen thus fixed goes back to the soil through annual decomposition of the plant material and through root exudates, both important for growth of associated plant communities. The ecological impact of the actinomycete-nodulated alder and other actinorhizal plants is already known[12,14,16]. However, no mention has been made of the contribution to nutrient-cycling in the plant-soil system by these types of plants in respect to many other nutrients but nitrogen, and some other major elements[8,15]. Specific data on the nitrogen status of the plant through its growth season is available[3,7,11], but data on other nutrients are scarce. It was considered of interest in this direction, to evaluate the seasonal fluctuation of nutrient elements in leaves, roots, and nodules of *Alnus glutinosa,* and its probable significance in the nutrient recycling and soil fertility. Although this has been done for *Alnus incana, A. inokumai,* and *A. rubra*[5,9,13,19] information for *A. glutinosa,* a typical European species is lacking. In addition, the present report gives data on the presence of certain trace elements which might intervene in the actual nutritional status of the alder in nature, and on its seasonal evolution.

Material and methods

Leaf samples were obtained from about 40–50 year old spontaneous pure thickets of alder trees growing in four different locations along the banks of the Tormes river, near Salamanca.

Four trees were selected at random in each of the four locations and in each tree several branches were selected from different parts of the tree. In each branch, leaves in the middle portion were collected and mixed together into a composite sample from where four replicates were drawn for analysis. Fresh and healthy samples of roots and nodules were also collected by carefully excavating around the selected trees up to a depth of 10–20 cm. Roots and nodules samples collected from different parts of the root system were mixed before drawing out four samples for analysis. All of the plant samples were collected on 22nd January, 28th April, 12th July, and 2nd November for the winter, spring, summer, and autumn seasons respectively. The samples were taken to the laboratory in polyethylene bags, and their content of nitrogen, phosphorus, potassium, calcium, magnesium, iron, manganese, molybdenum and cobalt determined respectively. Nitrogen was determined by Kjeldahl method and an Orion Research Digital Ion Analyzer (Ionalyzer 801) equipped with a plastic electrode for ammonia. The other elements were determined on a 0.5-g sample (leaves, roots, or nodules) which had been previously calcinated, dried, ground, and homogenized. Four ml of 25% HCl were added to the resulting ashes for further heat digestion. Deionized water was then gradually added during 1–2 h and the sample allowed to cool. Then, four ml of 25% HCl were added to the samples before being filtered into a 25-ml volumetric flask. Iron, manganese, molybdenum, and cobalt content were determined directly in that solution in a Varian Techtron 1250 atomic absorption spectrophotometer. Calcium and magnesium contents were also determined by atomic absorption. To this purpose 1-ml of the above solution was taken into a 25-ml volumetric flask with deionized water plus 2-ml 10% $SrCl_2$. The same procedure was adopted for potassium and phosphorus determination, except that 1000 ppm Na and 5-ml of the vanadomolybdate reagent substituted for $SrCl_2$.

The organic matter content and pH of the soils were determined along the four seasons. In each location, soil samples were drawn from five spots at a depth of 10–20 cm, depth at which nodulation was observed. These samples were mixed well and four replicates were drawn for analysis. The soil samples were transferred to the laboratory in polyethylene bags and passed through a 2 mm sieve prior to analysis.

All the data were subjected to statistical analysis adopting Snedecor "F" and Student's "t" tests. The variations have been expressed as S.E.

Results and discussion

The four soils from where alder samples were taken can be classified into three groups based on their pH characteristics (Fig. 1). Soil number 2, collected in Torrejon, has a pH which falls close to the limit given by Wheeler et al.[18] who reported that soils more acid than 4.5 give rise to nodules of lower specific activity due to less effective forms of the endophyte. The seasonal variation of the organic matter in the four locations follows a similar trend (Fig. 1). Maximum values were observed during winter and minimum values in summer. The large variation in organic matter between winter and summer could be attributed to the sandy nature of soil and high temperatures of about 34–38°C in summer which cause rapid oxidation of organic matter.

The data obtained are presented successively in Tables 1, 2, and 3, where it can be seen that macronutrients N, P, K, Ca, remain in the leaves at a higher level than in correspondent roots and nodules. Magnesium contents in leaves and nodules were rather similar, and slightly above those present in roots. Macronutrients N, P, K, Mg, decreased along the growth season, whereas calcium reaches its highest value in

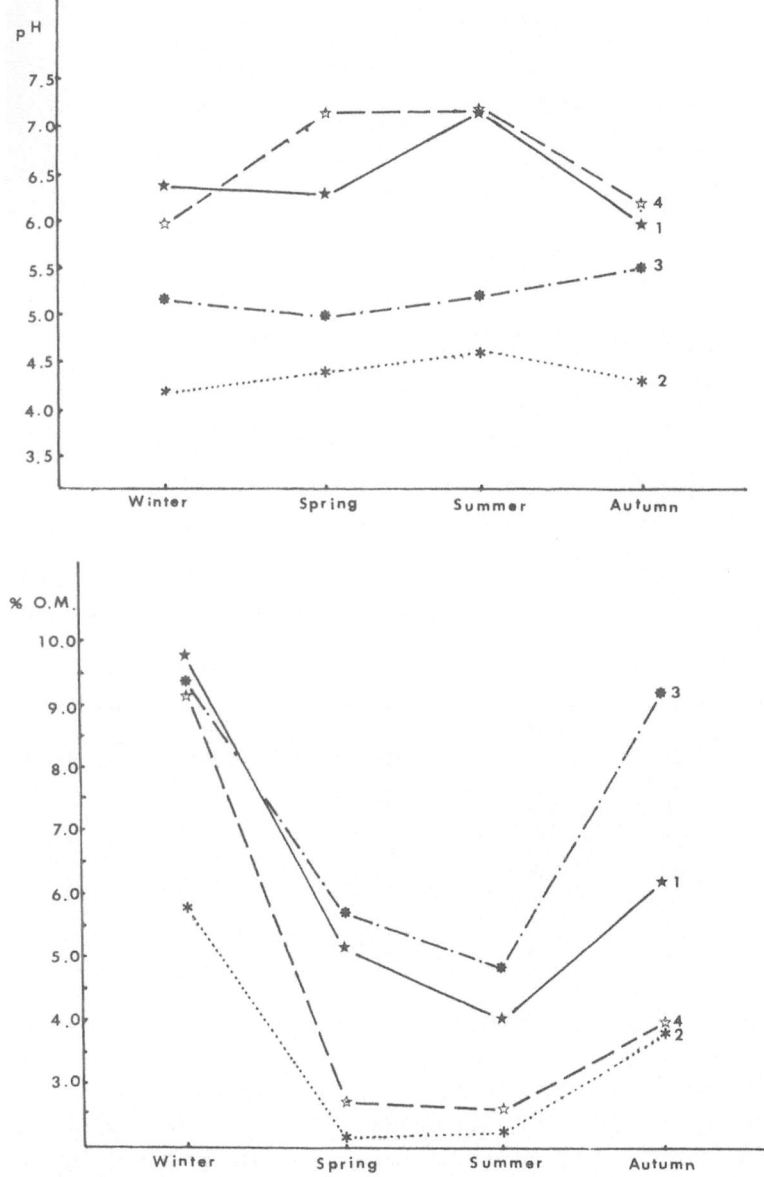

Fig. 1. Seasonal evolution of pH and organic matter content of the soils: 1 (La Serna), 2 (Torrejón), 3 (La Maya) and 4 (Zorita).

autumn, in the leaves. Iron content of roots and nodules was much higher than that of leaves. Manganese in the leaves showed its highest value in autumn, and was lowest in the nodules at any season. Molybdenum decreased from spring to autumn in leaves, roots, and nodules respectively, the latter giving higher figures. However, the course of

Table 1. Mineral concentration in the leaves of *Alnus glutinosa*. Mean values of four replicates

Element	Winter	Spring	Summer	Autumn	S.E.
N%	–	3.84	2.68	2.30	0.08
P%	–	0.36	0.16	0.10	0.03
K%	–	1.11	0.98	0.53	0.14
Ca%	–	0.85	0.63	1.24	0.08
Mg%	–	0.37	0.31	0.29	0.02
Fe ppm	–	314	300	404	35.4
Mn ppm	–	641	269	1014	236.3
Mo ppm	–	1.63	0.75	0.48	0.37
Co ppm	–	5.75	7.38	7.25	0.39

Table 2. Mineral concentration in the roots of *Alnus glutinosa*. Mean values of four replicates

Element	Winter	Spring	Summer	Autumn	S.E.
N%	0.94	1.58	1.01	1.34	0.21
P%	0.15	0.16	0.17	0.11	0.01
K%	0.42	0.50	0.37	0.26	0.03
Ca%	0.52	0.55	0.63	0.40	0.03
Mg%	0.23	0.20	0.23	0.18	0.02
Fe ppm	5110	3370	1685	2936	596.7
Mn ppm	223	154	165	137	19.2
Mo ppm	0.48	1.50	1.25	0.73	0.17
Co ppm	15.25	10.75	8.88	8.50	1.12

Table 3. Mineral concentration in the root-nodules of *Alnus glutinosa*. Mean values of four replicates

Element	Winter	Spring	Summer	Autumn	S.E.
N%	1.59	1.87	1.44	1.50	0.16
P%	0.15	0.16	0.11	0.13	0.01
K%	0.50	0.52	0.30	0.47	0.05
Ca%	0.39	0.49	0.39	0.26	0.04
Mg%	0.34	0.43	0.34	0.21	0.04
Fe ppm	2368	3504	1713	2341	582.1
Mn ppm	77	87	48	58	13.2
Mo ppm	1.00	2.90	1.90	0.60	0.37
Co ppm	17.30	20.80	13.30	10.40	1.68

events in winter for roots was contrary to that of nodules. Thus, nodule Mo content increased whereas that of roots decreased suggesting that Mo accumulated in nitrogen-fixing sites is not being used at a rate which could otherwise imply nitrogenase activity, and in agreement with other authors[1] who hardly detected fixation in alder nodules from the field in their November sampling. Again, Mo-containing nitrate reductase remains active in the host parts of the nodules and in root at that time of the season[10]. Cobalt figures increased slightly from spring to autumn in leaves, and decreased in roots and nodules, the figures for the latter being somehow higher. It should be noticed that cobalt figures obtained for roots and nodules in winter samples were higher than those for autumn. Despite the evolutionary trend observed in the data presented, the reader is addressed to Table 4 which includes levels

Table 4. Statistical analysis of data presented in Tables 1, 2, and 3

Element/ plant part	Level of significance	Statistical differences among seasons
Leaves		
N	99%	S > Su; S > A
P	99%	S > Su; S > A
Ca	99%	A > S; A > Su
Mg	99%	S > A
P	95%	Su > A
K	95%	S > A; Su > A
Ca	95%	S > Su
Mg	95%	S > Su
Mo	95%	S > A
Roots		
K	99%	S > Su; S > A; W > A
N	95%	S > Su; S > W
P	95%	S > A; Su > A
K	95%	Su > A
Fe	95%	Su > W
Mo	95%	Su > W; S > A
Nodules		
Mg	99%	S > A
Mo	99%	S > W; S > A
N	95%	S > Su
P	95%	S > Su
K	95%	S > Su; W > Su
Ca	95%	S > A
Mg	95%	W > A; S > A
Fe	95%	S > Su
Mo	95%	Su > A

Note: S, Su, A, and W stand for Spring, Summer, Autumn and Winter respectively.

of significance for the values obtained in Tables 1, 2, and 3. Also, in Fig. 2, the correlation indices between the nitrogen content and the rest of the elements studied in leaves, roots, and nodules are given.

On the whole, the nitrogen content in leaf samples was high, similar to what was reported for other alder species, as well as legumes, and again higher than that present in non-nitrogen-fixing trees[9,11,19]. However, although other elements are present in the latter group, in rather similar amounts, there is a trend for higher values in the nitrogen-fixing alder plant. In this context Mn and Fe, were rather high in alder, even higher than that reported in other alder species, although this can not be said precisely as the sampling season was not provided by the authors.

All the data agree with the acknowledged economical importance of alder and show the inocuous effect of high Mn and Fe levels for the alder stand[19]. Macronutrients in leaves are in higher amounts than in roots and nodules, whereas the latter contain higher levels of micro-nutrients. The seasonal fluctuations in leaves for each of the elements N, P, K, Mg, and Mo, indicates that minimum contents are observed in autumn with plant senescence, and maximum values are produced in

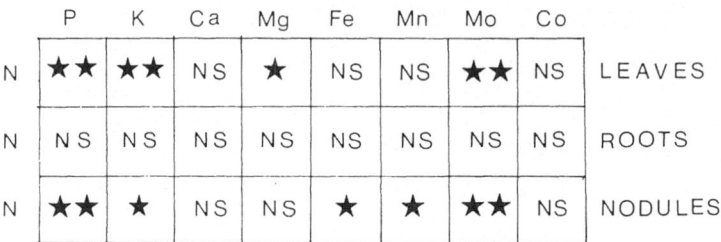

Fig. 2. Correlation between nitrogen content and other elements in different plant parts of *Alnus glutinosa* from the field.

spring. In this context[11] it has been found that N per cent in alder leaves in the field remains almost constant falling to the ground with a substantial amount of nitrogen, the values given being always higher than those found for other non-nitrogen-fixers like *Quercus, Ulmus,* and *Tilia,* although for the latter genus a different position has been reported[6]. The other elements, mainly Ca, Mn, Fe, and Co, increase in autumn, being minimum in summer. In nodules, maximum values are found in spring when higher demands for fixation are to be expected.

A positive correlation in leaves between N-P, and N-K is shown in Fig. 2, a result which has been observed in legumes at the family level[4]. Other positive correlations in Fig. 2 are those between N-Mo and N-Mg, in leaves. In roots no correlation between N and other elements studied was found, while in nodules a positive correlation was obtained between N and P, K, Fe, Mn and Mo respectively. Molybdenum is essential for the nitrogen-fixing plant, both in a more efficient function of the microsymbiont, and also in the accumulation of nitrate-nitrogen by the host. It should be pointed out that Mo requirements in *Rhizobium* are greater than in the host plant[2]. Molybdenum deficiency also results in a reduction of both soluble nitrogen and protein nitrogen, whereas nitrate accumulates in the plant tissues due to less nitrate reductase, which contains Mo as a structural constituent.

The present study, in revealing the pattern of accumulation and variation of some major and trace elements in alder through a growing season also enables the discussion of its relevance in nutrient-cycling and ultimate contribution to soil fertility. As in other nitrogen-fixing plants, owing to the presence of a dual nitrogen source, the leaves of alders contain a relatively high proportion of nitrogen. Besides, the nitrogen content in alder leaves undergo very little seasonal variation as compared to *Ulmus, Tilia* and *Quercus*[11]. Most of the winter deciduous trees conserve nitrogen by mobilizing leaf nitrogen in autumn and storing it in twigs and branches, from where it is made available for growth the following spring. This phenomenom is not observed in alders probably owing to the fact that they can fix atmospheric nitrogen and obtain lesser benefits from such conservation mechanisms[3]. This could perhaps explain why so much of nitrogen is present in the leaves at leaf fall. It is of particular interest from the standpoint of soil fertility that so much of the leaf nitrogen should still remain in the leaf at leaf fall. As a result, the addition of newly fixed nitrogen to the soil is a more continuous process than if most of the nitrogen remained locked up in the permanent parts of the tree until its death, which, for instance, is the situation in other trees mentioned earlier.

The percent retention of nutrients in leaves in autumn at leaf fall, as compared with that present in summer, is fairly high for N (86%), Mg (94%), and Co (98%), and slightly lower in the case of P(63%), K (54%) and Mo (64%), while in the case of Ca (197%), Fe (135%) and Mn (377%), a further accumulation in leaves was observed prior to leaf fall. These retention/accumulation values are comparable to those reported in the bibliography[17] for N, P, K, Ca, and Mg in *Alnus glutinosa*. The high retention of nutrients in the leaves at leaf-fall also obviously has a great significance in contributing to soil fertility through litter decomposition.

References

1 Akkermans A D L and van Dijk C 1975 The formation and nitrogen-fixing activity of the root-nodules of *Alnus glutinosa* under field conditions. *In* Symbotic N-Fixation in Plants. Ed. P S Nutman. IBP Vol. 7. Cambridge Univ. Press 511−520.

2 Andrew C S 1962 Influence of nutrition on nitrogen fixation and growth of legumes. *In* A review of nitrogen in the tropics with particular reference to pastures. Ed. Comm. Bur. Past. Field Crops Bull. 46, 130−146.

3 Dawson J O and Funk D T 1981 Seasonal change in foliar nitrogen concentration of *Alnus glutinosa*. Forest Sci. 27, 239−243.

4 Duque F 1970 Estudio quimico de suelo y especies pratenses de comunidades seminaturales de la provincia de Salamanca. Ph. D. Thesis. Univ. Salamanca. 476 p.

5 Hughes D R, Gessel S P and Walker R B 1968 Red alder deficiency symptoms and fertilizer trials. *In* Biology of Alder. Ed. Trappe *et al.* USDA Forest Service. Portland, Or. 225−237.

6 Insley H, Boswell R C and Gardiner J B H 1981 Foliar macronutrients (N, P, K, Ca and Mg) in lime (*Tilia* spp.). II. Seasonal variation. Plant and Soil 61, 391−401.

7 Miller H G 1982 Nutrient cycling in alder. Report Int. Energy Agency. JAB−21, 54 p.

8 Mun H T, Kim C M and Kim J H 1977 Distributions and cyclings of nitrogen, phosphorus and potassium in Korean alder and oak stands. Korean J. Bot. 20, 109−118.

9 Ovington J D 1956 Decomposition of tree leaves. Forestry 29, 22−28.

10 Pizelle G and Thiery G A 1977 Variations saisonnières des activités nitrogénase et nitrate-reductase chez l'Aune glutineux (*Alnus glutinosa* L. Gaertn.). Physiol. Veg. 15, 333−342.

11 Rodriguez-Barrueco C 1968 Elemental nitrogen as a starting point for protein synthesis in

plants. Proc. VII Int. Symp. Agrochimica. Salamanca, 354–360.

12   Silvester W B 1974 Ecological and economic significance of the non-legume symbioses.
      In Proc. 1st Int. Symp. N-fixation. Ed. W E Newton and C J Nyman. Wash. Sta. Univ.
      Press. 489–506.

13   Tarrant R F 1961 Stand development and soil fertility in a Douglas-fir-Red alder plan-
      tation. Forest Science 7, 238–246.

14   Tarrant R F and Trappe J M 1971 The role of *Alnus* in improving the forest enviroment.
      Plant and Soil Special Vol. 335–348.

15   Turner J, Cole D W and Gessel S P 1976 Mineral nutrient accumulation and cycling in a
      stand of red alder (*Alnus rubra*). J. Ecology 64, 965–974.

16   Uemura S 1971 Non-leguminous root-nodules in Japan. Plant and Soil Special Vol.
      349–360.

17   Viro P J 1955 Investigations on forest litter. Metsätiet Tutkimuslait Julk. 45, 1–63.

18   Wheeler C T, McLaughlin M E and Steele P 1981 A comparison of symbiotic nitrogen
      fixation in Scotland in *Alnus glutinosa* and *Alnus rubra*. Plant and Soil 61, 169–188.

19   Yamaya K 1962 Relationship between properties of the "Hiba" (*Thujopsis dolabrata*
      var. Hondai MAKINO) forest soils and forest growth in the Tsugaru and Shimokita
      Penninsulas. Forest Soils of Japan 12, 155 p.

Plant and Soil 78, 209–220 (1984).   Ms. Fr 23
© 1984 *Martinus Nijhoff/Dr W. Junk Publishers, The Hague.*

Growth, nitrogen accumulation, and symbiotic dinitrogen fixation in pure and mixed plantings of hybrid poplar and black alder

B. COTÉ and C. CAMIRE
Départment d'Ecologie et de Pédologie, Faculté de Foresterie et de Géodésie, Université Laval, Québec, Canada G1K 7P4

Key words Black alder   Hybrid poplar   Interactions   Natural ^{15}N dilution   N$_2$ fixation   N accumulation   Short-rotation plantation   Tree biomass

Summary Growth and N accumulation were assessed in pure and mixed plantings (2 years old) of hybrid poplar and black alder in southern Québec. Symbiotic dinitrogen fixation was evaluated by natural ^{15}N dilution. Growth of hybrid poplar plants and N accumulation in their tissues increased with their decreasing contribution to species ratio whereas no differences among treatments were measured for black alder. Yield and N content per hectare of aboveground components increased with the proportion of black alder in the plantation. Symbiotic dinitrogen fixation was estimated at 68% of alder nitrogen in both pure and mixed treatments. The maximum rate of N-fixation was 53 kg ha^{-1} yr^{-1} in pure alder plots. The amount of nitrogen accumulated in entire plants of black alder from symbiotic fixation could be sufficient to balance the N export in harvested stems and branches of short-rotation plantations containing at least 33% of alder.

Introduction

The concept of short-rotation intensively cultured plantations has rapidly gained interest since the study of McAlpine *et al.*[25]. Rates of nitrogen uptake by young plantations of hybrid poplar may be very high on highly productive sites[17]. The possibility of using a nitrogen-fixing species to supply nitrogen to a companion tree has shown promising results[10,27,29,40]. Mixed plantings with alder can increase height and diameter growth of hybrid poplars[18,44], as well as total yield[13]. However, alder growth may be reduced by shading of hybrid poplar[12]. Younger and Kapustka[45] noted a reduced rate of nitrogen fixation by *Alnus rugosa* in mixed stands with *Populus tremuloides*, and Jobidon and Thibault[23] observed an inhibitory effect of leachates of *Populus balsamifera* on nodulation and N$_2$ fixation by *Alnus crispa*.

Pure and mixed plantings of black alder (*Alnus glutinosa* Gaert.) and hybrid poplar cv. Roxbury (*Populus nigra* L. x *P. trichocarpa* Torr and Gray) were established in spring of 1981. Our purpose was to evaluate possible interactions between both species on their respective biomass production and nitrogen accumulation, and to assess the contribution of symbiotic nitrogen fixation on the economy of this nutrient in short-rotation plantations in southern Québec, Canada.

Materials and methods

Site description

The study area is located on the campus of Laval University, Québec, Canada (46°41′N, 71°16′W, elevation = 90 m). The climate of the region is cold: average January and July temperatures are − 12 and 19°C, respectively, with an average frost-free period of 170 days. Annual precipitation averages 1,200 mm with 650 mm falling between May and October. The study area is included in the L-3 forest region[35]. The soil is an acid loam (pH 4.1, 0.01 M $CaCl_2$), derived from marine sediments, with a good drainage and a medium fertility level (0.M. 5.5%, N 0.25%, P-Truog 19 ppm, extractable K, Ca, and Mg, 90, 625, and 57 ppm, respectively; and C.E.C.: 20 meq 100 g^{-1} (1 N NH$_4$OAc, pH 7.0). It is classified as an orthic dystric brunisol[9] (typic dystrochrepts[38]). This abandoned agricultural site was ploughed and harrowed in the spring of 1981 prior to plantation. No fertilizer nor herbicide was applied.

Plot establishment and maintenance

Twelve (5 × 4 m) plots were delineated within the study site. Four combinations of mixing alder and poplar were tested: pure alder (3A), two thirds alder for one third hybrid poplar (2A1P), one third alder for two thirds hybrid poplars (1A2P), and pure hybrid poplar (3P). Trees were planted systematically at a spacing of 33 cm × 33 cm (90,000 plants ha^{-1}) in alternative spots within rows, according to their relative ratios. The experimental design was completely randomized.

Plantation was made in May of 1981. Hybrid poplar cuttings, 40 cm long, were planted at 30 cm depth. Nodulated bareroot black alder seedlings, of unknown origin, were 1−2 years old and 20−40 cm tall; they were gathered from an old experimental plantation of the immediate region just before plantation. During the first growing season, alder survival was good (87%) but mortality was high for hybrid poplar (41%), resulting probably from the combined effects of the late date of plantation[30] and the drought that followed. Both species were replanted in summer of 1981 with plants from unused nearby plots. During the 2 years of experimentation, plots were manually hoed to control weed competition.

Measurements of growth and biomass

In mid-September 1982, before leaf fall, plants were measured for height (cm) and collar diameter (mm). To eliminate the border effect of such small plots[17], only the 49 interior plants were measured. Previously, hybrid poplar shoots that were killed by frost during the 1981 winter were cut, and the length and the base diameter were measured in May of 1982. Twelve frost-killed tops and fifteen plants of each species, of which five root systems were excavated, were randomly chosen and harvested at the time of measurements. Each component (leaves, stem + branches, and roots) was oven-dried at 65°C for 96 hours and weighed.

Plant nitrogen status

Each sample was ground to pass through a 40 mesh sieve and analyzed for total nitrogen content by the standard micro-Kjeldahl technique on a Kjeltec Auto System (Digestion System 40 and Kjeltec Auto 1030 Analyzer).

Non-fixing system (nfs)

The non-fixing control plants were grown in the greenhouse during the summer of 1982. Sixteen seeds of black alder were sown in a water tight plexiglass box (60 × 60 cm) containing 85 kg (dry weight basis) of soil (0−15 cm depth) removed from the study area and sieved to 6 mm. Four boxes (replicates) were placed in a water bath to maintain the soil temperature between 17 and 21°C during the course of the experiment. Deionized water was added periodically ($\approx$ 0.26 mm every 1.4 day) to maintain soil moisture at field capacity. Excess water, accounting for 4% of the watering, was continuously drained by 2 tension lysimeter cups (6 cm O.D.; Soil Moisture Equipment, Santa Barbara, California), maintained with a vacuum of 0.3 atmosphere. After 4 months, the entire plants were harvested, oven-dried at 65°C for 72 hours and ground to 40 mesh. The non-fixing capacity was ensured by the total absence of root nodules.

15N determination

Four seedlings in the greenhouse and total aboveground tissues of 6 black alders in the field (2 per treatment) were analyzed in bulk for ^{15}N concentration. Samples were digested in the Kjeltec System, using the micro-Kjeldahl technique, modified to include nitrite and nitrate with the salicylic acid pretreatment[8]. Steam distillation with NaOH (40%) was performed in a complete glass distillation unit with 250 ml sample flasks. Ammonia was recovered in 2 ml of 4% boric acid without indicator over the 40 ml mark of a 100 ml beaker. Cross contamination was avoided by distilling an additional 20 ml of 95% ethanol between samples. Distillates were immediately acidifed with 2 drops of $1 N$ H_2SO_4 and dried in a ventilated oven at 85°C. They were then transferred with 2 ml of distilled water in a disposable culture tube and dried again to be analyzed by a mass spectrometer (Micromass 602E; Iso-Mass Consulting Ltd., Saskatoon, Canada). The percent N in the plant derived from the atmosphere (% Ndfa) was assessed by the natural ^{15}N dilution technique[2,3,4,19,24] and calculated as follows[31]:

$$\% \; Ndfa \; = \; 1 - \left(\frac{at \; \% \; ^{15}N \; ex \; (fs)}{at \; \% \; ^{15}N \; ex \; (nfs)} \right) \times 100$$

where fs is the fixing system and nfs is the non-fixing system. The at. % ^{15}N ex is atom percent ^{15}N excess[33] where the control is N-fixing black alder seedlings grown in a growth chamber for 3 months in washed quartz using a N-free nutrient solution.

Statistical analysis

Biomass and N content of aboveground plant tissues (leaves, stem + branches, leaves + stem + branches) were evaluated by regression equations $(Y = aX^b c^X)$ where X is the combined variable D^2H (D = collar diameter (mm) and H = height (cm)). R^2 varied from 0.92 to 0.99. Biomass and N content per living plant and per hectare, Ndfa, N in the plant derived from the soil (Ndfs), and % Ndfa were compared using a one-way analysis of variance and Duncan's new multiple range test. SAS[21] software was used to run these statistics.

Results

Biomass and nitrogen content per living plant

Dry weight of leaves, stem + branches and total aboveground tissues per living plant of hybrid poplar decreased with increasing its density. (Fig. 1) Aboveground biomasses were 160, 110 and 59 g plant^{-1} for 2A1P, 1A2P and 3P, respectively. Black alder showed no trend in relation to treatments. (Fig. 1) Height of both species did not differ after 2 years in plantation because the frost-killed part of hybrid poplar was removed before the second growing season. At the time of measurement, mortality accounted for 33, 35 and 21% of hybrid poplar plants in 2A1P, 1A2P, and 3P treatments, respectively, but was very low for black alder (*ca* 4%).

The effects of species ratio in plantation on nitrogen content of the two species were similar to those observed for biomass. (Fig. 2) Total aboveground N content of hybrid poplar was 950, 400 and 370 mg plant^{-1} for 2A1P, 1A2P and 3P, respectively.

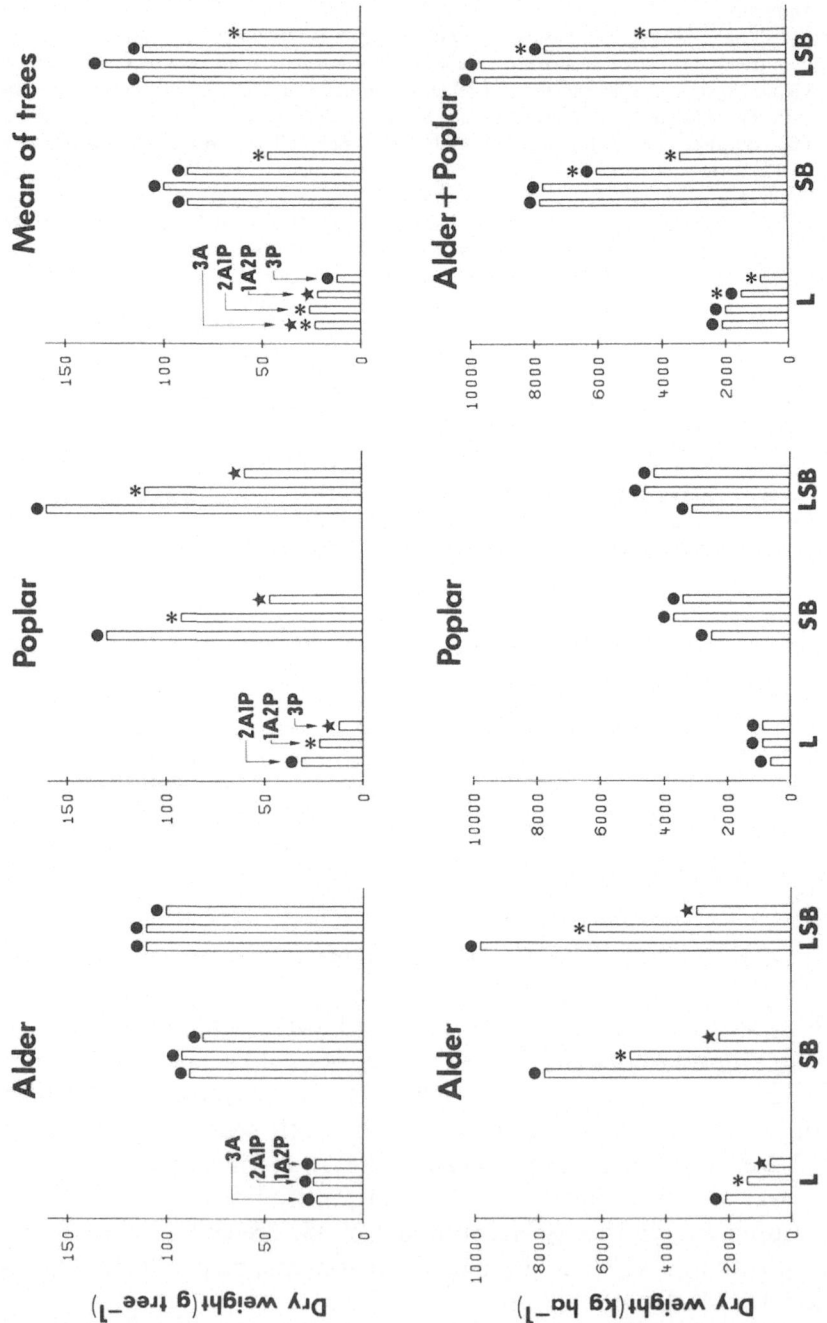

Fig. 1. Effect of species ratio (A = alder; P = poplar) on dry weight per plant and per hectare of leaves (L), stem + branches (SB), and leaves + stem + branches (LSB) for each species. Columns with the same symbol within a component are not significantly different at the 5% level.

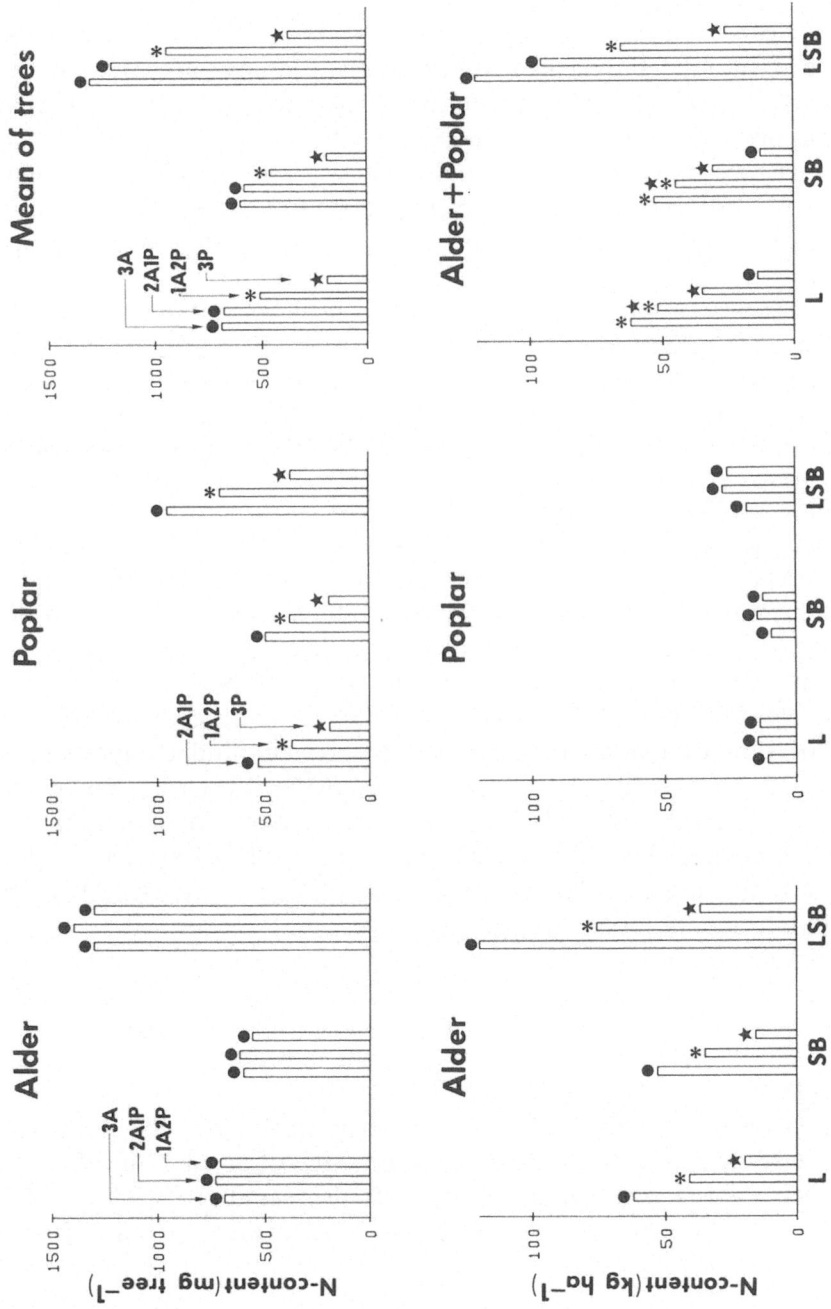

Fig. 2. Effect of species ratio (A = alder; P = poplar) on N-content per plant and per hectare of leaves (L), stem + branches (SB), and leaves + stem + branches (LSB) for each species. Columns with the same symbol within a component are not significantly different at the 5% level.

Biomass and nitrogen content per hectare

On a per hectare basis, dry weight of leaves, stem + branches, and total aboveground biomass of hybrid poplar did not differ significantly among the 2A1P, 1A2P and 3P treatments, while that of black alder showed marked increases with increase in its density (Fig. 1). Over two years, average rates of aboveground biomass production decreased with the density of hybrid poplar, averaging 4 900, 4 800, 3 850, and 2 150 kg ha^{-1} yr^{-1} for treatments 3A, 2A1P, 1A2P, and 3P, respectively.

Species ratio affected the N content of both species in a way similar to that of biomass. Rates of N accumulation decreased with the increasing ratio of hybrid poplar, accounting to 60, 48, 33 and 13 kg ha^{-1} yr^{-1} for the 3A, 2A1P, 1A2P and 3P treatments, respectively. For both species, N in foliage averaged 55% of the total aboveground N.

Symbiotic dinitrogen fixation (Ndfa)

Since no significant difference was observed between treatments (3A, 2A1P, 1A2P) in dinitrogen fixation (% Ndfa) by black alder, as determined by natural ^{15}N dilution technique, we assessed the average % Ndfa to 68% (Table 1) of the total aboveground nitrogen content of black alder plants. Annual rates of N-fixation were respectively 41, 26, and 13 kg ha^{-1} yr^{-1}, corresponding to 68, 55, and 40% of the total annual N demand of aboveground components in 3A, 2A1P, and 1A2P plots. Roots averaged 35% of the total dry weight of an alder plant and their N concentrations were similar to those of stem + branches. Therefore, total annual rates of N-fixation by alder were estimated to 53, 34, and 16 kg ha^{-1} yr^{-1} for the 3A, 2A1P and 1A2P treatments, respectively. The N content in the 3P treatment was estimated to 37 kg N ha^{-1} by linear extrapolation from the 2A1P and 1A2P data; corresponding % Ndfa of black alder were assessed to 66, 76, and 69% for the 3A, 2A1P, and 1A2P treatments, respectively.

Nitrogen derived from soil (Ndfs)

Based on the % Ndfa (68%) estimated in black alder for 3A, 2A1P, and 1A2P treatments, % Ndfs averaged 32% of total N in all the 3 treatments; on a per hectare basis, Ndfs decreased from 3A to 1A2P (Table 2). However, mixed treatments gave higher overall Ndfs for all components than pure treatments.

Discussion

Biomass and N content per plant of black alder were independent of the species ratio, and therefore independant of the competition

Table 1. Effect of the species ratio on % Ndfa and Ndfa (kg ha^{-1}) in 2 yr-old plantations (aboveground parts only)

Species ratio[1]	% Ndfa		Ndfa[2]
	X̄	SD	
3A	66a	6.7	82b
2A1P	64a	17.9	52c
1A2P	75a	14.9	25d
Total	68a	6.7	–

[1] 3A, 3 alders; 2A1P, 2 alders for 1 poplar; 1A2P, 1 alder for 2 poplars; 3P, 3 poplars.
[2] Using % Ndfa = 68%.
[3] Treatment means within a column followed by a common letter are not significantly different at the 5% level.

Table 2. Effect of the species ratio on Ndfs (kg ha^{-1}) of black alder and hybrid poplar in 2 yr-old plantations (aboveground parts only)

Species ratio[1]	Species		
	Alder[2]	Poplar	Alder + poplar
3A	38a[3]	–	38e
2A1P	24b	19d	43e
1A2P	12c	28d	40e
3P	–	26d	26f

[1] 3A, 3 alders; 2A1P, 2 alders for 1 poplar; 1A2P, 1 alder for 2 poplars; 3P, 3 poplars.
[2] Using % Ndfs = 32%.
[3] Treatment means within a column followed by a common letter are not significantly different at the 5% level.

of hybrid poplar for light, water, and nutrients. DeBell and Radwan[13] reported similar results for *Alnus rubra* and *Populus trichocarpa*. The competition for light by hybrid poplar could have been more pronounced on alder[18], if it had not been partly frost-killed after the first growing season. This does not mean that competition did not occur in mixed plantings, but rather suggests that hybrid poplar competed as intensively as did black alder. On the other hand, hybrid poplar showed increased values when decreasing its relative density, therefore benefiting from the presence of black alder. Hansen and Dawson[18] explained the significant height increase of hybrid *Populus* in mixed culture with *Alnus* by a positive effect of the latter rather than a reduced competition of alders. Our study cannot rule out this possibility, but the patterns of growth and N accumulation per plant of both species, and the decreasing Ndfs per hectare of black alder from 3A to 1A2P treatments associated with the increasing values of hybrid poplar in mixed treatments, favor the hypothesis of a reduced N competition of alder.

The natural ^{15}N dilution technique provides an integrated value for the % Ndfa. Results obtained in this study are for a period of 2 years, but values for the first and the second growing seasons could differ significantly in response to variations of environmental factors, as affected by the growth of both species. Percentages of Ndfa observed for black alder were relatively high but in agreement with the results of Rennie et al.[32], Ruschel et al.[36], and Kohl et al.[24] for leguminous plants cultivated in soils without addition of N-fertilizer. Miller[26] stated that alder could fix N in significant amounts only on sites that are incipiently or strongly N-deficient. Since our soil cannot be considered strongly N-deficient (total N 0.25%), one can assume that the high density of the plantation used in this study created a N deficiency that stimulated the symbiotic dinitrogen fixation. However, our results are not the first to show high rates of symbiotic dinitrogen fixation on relatively fertile sites. Binkley et al.[5] with Ceanothus velutinus, Franklin et al.[16], and DeBell and Radwan[13] with Alnus rubra assessed soil nitrogen accretion to 100, 130, and 80 kg ha^{-1} yr^{-1}, respectively, on productive soils. Accumulation of the fixed N in trees of the 3A treatment (53 kg ha^{-1} yr^{-1}) is comparable to the N-fixation rates observed by Akkermans et al.[1] and Tripp et al.[42] for Alnus glutinosa and Alnus rubra, using the acetylene reduction technique. It should be emphasized that the natural ^{15}N dilution technique provides an integrated assessment of the symbiotic N immobilized in the N fixing tree only, and that it does not take into account the soil nitrogen accretion that can occur. Therefore, care must be taken in comparing results of N-fixation estimated with other techniques.

This study did not show a negative interaction of hybrid poplar on growth and N-fixation by black alder. Jobidon and Thibault[23] observed an inhibitory effect of leachates of balsam poplar on the N-fixation by green alder. Such allelochemic properties are apparently missing in Populus nigra x P. trichocarpa. Black alder susceptibility to these inhibitors, if present, can also be lower than of green alder.

Higher amounts of Ndfs per hectare in the total aboveground tissues of mixed treatments over pure stands indicate a better utilization by the former of soil nitrogen. The longer growing season observed for hybrid poplar over black alder, and a greater volume of soil explored in mixed plots due to enhanced growth of Populus[15,20,27,39] could possibly explain these differences. However, increasing values of Ndfs from 1A2P to 2A1P treatments, and the corresponding higher accretion of 3A over 3P support the hypothesis of an increased rate of soil nitrogen mineralization under alder plantations as measured by Bollen and Lu[6].

The highest biomass per plant obtained in mixed treatments resulted primarily from enhanced growth of hybrid poplar, as previously noted by DeBell and Radwan[13]. However, yields of mixed plots did not differ significantly from that of 3A mainly because of mortality of hybrid poplar. The higher yield of 3A over 3P is indicative of the high N requirements of hybrid poplar to realize its growth potential, and of the capacity of black alder to symbiotically fix dinitrogen. The symbiotic N-fixation could possibly explain the highest N accumulation per plant and per hectare obtained in 3A and 2A1P.

The natural ^{15}N dilution technique used to evaluate the % Ndfa has given good results in agriculture where appropriate nfs controls exist. Thus far, the absence of an adequate nfs control for alder has prevented the use of this technique to assess the % Ndfa of this species. It could be argued that the present control alder (nfs) and alder in the field (fs) did not have access to the same soil N pool, since the former did not nodulate. In fact, despite the same soil and the same temperature gradient in both systems, N accumulation was enhanced by sieving the soil, and by the higher soil moisture content[37] maintained during the course of the greenhouse experiment, as shown by the high concentration of N-NO$_3$ in the soil percolates ($\bar{X} \geqslant 800$ ppm). Enhanced soil nitrogen mineralization, which exceeded by far N demands by black alder seedlings may explain why alder plants did not nodulate[28]. Righetti and Munns[34] observed no nodulation on *Purshia* in a greenhouse experiment using a soil with the same N content as ours. The sampling of alder trees for ^{15}N analyses is also questionable since we chose average weight trees, rather than performing a randomized sampling, in order to reduce the number of analyses. Dawson and Gordon[12] noted significant correlations between leaf area and acetylene reduction. Thus, we can assume that the average dry weight tree corresponds to the average N-fixing tree. However, the natural ^{15}N dilution technique and the comparison of plots with and without black alder gave similar assessments of the N-fixation. We used a linear extrapolation of the N content (kg ha^{-1}) of hybrid poplar in 2A1P and 1A2P treatments to evaluate the Ndfs in 3P plots because the presence of alder can increase the N mineralization[6]. This procedure should give a better estimate of the availability of soil nitrogen under plantations containing alder than the measured N content of pure hybrid poplar plantations. Results obtained in this study, using the natural ^{15}N dilution technique, were only semi-quantitative but a more intensive sampling could increase the precision of the technique, and therefore its usefulness.

Rates of aboveground biomass production measured in this study for mixtures of alder and poplar hybrid showed the same compatibility and almost the same productivity as the plantations studied by DeBell and Radwan[13] and by Tarrant and Trappe[41]. The lower yields obtained in this experiment can be due to the elimination of the positive border effects[17], to a less intensive site preparation and maintenance, and to a less productive study area. Results of N accumulation and N-fixation per hectare showed that the harvest of stem and branches only would create a N output that would be balanced by the Ndfa contained in entire plants of alder of the 1A2P treatment; 3A and 2A1P plots would increase their N content, and 3P plantations would be N-depleted. Harvesting the total aboveground biomass would necessitate a N-fertilization unless a sufficient soil nitrogen accretion occurred in plantations as reported by many workers[6,7,11,13,40]. These previsions are based on the assumption that the % Ndfa is similar for leaves, stem + branches, and roots, which is most likely to be an approximation[33], but errors should be minimal[36].

Conclusion

Plantations of very dense spacings, such as 33×33 cm, are not justifiable from an economic point of view[14]. Nevertheless, they allowed, in a short period of time, the demonstration of beneficial effects of black alder in mixed plantings on biomass production and N assimilation by hybrid poplar, and this, without adverse effect on the N-fixing tree. Our findings reveal that the nitrogen derived from symbiotic fixation immobilized in tissues of entire plants of black alder could provide sufficient N to balance the N exportation in harvested stems and branches of short-rotation plantations containing at least 33% of alder.

More complete studies are needed to understand the patterns of utilization of the fixed nitrogen by companion species and soil microorganisms, and the long term effects of these plantations on the full nutrient cycle.

Acknowledgements This research was supported by ENFOR program of Canadian Forest Service (project P-78), Major program No. 21 of Formation de chercheurs par action concertée, province de Québec, and Natural Sciences and Engineering Research Council of Canada (grant No. A7817).

References

1 Akkermans A D L and van Dijk C 1976 The formation and nitrogen fixing activity of the root nodules of *Alnus glutinosa* under field conditions. *In* Symbiotic Nitrogen Fixation in Plants. Ed. P S Nutman. pp 511–520. Cambridge University Press, Cambridge.

2 Amarger N, Mariotti A and Mariotti F 1977 Essai d'estimation du taux d'azote fixé symbiotiquement chez le lupin par le traçage isotopique naturel (^{15}N). C. R. Acad. Sci. Paris 284, 2179–2182.

3 Amarger N, Mariotti A, Mariotti F, Durr J C, Bourguignon C and Lagacherie B 1979 Estimate of symbiotically fixed nitrogen in field grown soybeans using variations in ^{15}N natural abundance. Plant and Soil 52, 269–280.

4 Bardin R, Domenach A M and Chalamet A 1977 Rapports isotopiques naturels de l'azote. II. Application à la mesure de la fixation symbiotique de l'azote in situ. Rev. Ecol. Biol. Sol 14, 395–402.

5 Binkley D, Cromack Jr K and Fredriksen R L 1982 Nitrogen accretion and availability in some snowbrush ecosystems. Forest Sci. 28, 720–724.

6 Bollen W B and Lu K C 1968 Nitrogen transformations in soils beneath red alder and conifers. In Biology of alder. Eds. J M Trappe, J F Franklin, R F Tarrant and G M Hansen. pp 141–148. Pacific Northwest Forest and Range Experiment Station, Portland, Oreg., 292 p.

7 Bormann B T and DeBell D S 1981 Nitrogen content and other soil properties related to age of red alder stands. Soil Sci. Soc. Am. J. 45, 428–432.

8 Bremner J M 1965 Total nitrogen. In Methods of Soil Analysis. Ed. C A Black. Part 2, pp 1149–1178. Am. Soc. Agron., Madison, Wis., U.S.A.

9 Canada Soil Survey Committee, Subcommittee on soil classification 1978 The Canadian system of soil classification. Can. Dep. Agric. Publ. 1646. Supply and Services Canada, Ottawa, Ont. 164 p.

10 Dale M E 1963 Interplant alder to increase growth in strip-mine plantations. U.S.D.A. Forest Serv. Res. Note CS-14, 4 p.

11 Daly G T 1966 Nitrogen fixation by nodulated Alnus rugosa. Can. J. Bot. 44, 1607–1621.

12 Dawson J O and Gordon J C 1979 Nitrogen fixation in relation to photosynthesis in Alnus glutinosa. Bot. Gaz. 140 (Supp.), S70–S75.

13 DeBell D S and Radwan M A 1979 Growth and nitrogen relations of coppiced black cottonwood and red alder in pure and mixed plantings. Bot. Gaz. 140 (Supp.), S97–S101.

14 Ek A R and Dawson D H 1976 Actual and projected growth and yields of Populus tristas #1 under intensive culture. Can. J. For. Res. 6, 132–144.

15 Ellern S J, Harper J L and Sagar G R 1970 A comparative study of the distribution of the roots of Avena fatua and A. strigosa in mixed stands using a ^{14}C-labelling technique. J. Ecol. 58, 865–868.

16 Franklin J F, Dyrness C T, Moore D G and Tarrant R F 1968 Chemical soil properties under coastal Oregon Stands of alder and conifers. In Biology of alder. Eds. J M Trappe, J F Franklin, R F Tarrant and G M Hansen. pp 157–172. Pacific Northwest Forest and Range Experiment Station, Portland, Oreg., 292 p.

17 Hansen E A and Baker J B 1979 Biomass and nutrient removal in short rotation intensively cultured plantations. In Impact of intensive harvesting on forest nutrient cycling. pp 130–151. Proc. Symp. Syracuse N.Y., 421 p.

18 Hansen E A and Dawson J O 1982 Effect of Alnus glutinosa on hybrid Populus height growth in a short-rotation intensively cultured plantation. Forest Sci. 28, 49–59.

19 Hauck R D 1973 Nitrogen tracers in nitrogen cycle studies – Past use and future needs. J. Environ. Quality 2, 317–327.

20 Hauck R D and Bremner J M 1976 Use of tracers for soil and fertilizer nitrogen research. Adv. in Agron. 28, 219–266.

21 Helwig J T (ed) 1979 SAS user's guide, 1979 edition. SAS Institute Inc., Cary, N.C., 495 p.

22 Jansson S L 1971 Use of ^{15}N in studies of soil nitrogen. In Soil Biochemistry. Eds. A Douglas McLaren and J Skujins. Vol. 2, pp 129–166. Marcel Dekker Inc, New York, U.S.A.

23   Jobidon R and Thibault J R 1982 Allelopathic growth inhibition of nodulated and un-
      nodulated *Alnus crispa* seedlings by *Populus balsamifera.* Amer. J. Bot. 69, 1213–1223.

24   Kohl D H, Shearer G and Harper J E 1980 Estimates of N_2 fixation based on differences
      in the natural abundance of ^{15}N in nodulating and nonnodulating isolines of soybeans.
      Plant Physiol. 66, 61–65.

25   McAlpine R C, Brown C L, Herrick A M and Ruark H E 1966 Silage sycamore. Forest
      Farmer 26, 6–7.

26   Miller H G 1983 Nutrient cycling in alder. IEA Report. National Swedish Board for
      Energy Source Development, 54 p.

27   Miller R E and Murray M D 1978 The effects of red alder on growth of Douglas-fir. *In*
      Utilization and management of alder. Eds. D G Briggs, D S DeBell and W A Atkinson.
      pp 283–306. Pacific Northwest Forest and Range Experiment Station Forest Service,
      U.S.D.A., Portland, Oreg., 379 p.

28   Pizelle G 1966 L'azote minéral et la nodulation de l'aulne glutineux (*Alnus glutinosa*).
      II. Observations sur l'action inhibitrice de l'azote minéral à l'égard de la nodulation.
      Ann. Inst. Pasteur, (Supp.), 259–264.

29   Plass W T 1977 Growth and survival of hardwoods and pine interplanted with European
      Alder. U.S. Dep. Agr. Forest Service Res. Paper NE-376, 10 p.

30   Raitenen W E 1980 Farming fast growing hardwoods for energy in Ontario, Canada.
      Second Western Hemisphere Energy Symposium, Rio de Janeiro, 30 p.

31   Rennie R J 1982 Quantifying dinitrogen (N_2) fixation in soybeans by ^{15}N isotope di-
      lution: the question of the nonfixing control plant. Can. J. Bot. 60, 856–861.

32   Rennie R J, Dubetz S, Bole J B and Muendel H-H 1982 Dinitrogen fixation measured
      by ^{15}N isotope dilution in two Canadian soybean cultivars. Agron. J. 74, 725–730.

33   Rennie R J, Rennie D A and Fried M 1978 Concepts of ^{15}N usage in dinitrogen fixation
      studies. *In* Isotopes in Biological Dinitrogen Fixation. IAEA, Vienna, pp 107–133 Unipub.
      New York, U.S.A.

34   Righetti T L and Munns D N 1982 Nodulation and nitrogen fixation in *Purshia:* inocu-
      lation responses and species comparisons. Plant and Soil 65, 383–396.

35   Rowe J S 1972 Les régions forestiè res du Canada. Ministè re de l'environnement, Service
      canadien des Forêts, Ottawa. Publication no 1300 F.

36   Ruschel A P, Vose P B, Matsui E, Victoria R L and Tsai Saito S M 1982 Field evauation
      of N_2-fixation and N-utilization by *Phaseolus* bean varieties determined by ^{15}N isotope
      dilution. Plant and Soil 65, 397–407.

37   Sarathchandra S U and Upsdell M P 1981 Nitrogen mineralization and the activity and
      populations of microflora in a high producing yellow-brown loam under pasture. N.Z.
      J. Agric. Res. 24, 171–176.

38   Soil Survey Staff 1975 Soil Taxonomy. U.S. Dep. Agric. Handbook No. 436. U.S. Govern-
      ment Printing Office, Washington, D.C. 754 p.

39   Sorensen L H 1982 Mineralization of organically bound nitrogen in soil as influenced
      by plant growth and fertilization. Plant and Soil 65, 51–61.

40   Tarrant R F 1961 Stand development and soil fertility in a Douglas-fir-red alder plan-
      tation. Forest Sci. 7, 238–246.

41   Tarrant R F and Trappe J M 1971 The role of *Alnus* in improving the forest environ-
      ment. Plant and Soil, Spec. Vol. 1971, 335–348.

42   Tripp L N, Bezdicek D F and Heilman P E 1979 Seasonal and diurnal patterns and rates
      of nitrogen fixation by young red alder. Forest Sci. 25, 371–380.

43   Van Cleve K, Viereck L A and Schlentner R L 1971 Accumulation of nitrogen in alder
      (*Alnus*) ecosystems near Fairbanks, Alaska. Arctic and Alpine Res. 3, 101–114.

44   Van der Meiden H A 1961 De els in populierenbeplantingen. (Alder in mixture with
      poplar). Nederlands Bosbouwk. Tÿdschrift 33, 168–171.

45   Younger P D and Kapustka L A 1983 $N_2(C_2H_2)$ ase activity by *Alnus incana* ssp. *rugosa*
      (Betulaceae) in the northern hardwood forest. Am. J. Bot. 70, 30–39.

Plant and Soil 78, 221–233 (1984).
© 1984 *Martinus Nijhoff/Dr W. Junk Publishers, The Hague.*

N$_2$ fixation by red alder (*Alnus rubra*) and scotch broom (*Cytisus scoparius*) planted under precommercially thinned Douglas-fir (*Pseudotsuga menziesii*)

O. T. HELGERSON,[1] J. C. GORDON[2] and D. A. PERRY[1]

[1] *Department of Forest Science, Oregon State University, Corvallis, OR, 97331, USA*
[2] *School of Forestry and Environmental Studies, Yale University, New Haven, CT 06511, USA*

Key words *Alnus rubra* *Cytisus scoparius* Growth N$_2$ fixation Precommercial thinning *Pseudotsuga menziesii*

Summary From acetylene reduction assays over a 10-month period starting in April 1979, nodule activities averaged 18.78 (se 4.67) μmoles C_2H_4 g nodule dw^{-1} h^{-1} for *Alnus rubra* and 59.95 (se 12.14) μmoles C_2H_4 g nodule dw^{-1} h^{-1} for *Cytisus scorparius*. Plant rates were 1.91 (se .47) μmoles C_2H_4 plant^{-1} h^{-1} for *A. rubra* and 0.55 (se .17) μmoles C_2H_4 plant^{-1} h^{-1} for *C. Scoparius*. Plant activity and total leaf N were strongly correlated with the dw of other plant parts, but nodule activity and percent leaf N were not. Plant and nodule activities were not associated with temperature, moisture stress, precipitation events or percent light for either species over the growing season nor for 54 *A. rubra* sampled in mid-season 1979 on one replication. After 5 to 6 growing seasons, 14 *A. rubra* on the same site ranged from 30 to 332 cm in height and showed strong correlation between nodule dw, leaf dw, plant size and total leaf N. Results from this study and others indicate logistic equations may be modified to predict the effect of adding a N$_2$ fixing plant to a population of non N$_2$ fixing trees.

Introduction

The use of a symbiotic N$_2$ fixing plant species (N$_2$ fixer) to enhance the yield of another non N$_2$-fixing tree species is based on the premise that allocation of site resources, temporarilly or spatially, to the N$_2$ fixer will increase net site productivity by making N less limiting[31]. Addition of the N$_2$ fixer may enhance energy storage or lower entropy[7], or increase the efficiency of photosynthetic area[9]. Although nitrogen may become less limiting, the N$_2$ fixer may also cause another growth factor to become more limiting and decrease yield (Fig. 1).

In the Pacific Northwest, *Alnus rubra* Bong. (red alder) has enhanced the growth and yield of *Pseudotsuga menziesii* (Mirb.) Franco. (Douglas-fir)[5,21,31] especially on sites low in productivity and nitrogen capital[5]. N$_2$ fixers that grow more slowly than non N$_2$ fixers can be beneficial[4], but N$_2$ fixers such as alder with fast juvenile growth are easily capable of suppressing associated species with slower juvenile growth such as Douglas-fir[21,22,36]. Growth reductions in Douglas-fir have been related to water use by the N$_2$ fixers *Ceanothus velutinus velutinus* Dougl. (snowbrush)[24] and red alder[39]. Reductions in P have also been associated with stands of red alder[5]. Conversely, fast-growing trees

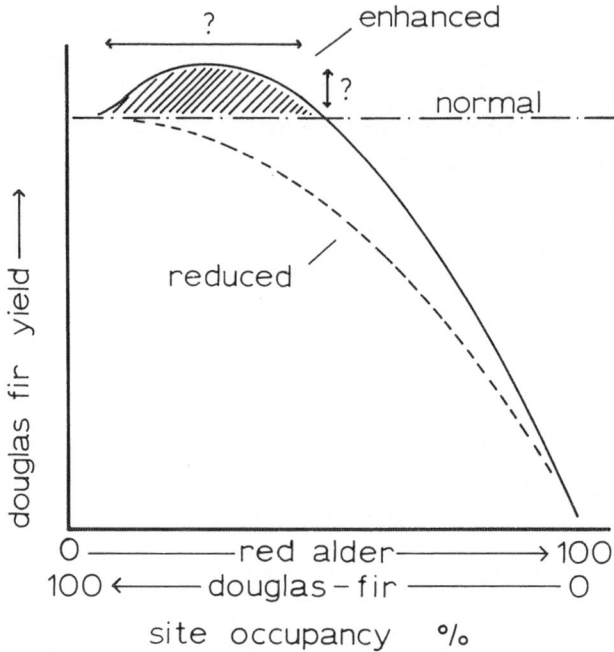

Fig. 1. Diagram of possible yield enhancement or decrease of a heterotroph (Douglas-fir) created by adding an autotroph (red alder) in space or time (after[21]).

such as *Populus* sp. are capable of faster juvenile growth than associated *Alnus* sp.[14,15,16,36,37], and non N_2 fixing trees with fast juvenile growth show increased height growth as distance to the N_2 fixer decreases[38].

One way to minimize competitive effects is to introduce the N_2 fixer after the associated species is well established, but also while site resources are still adequate for nitrogen fixation. Precommercial and commercial thinning of timber stands appear to offer this opportunity[3,29,30]. Many stands of Douglas-fir in the Pacific Northwest are near or at the precommercial thinning stage, thus opportunity may exist to enhance the yield of these stands by adding a N_2-fixing plant. The objectives of this study were to see whether *Cytisus scoparius* L. (scotch broom), a perennial woody legume, and red alder could survive, grow, and fix nitrogen in the environment created by precommercial thinning of Douglas-fir in the central portion of the Cascade Mountains in Oregon as a step toward determining whether biological N_2 fixation can enhance the growth of such stands.

Methods

The study area is 80 km east of Eugene Oregon, USA. In the spring of 1978, six precommercially thinned stands of 18- to 20-year old Douglas-fir were underplanted with alder and broom in a randomized block design with the stands as replicates. These six sites ranged across Douglas-fir productivity classes III and IV[18], north and south aspects, and elevations of 550 to 1000 m. The stands were approximately 4 to 6 m in height and were thinned to approximately 555 trees ha[-1]. The propagules of broom and alder were one to three year old wildlings from the Oregon Coast Range and were planted in late March, 1978. To replace mortality, alder wildlings and container grown seedlings, broom wildlings and snowbrush wildlings were planted in April, 1979. The replacement broom was also from the Coast Range source. The alder and snowbrush wildlings came from sites within 15 km of the study area in the Cascades. The container grown alder were 1-year-old 66 ml plugs grown in Nisqually, Washington, USA from a low elevation Washington seed source.

A total of 39 broom and 40 alder were assayed for nitrogenase activity across the six replications, on eight dates between April and November, 1979 with an acetylene reduction assay[19]. The sites were sampled on 2 consecutive days in August and in September. The six sites were visited in a random order on each sampling date with an attempt made to sample one plant of each species from each site. Plants were sampled from two low elevation sites that were free of snow in February, 1980.

Nodules were excised from sample plants on intact root systems and incubated within glass vessels containing 10 percent (vv) acetylene for 40 minutes in the soil. The volume and temperature of acetylene placed in each vessel, the temperature of adjacent soil, and the barometric pressure were recorded. At the end of incubation, gas samples were drawn into partially evacuated 13 ml glass tubes (Vacutainer, T.M.). To ascertain the base level of ethylene, control samples were also drawn from incubation vessels that were treated similarly but that contained only acetylene.

From each Vacutainer, an 0.1 ml sample was injected into a Hewlett-Packard 5830 gas chromatograph fitted with a 6 × 1000 mm column filled with Porapak N. A programmable electronic integrator coupled to a flame ionization detector calculated the molar proportion of ethylene in the sample. The carrier gas was N_2. A pressure bomb was used to measure moisture stress at the time of sampling as xylem pressure potential of the sampled plants[26, 33].

On two dates in the middle of the 1979 growing season (June 20 and July 5) a total of 54 alders were sampled on one replicate to attempt to better define environmental effects on nitrogen fixation. Acetylene reduction, moisture stress, available light and the basal area of surrounding Douglas-fir were measured. Available light was measured as a percent of total sunlight with pads of Ozalid T.M. paper[13] exposed adjacent to individual plants. Basal area was measured with a 10 factor (American) prism using sample plants as center points.

The plots were surveyed for mortality in September 1978 and October 1979. A minimum of 30 plants (alive plus dead) were tallied on each date. Browsing damage was also assessed in 1979. The site where 54 alder were sampled in midsummer 1979 was revisited in July, 1983. This time, 14 alders were measured for individual size, basal area of surrounding Douglas-fir using a 10 factor prism, and plant cover in a 1 m radius around each alder above and below its top.

Results and discussion

After one growing season, survival rates for the broom and plug alder planted in 1979 were significantly better than the other propagules and other planting dates (Table 1). Deer or elk browsing damaged some broom and alder, and most snowbrush were clipped off near ground level by rodents. The surviving snowbrush were judged to be too few and too damaged to be assayed for nitrogenase activity.

The mean values for nodule activity and plant activity (nodule activity × nodule dw) showed considerable variation both on individual sampling dates and over the growing season for both species, with

Table 1. Percent survival for underplanted alder and broom. Mean above, standard error below

	Survival (percent)	
Year of planting and propagule	September 1978	October 1979
1978	37	20
Alder wildlings	8	8
1978	35	23
Broom wildlings	11	11
1979		62*
Plug alder		11
1979		24
Alder wildlings		12
1979		65*
Broom wildlings		9

* The 1979 plug alder and broom showed significantly greater survival after one year ($p = 0.05$) using arc sin percent survival^{-2}.

Table 2. Mean acetylene reduction rates and allometric values (standard error below) for red alder and broom sampled over 10 months and sampled in mid-1979 and mid-1983

	μ moles C_2H_4 g nodule dw^{-1} h^{-1}	μ moles C_2H_4 plant^{-1} h^{-1}	Leaf dw g	Nod dw g	Top dw g
Averaged over 1979					
Red alder	18.78	1.91	1.07	0.18	7.50
n = 40	4.67	0.47	0.32	0.03	0.99
Broom	58.95	0.55	0.23	0.02	5.20
n = 39	12.24	0.17	0.08	0.006	1.08
Midseason 1979					
Red alder	25.78	6.13	1.71	0.20	5.57
n = 54	2.89	1.33	0.27	0.03	0.88
Midseason 1983					
Red alder			22.31	1.71	91.44
n = 14			7.13	0.52	33.15

differences appearing between sampling dates 1 day apart in September for nodule activity (Figs. 2, 3). Averaged over all sampling dates, nodule activity for broom was approximately three times greater than for alder. For plant activity, however, the positions were reversed, with alder having about 3.5 times more activity than broom (Table 2). The large differences in nodule activity appear to be offset by differences in the amount of nodules per plant. For alder, nodule dw averaged 2.4 percent of top dw (leaf plus stem dw), whereas for broom, nodule dw averaged just 0.38 percent of top dw. The tendency for nodule weight to offset unequal nodule activities has also been observed between red alder and *Alnus sinuata* (Reg.) Rybd.[8].

Using the ratio of the means, broom had a much higher ratio of leaf-to-nodule dw (11.5) compared to alder 5.94), but broom leaves

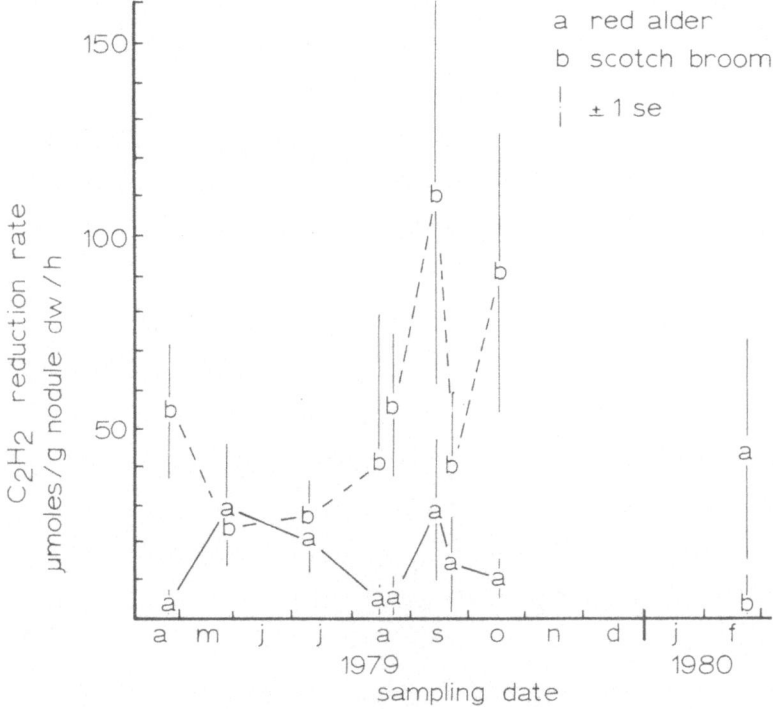

Fig. 2. Nodule activity for alder and broom sampled over 10 months.

made up a far smaller portion of top dw (4.4 percent) than alder (14.3 percent). Averaged plant activity expressed per unit leaf dw was 1.78 μmoles C_2H_4 h^{-1} g leaf dw^{-1} for alder and 2.39 μmoles C_2H_4 h^{-1} g leaf dw^{-1} for broom. Broom had far less of its top dw as foliage and a far greater leaf-to-nodule ratio than did alder. However, broom's nodule activity expressed as leaf dw was greater than alder's. This suggests that broom leaves may be more efficient than alder at supplying photosynthate to nodules or that broom's young green stems can also supply photosynthate to root nodules. The importance of broom's stems is suggested by the activity recorded in February when it had no leaves, and by an observed cessation of nodule activity when broom's young green stems were killed by frost[34].

The activity observed in February in the alder occurred in leafless plants sampled on south facing lower elevation sites free of snow with soil temperatures of 6°C. The rates are higher than reported in a summary of other studies of N₂ fixation during plant dormancy[35]. The observed activity may indicate that appreciable N₂ fixation may occur under some wintertime conditions although the rates may have resulted from unnoticed ethylene contamination.

For alder sampled mid-season in 1979, nodule and plant activities

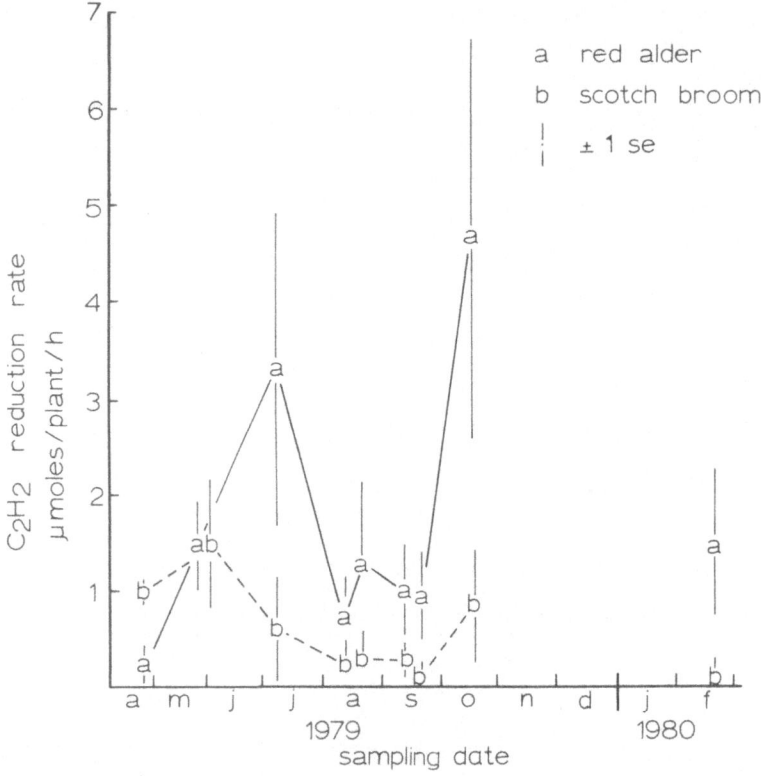

Fig. 3. Plant activity for alder and broom sampled over 10 months.

were greater than the 10 month averages. The leaf-to-top dw ratio at this time was 0.31, compared to 0.14 for the 10 month average. The increase in acetylene reduction coupled with an increase in leaf mass is consistent with previously observed relationships between photosynthesis and N_2 fixation[6,10,14].

Over the growing season, neither nodule nor plant activities for either species were associated with precipitation events, soil temperature or moisture stress. Examination of plotted values did not reveal any obvious relations with these variables or for diurnal trends of N-fixation. Soil temperature and plant moisture stress were unrelated to acetylene reduction on individual sampling dates through the season or for the 54 alders sampled in mid 1979. Nor was basal area of surrounding Douglas-fir or percent sunlight related to nitrogen fixation or plant size. The lack of diurnal trends or any relations between nitrogen fixation and soil temperature, light, plant moisture stress or surrounding basal area is puzzling given the considerable research[23,35] that shows fixation rates to be often related to those variables. It is possible that the effects of these variables masked each

Table 3. Linear correlations between acetylene reduction and allometric variables for under-planted alder and broom sampled over 10 months. Correlation coefficients above, probability of larger r below

	Nodule activity	Plant activity	Nodule dw	Leaf dw	Top dw
Red alder					
Plant activity	0.13				
	0.40				
Nodule dw	− 0.29	0.54			
	0.06	0.0003			
Leaf dw	0.02	0.635	0.72		
	0.89	0.0001	0.0001		
Top dw	− 0.17	0.30	0.25	0.30	
	0.31	0.08	0.13	0.07	
Leaf N %	0.06	0.01	− 0.21	− 0.33	− 0.30
	0.44	0.96	0.29	0.09	0.12
Total leaf N	0.18	0.73	0.79	0.99	0.22
	0.40	0.0001	0.0001	0.0001	0.26
Scotch broom					
Plant activity	0.18				
	0.28				
Nodule dw	− 0.22	0.47			
	0.17	0.002			
Leaf dw	− 0.10	0.48	0.45		
	0.60	0.006	0.009		
Top dw	− 0.21	0.52	0.83	0.74	
	0.25	0.003	0.001	0.001	
Leaf N %	0.38	0.38	0.03	0.12	− 0.14
	0.12	0.12	0.90	0.63	0.55
Total leaf N	0.07	0.44	0.40	0.95	0.41
	0.80	0.07	0.10	0.0001	0.08

Table 4. Linear correlations between acetylene reduction and allometric variables for alder sampled on June 20 and July 6, 1979. Correlation coefficient above, probability of larger r below

	Nodule activity	Plant activity	Nodule dw	Leaf dw	Top dw
Plant activity	0.56				
	0.0001				
Nodule dw	0.21	0.87			
	0.13	0.0001			
Leaf dw	0.38	0.90	0.92		
	0.005	0.0001	0.0001		
Top dw	0.26	0.63	0.71	0.89	
	0.06	0.0001	0.0001	0.0001	
Leaf N %	0.08	0.04	− 0.004	− 0.06	− 0.04
	0.56	0.79	0.97	0.67	0.76
Total leaf N	0.40	0.91	0.91	0.99	0.88
	0.003	0.0001	0.0001	0.0001	0.0001

other or were masked by some other factor such as phenological development, planting shock, or an inadequate sample size.

Allometric correlations between nodule activity and the five variables — dw of nodules, leaves and tops, and leaf and stem N contents — tended to be insignificant for either broom or alder over the growing season. However, the correlations between plant activity and these variables tended to be strongly significant for both species (Table 3). Allometric correlations were even stronger for the 54 alders measured in mid-1979. These alders showed particularly strong associations between leaf dw and nodule activity, plant activity and nodule dw, and between total leaf N (percent leaf N x leaf dw) with plant activity, nodule and leaf dw (Table 4).

The 14 alders measured in mid 1983 ranged from 30 to 332 cm in height and 4.4 and 473 g top dw. The allometric correlations observed in the first year (1979) between dry weights and total leaf N remained strong. Nodule dw was strongly associated with height, root collar diameter and stem volume (diameter^2 x height). The association of percent leaf N with other variables remained weak. Compared to the plants sampled in mid-1979, the alder sampled in 1983 showed significant negative correlations between plant size and weight and the basal area of surrounding Douglas-fir. Estimated cover was not related to alder size (Table 5). The negative associations between plant size and the surrounding basal area of Douglas-fir are similar to other density dependent changes in alder size and imply that similar changes in nitrogen fixation exist[6, 27, 29, 30]. Measurements on more plants are, however, necessary to better define this relationship to Douglas-fir basal area. Numerous live broom are also noted, but lack of time prevented measurement of this species.

The ratio of nodule-to-top dw was 1.87 percent in 1983 compared to 3.5 percent for the alders sampled in mid-1979, a decrease similar to that reported for *A. glutinosa* (L.) Gaertn.[2] The leaf-to-nodule dw ratio increased from 8.58 in 1979 to 16.14 in 1983. Because the nodules were not assayed for nitrogenase activity in 1983, it cannot be established that the decreased proportion of nodules represents a lessening of photosynthate allocation to nitrogen fixation.

This study and others cited show a relationship between plant size, surrounding stand density, available nitrogen and N_2 fixation. These results indicate that the Verhulst-Pearl logistic equations that describe competition between two organisms[17, 25] offer a convenient starting point for predicting whether the addition of a N_2 fixing plant will enhance the yield of a stand of non N_2 fixers. They appear to be able to incorporate factors that may affect N_2 supply: the

Table 5. Correlations between allometry and Douglas-fir basal area for underplanted alder five to six years after planting. Correlation coefficient above, probability of greater r below

	Height	Diameter	Stem volume	Nodule dw	Stem dw	Leaf dw	Top dw	Leaf % N	Leaf total N
Diameter	0.83								
	0.002								
Stem	0.90	0.89							
volume	0.0001	0.0001							
Nodule	0.98	0.81	0.93						
dw	0.0001	0.0004	0.0001						
Stem	0.94	0.80	0.96	0.97					
dw	0.0001	0.0006	0.0001	0.0001					
Leaf dw	0.95	0.83	0.96	0.97	0.99				
	0.0001	0.0003	0.0001	0.0001	0.0001				
Top dw	0.94	0.80	0.96	0.97	0.99	0.99			
	0.0001	0.0005	0.0001	0.0001	0.0001	0.0001			
Leaf	0.031	0.20	0.03	− 0.009	− 0.009	0.04	0.02		
% N	0.92	0.52	0.93	0.98	0.98	0.89	0.99		
Total	0.94	0.85	0.95	0.95	0.97	0.99	0.98	0.12	
leaf N	0.0001	0.0003	0.0001	0.0001	0.0001	0.0001	0.0001	0.69	
Douglas-	0.61	− 0.29	− 0.57	− 0.64	− 0.70	− 0.67	− 0.69	0.31	− 0.64
fir BA	0.06	0.41	0.08	0.04	0.02	0.03	0.03	0.41	0.06

percentage of N_2 fixers on the site, the N_2 fixation rate, and the availability of N to associated trees[15]. For a single species, one form of the logistic equation is

$$\frac{dP}{dt} = aP \frac{a/b - P}{a/b} = P(a - bP)$$

where (P) denotes population size or total biomass; (a/b) is the upper limit on population growth established as a number of individuals or biomass that a site can support; and the parameter (a) represents an intrinsic growth rate based on current site resources. The population growth rate slows as (P) approaches the carrying capacity of the site (a/b). The plot of P *versus* time follows a sigmoid (S-shaped) curve.

Although originally developed for animal populations, the logistic equation can also describe the growth of individual plants in even-aged populations[1,12,27]. The volume yields of even-aged stands of Douglas-fir[18], other North American conifers[11] and red alder[22,28,36] tend to follow sigmoid patterns over time, with better sites tending to support greater rates of growth and standing volumes.

Describing a mixed stand of N_2 fixers and an associated species follows a model[25] used to describe the interaction between two uniformly distributed competing species. Here, competition between the N_2 fixers (A) such as red alder, *etc.* and a non N_2 fixer (H) such as

Douglas-fir, *etc.* may be expressed as

$$\frac{dH}{dt} = H(a_H - b_H H - b_{HA} A)$$

$$\frac{dA}{dt} = A(a_A - b_A H - b_A A)$$

In these two equations; a_H denotes the intrinsic growth rate of species H; b_H denotes the effect that the presence of H has on its own growth; and b_{HA} denotes the effect that the presence of species A has on the growth of H. Similarly, a_A denotes the intrinsic growth rate of species A, and b_{AH} denotes the effect that species H has on the growth of A, and b_A denotes the effect that the presence of A has on its own growth. These equations indicate that the growth rate of either species is slowed not only by the presence of its own members, but also by the presence of the second species.

The addition of the N_2 fixer, however, can increase the growth rate and standing biomass of the other species on nitrogen deficient sites. Thus the parameters (a) and (a/b) for the non N_2 fixer may be increased with greater nitrogen availability induced by the N_2 fixer, depending on its growth. As N availability increases, however, N_2 fixation rates probably will likely decline[5,35] as should the response of the non N_2 fixer to incremental nitrogen[5,20]. Following the logistic concepts of Mitscherlich and Spillman[32], the amount of nitrogen fixed (N_f) is proportional to the growth of the N_2 fixer $\left(\frac{dA}{df}\right)$, but may be moderated by the difference between its optimum level (N_{opt}) and what is available (N_{avail}), with availability in turn partially dependent on N_2 fixation as well as on rates of N release, turnover and loss.

$$\frac{dN_f}{dt} \propto \frac{dA}{dt} \left[\frac{N_{optA} - N_a}{N_{optA}} \right]$$

Likewise, for the non N_2 fixer its growth rate (a_H) and the carrying capacity of the site (a_H/b_H) may be similarly controlled. With this relation reflecting rates of N release and turnover:

$$\frac{da_H}{dt} \propto \frac{dN_{avail}}{dt} \left[\frac{N_{optA} - N_{avail}}{N_{optH}} \right].$$

Enhancement of growth of the associated species should occur when the competitive effect of the N_2 fixer is less than it's positive effect on the non N_2 fixers growth. This set of relationships appears to provide a conceptual framework that can be used to predict the effects of adding an N_2 fixer. The effects vary; yield may increase, decrease, or stay the same. The change induced depends on the starting amounts of each species, their intrinsic growth rates, maximum biomass potentials, N_2 fixation rates and N availability.

Conclusions

Broom and red alder are capable of survival and nitrogen fixation beneath precommercially thinned Douglas-fir on sites typical of those found in the central Oregon Cascade mountains. The difference in survival between stocktypes and planting year illustrates the need to use vigorous propagules. The increase in size of the red alder 5 to 6 years after planting, the strong relationships between size of the alder and total leaf N, and the negative relationship between size and surrounding basal area of Douglas-fir are further evidences that nitrogen fixation is proportional to plant growth and photosynthetic area and that it may be controlled by the presence of the surrounding conifer. If so, the logistic equations may indicate whether the addition of a nitrogen fixing plant species will increase the yield of a non N_2 fixer.

References

1 Aikman D P and Watkinson A R 1980 A model for growth and self thinning in even-aged monocultures of plants. Ann. Bot. 45, 419–417.

2 Akkermans A D L and Houwers A 1979 Symbiotic nitrogen fixers available for use in temperate forestry. *In* Symbiotic Nitrogen Fixation in the Management of temperate Forests. Eds. J C Gordon, C T Wheeler and D A Perry. pp 23–35, For. Res. Lab., Oregon State Univ., Corvallis, OR.

3 Berg A and Doerksen A 1975 Natural fertilization of a heavily thinned Douglas-fir stand by understory red alder. Res. Note No 56. Forest Res. Lab., Oregon State Univ., Corvallis, OR.

4 Bengston G W and Mays D A 1978 Growth and nutrition of loblolly pine on coal mine spoil as affected by nitrogen and phosphorus fertilizer and cover crops. For. Sci. 24, 389–409.

5 Binckley D 1982 Case studies of red alder and Sitka alder in Douglas-fir plantations: nitrogen fixation and ecosystem production. Ph. D. Diss. Oregon State Univ., Corvallis, OR 115 p.

6 Borman B T 1981 Stand density in young red alder plantations: production, photosynthate partitioning, nitrogen fixation and an optimal spacing model. Ph. D. Diss. Oregon State Univ., Corvallis, OR 62 p.

7 Beuter J 1979 Forest fertilization and the economics of perpetual motion machines. *In* Symbiotic Nitrogen Fixation in the Management of Temperate Forests. Eds. J C Gordon, C T Wheeler and D A Perry. pp 4–14. Forest Res. Lab. Oregon State University, Corvallis, OR.

8 Carpenter C V, Baribo L E, Robertson L R, Van DeBogart F and Onufer G M 1979 Acetylene reduction by excised root nodules from *Alnus rubra* and *Alnus sinuata*. *In* Symbiotic Nitrogen Fixation in the Management of temperate Forests. Eds. J C Gordon, C T Wheeler and D A Perry. p 475. For. Res. Lab., Oregon State Univ., Corvallis, OR.

9 Cromack K, Delwiche C and McNabb D H 1979 Prospects and problems of nitrogen

management using symbiotic nitrogen fixers. *In* Symbiotic Nitrogen Fixation in the Management of temperate Forests. Eds. J C Gordon, C T Wheeler and D A Perry. pp 210–223. For. Res. Lab., Oregon State Univ., Corvallis, OR.

10   Dawson J O and Gordon J C 1979 Nitrogen fixation in relation to photosynthesis in *Alnus glutinosa*. Bot. Gaz. (supplement) 140, 570–575.

11   Forbes R D 1961 Forestry Handbook. pp 3, 1–61. The Ronald Press, New York.

12   Fresco L F M 1973 A model for plant growth. Estimation of the parameters of the logistic function. Acta. Bot. Neerl. 22, 486–489.

13   Friend D T 1961 A simple method of measuring integrated light values in the field. Ecology 42, 577–580.

14   Gordon J C and Wheeler C T 1978 Whole plant studies on photosynthesis and acetylene reduction in *Alnus glutinosa*. New Phytol. 80, 179–186.

15   Hansen E A and Dawson J O 1982 Effect of *Alnus glutinosa* on hybrid *Populus* height growth in a short rotation intensively cultured plantation. For. Sci. 28, 49–59.

16   Harrington C A, DeBell D S and Strand R F 1979 An experiment in biomass production: results from three consecutive harvests of cottonwood and alder. *In* Proc. of Solar 79 Northwest. pp 363–366. Seattle, WA.

17   Harper J L 1977 Population biology of plants. Academic Press, New York 892 p.

18   McArdle R E, Meyer W and Bruce D 1961 The yield of Douglas-fir in the Pacific Northwest. Tech. Bull. No. 201. USDA, Wash. DC.

19   McNabb D H and Geist J M 1979 Acetylene reduction assay of symbiotic N$_2$ fixation under field conditions. Ecology 60, 1070–1072.

20   Miller R E and Fight R D 1979 Fertilizing Douglas-fir forests. Gen. Tech. Rep. PNW-83. Pac. Northwest For. and Range Exp. Stn. Portland, OR.

21   Miller R E and Murray M D 1978 The effects of red alder on the growth of Douglas-fir. *In* Utilization and Management of Alder. Eds. D G Briggs, D S DeBell and W A Atkinson. pp 283–306. Pac. Northwest For. and Range Exp. Stn. Portland, OR.

22   Newton M, El Hassan B A and Zavitkovski J 1968 Role of red alder in western Oregon forest succession. *In* Biology of Alder. Eds. J M Trappe, J F Franklin, R F Tarrant and G M Hansen. pp 73–84. Pac. Northwest For. and Range Exp. Stn. Portland, OR.

23   Perry D A, Wheeler C T and Helgerson O T 1979 Nitrogen fixing plants for silviculture; some genecological considerations. *In* Symbiotic Nitrogen Fixation in the Management of temperate Forests. Eds. J C Gordon, C T Wheeler and D A Perry. pp 243–252. For. Res. Lab., Oregon State Univ., Corvallis, OR.

24   Petersen T D 1981 First year response of Douglas-fir after release from snowbrush. M. S. Thesis. Oregon State Univ., Corvallis, OR, 49 p.

25   Pielou E C 1969 An introduction to mathematical ecology. Wiley-Interscience, New York, 286 p.

26   Scholander D F, Hammel E D, Bradstreet E D and Hemmingsen E A 1965 Sap pressure in vascular plants. Science 148, 339–346.

27   Smith N J 1983 Effects of density stress and soil productivity on size mortality and nitrogen fixation in artificial populations of seedling red alder (*Alnus rubra* Bong.). Ph. D. Diss. Oregon State Univ., Corvallis, OR, 146 p.

28   Smith J H G 1968 Growth and yield of red alder in British Colombia. *In* Biology of Alder. Eds. J M Trappe, J F Franklin, R F Tarrant and G M Hansen. pp 273–286. Pac. Northwest For. and Range Exp. Stn. Portland, OR.

29   Sprent, J I 1973 Growth and nitrogen fixation in *Lupinus arboreus* as affected by shading and water supply. New Phytol. 72, 1005–1022.

30   Sprent J I and Sylvester W B 1973 Nitrogen fixation by *Lupinus arboreus* grown in stands in the open and under different aged stands of *Pinus radiata*. New Phytol. 72, 991–1003.

31   Tarrant R F 1961 Stand development and soil fertility in a Douglas-fir-red alder plantation. For. Sci. 7, 238–246.

32   Tisdale S L and Nelson W L 1966 Soil fertility and fertilizers (2nd Ed.) The MacMillan Co., New York, 693 p.

33   Waring R H and Cleary B D 1967 Plant moisture stress: evaluation by pressure bomb. Science 15, 1248–1254.

34   Wheeler CT, Perry D A, Helgerson O and Gordon J C 1979 Winter fixation of nitrogen in scotch broom (*Cytisus scoparius* L.). New Phytol. 82, 697–701.

35   Wheeler C T and McLauglin M E 1979 Environmental modulation of nitrogen fixation in actinomycete nodulated plants. *In* Symbiotic Nitrogen Fixation in the Management of temperate Forests. Eds. J C Gordon, C T Wheeler and D A Perry pp 124–142. For. Res. Lab., Oregon State Univ., Corvallis, OR.

36   Williamson R L 1969 Productivity of red alder in western Oregon and Washington. *In* Biology of Alder. Eds. J M Trappe, J F Franklin, R F Tarrant and G M Hansen. pp 287–291. Pac. Northwest For. and Range Exp. Stn. Portland, OR.

37   Whittwer R F and Immel M J 1979 A comparison of five tree species for intensive fiber production. For. Ecol. Manage. 1, 249–254.

38   Zavitkovski J, Hansen E A and McNeel H A 1979 Nitrogen fixing species in short rotation systems for fiber and energy production. *In* Symbiotic Nitrogen Fixation in the Management of temperate Forests. Eds. J C Gordon, C T Wheeler and D A Perry. pp 388–402. For. Res. Lab., Oregon State Univ., Corvallis, OR.

39   Zedaker S M 1981 Growth and development of young Douglas-fir in relation to intra- and inter-specific competition. Ph. D. Diss. Oregon State Univ. 175 p.

Plant and Soil 78, 235–243 (1984).
© 1984 *Martinus Nijhoff/Dr W. Junk Publishers, The Hague.*

Alder–*Frankia* interaction and alder–poplar association for biomass production

E. TEISSIER du CROS,
INRA, Station d'Amélioration des Arbres Forestiers, Ardon, F-45160 Olivet, France

G. JUNG
Rhône-Poulenc, Centre de Recherches de la Croix de Berny, 182–184, avenue Aristide Briand, F-92160 Antony, France

and M. BARITEAU
INRA, Service technique biomasse, Ardon, F-45160 Olivet, France

Key words *Alnus Frankia* N_2-fixation Mixed stands *Populus*

Summary Alders have an important role to play in biomass producing stands because of their N_2-fixing ability and their capacity to withstand soils having an excess of moisture. The objectives of preliminary trials were (1) to find if there is any alder-genotype × *Frankia*-strain interaction when the effect of inoculating the bacteria was compared to no inoculation in seed beds of different species and provenances of alder, (2) to measure the possible effect of black alders interplanted in poplars compared to pure poplar plots. Two trials were laid out to study the alder-*Frankia* interaction. Both produced interaction. In the first one the inoculation had a favorable effect on *Alnus glutinosa* at age 2 years and *A. cordata* at age 1 and 2 and no effect on *A. rubra*. In the second one the inoculation had a depressive effect at age 1 on 2 of 3 provenances of *A. rubra* and no effect on 1 *A. rubra*, 3 *A. glutinosa* and 3 *A. cordata* provenances.

A closely spaced field trial associating one black alder provenance and the poplar clone UNAL gives no superiority of mixed plots compared to pure plots. The results suggest that the N_2-fixation of alders is not profitable to poplars at age 3 with a 1.5 × 2 m spacing.

Introduction

Among the species which could be used for short term biomass production in forest land, alders have an important role to play for several reasons. Their ability to fix atmospheric nitrogen offers them large possibilities in poor or reclaimed soils and also in species mixtures. Their natural disposition to withstand moist conditions leads foresters to consider them in the reforestation of low potential hydromorphic sites.

But in these conditions of utilization where the main objectives are adaptation to sites and high juvenile growth, three questions are raised which need special emphasis:

– improvement programmes are usually faced with transfers of species and provenances from their natural ranges to new sites. The question is whether these genotypes will find the right or even the best *Frankia* endophyte in these new locations.

– a similar question is raised when tree breeders deal with inter-specific hybrids.

– species mixtures associating alders have proven efficient in certain cases. Is this general?

The present paper gives information on two of these questions: A – Nursery study of a host-endophyte interaction with alders and *Frankia* and B – Field study of a mixture associating a poplar clone and a black alder provenance (*Alnus glutinosa*).

Method

A. *Alder-*Frankia *interaction*

The objective was to apply a *Frankia* strain on a *Frankia*-free seed bed when sowing alder seeds.

The *Frankia*-strain used was ARbN4$_b$ from M. Lalonde (Canada). It was cultivated in a modified Qmod medium as described by Lalonde in 1979[7]. The contents of 32 test tubes of a 40 ml single volume, obtained after 2 months incubation at 28°C was centrifuged for 5 minutes at 5000 rpm. The residue was crushed (Ultra-Turrax) to burst sporangia, and re-suspended into 500 ml water. The suspension was used to prepare 1500 ml of gel including Xanthane (Rhodopol 23 by Rhône-Poulenc) and carob gum (by François). To the gel was added 40% of its weight of precipitated semiporous silica (S 1069 by Rhône-Poulenc), mixed (Küstner) then dried to a final water content between 150 and 200% dry matter. The final powder was weighed, divided in equal quantities, spread on the seed beds and mixed with the substrate by slight raking. Non-inoculated controls were kept to observe the natural nodulation in the sowing conditions applied in our experiments.

The sowing substrate was a mixture of 15% peat and 85% grinded, composted and fertilized Maritime pine bark. A 15-cm-thick layer was spread onto the nursery soil and divided by plastic frames in 0.20 m² surfaces which contained the experimental single plots. The substrate was *Frankia*-free before use in our experiments as shown by Rhône-Poulenc Research Laboratory after having sown alder seeds in controlled laboratory conditions and having obtained no nodulation.

Alder species were *Alnus glutinosa*, *A. cordata* and *A. rubra* with one provenance each in the 1981 trial and three provenances each in the 1982 trial. The amount of seeds sown was estimated to give roughly 30 seedlings per single plot.

The experimental lay-out was factorial and had four replications.

The efficiency of the inoculation was estimated from the total height of the seedlings after one vegetation period for both trials and after a second vegetation period following transplanting without any new inoculation, for the 1981 trial.

The data were interpreted according to the linear model of unbalanced multivariate analysis[14] in which means adjustment may have led to a "mean effect" slightly different from the arithmetic mean calculated with the figures given in Table 2.

B. *Poplar-alder association*

The objective was to compare the total production of mixed plots of 50% poplar (Belgian clone "UNAL": *P. trichocarpa* × *P. deltoides*) and 50% black alder provenance, and pure plots of each species. The site had a high fine particle content and was poor (Table 1). Since it was known from previous trials in the vicinity that some poplar clones not only withstand such sites but grow quickly, these conditions were considered good for putting to evidence a significant effect of a nitrogen fixing species on a highly productive species. The 4 replications were distributed according to an ecological survey. Each plot covered 72 m², the spacing was 2 × 1.5 m, the whole experiment had an area of 864 m². It was laid out in Spring 1980 with sets for "UNAL" and 2-years-old seedlings for the alder.

Table 1. Physical and chemical analyses of study site (means of 9 samples)

Physical analyses		Chemical analyses
0–0.002 mm	59%	Total N (Kjeldahl) 0.41%
0.002–0.02 mm	20%	N-NO$_3$ 11.4 ppm
0.02–0.05 mm	2%	N-NH$_4$ 9.2 ppm
0.05–0.2 mm	6%	P (fluorhydric acid) 0.09%
0.2–2 mm	13%	K (fluorhydric acid) 1.2%

Table 2a. Total heights (cm) of alder seedlings at age 1 and 2 years in 1981 trial

	A. glutinosa		*A. cordata*		*A. rubra*		Mean effect	
	1	2	1	2	1	2	1	2
Control	29.8	105.8	28.9	92.2	16.2	83.1	26.0	93.4
Inoculated	30.3	123.7	33.1	97.4	17.8	83.9	27.7	101.1
Significance level	NS	1%	1%	1%	NS	NS	5%	1%

Species × inoculation interaction: 1 year $P = 0.05$, 2 years $P = 0.01$

Table 2b. Total heights (cm) of alder provenances at age 1 year in 1982 trial

	A. glutinosa				*A. cordata*				*A. rubra*				Mean
	1	2	3	Mean	1	2	3	Mean	1	2	3	Mean	effect
Control	17	15	19	17	24	20	20	21	40	25	31	32	23
Inoculated	18	13	19	17	25	19	18	21	34	23	23	27	22
Sign. levels	NS	NS	NS	NS	NS	NS	NS	NS	5%	NS	1%	1%	1%

Provenance × inoculation interaction: NS
Species × inoculation interaction: 1%

Results

A. Alder-Frankia *interaction*

Independent of the inoculation, all alder seedlings bore nodules and it was difficult to notice any differences in their shape, abundance or localization. Observations were concentrated on seedling total heights at age 1 and 2 for the 1981 trial and age 1 for the 1982 trial (Table 2).

A slight interaction appears at age 1 for the 1981 trial. It is due to the favorable effect of the inoculation on *Alnus cordata* (+ 14% in height). The interaction grows stronger at age 2. The effect of the inoculation is still felt for *A. cordata* (+ 8%), but the interesting fact is that it also appears for *A. glutinosa* (+ 15%).

For the second trial, the results seem fairly different. Although a strong "species × inoculation" interaction exists, the general effect of the inoculation is significantly depressive. This general unfavorable effect is due to two provenances of *Alnus rubra*, whereas for the other species the inoculation of *Frankia* has no effect.

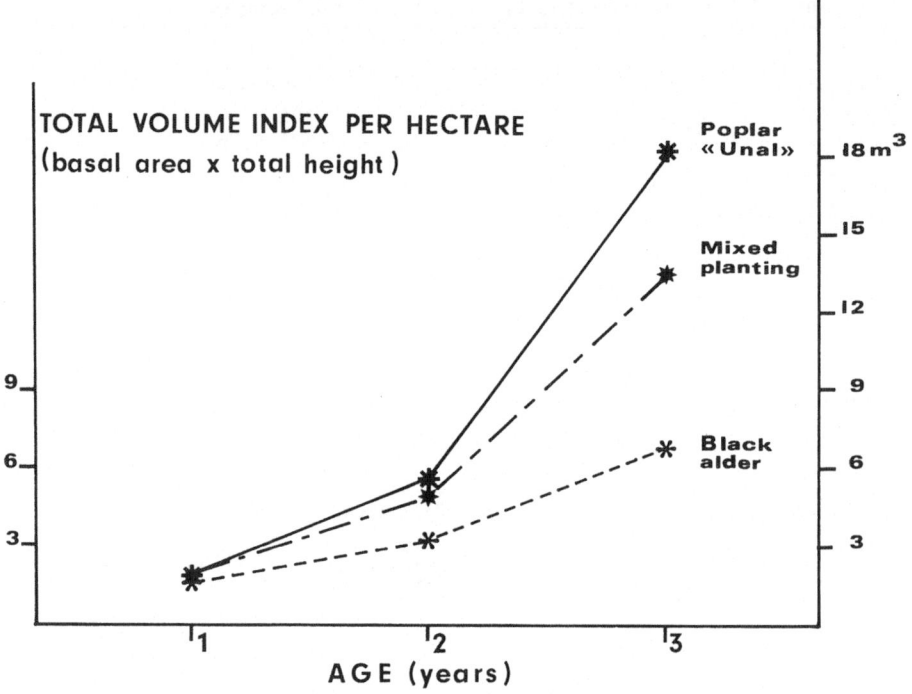

Fig. 1. Association of poplar and alder compared to pure plots of both species.

B. *Poplar-alder association*

Stem volumes have been estimated by multiplying the tree basal area by its total height. Thus, values obtained can only be used to compare the production of the different types of plantings.

Figure 1 shows clearly that the total volume production of mixed plots is lower than that of pure poplar plots. This inferiority is mostly due to the number of poplar stems per unit area which is half in mixed plots than in pure plots. Except for poplars at age 2, the single-stem-volume is not affected by the type of planting (Table 3). This means that no competitive or cooperative effects between the species have yet appeared.

Discussion

A. *Alder*-Frankia *interaction*

Different research groups suggest that the genetic improvement of alder should take in account its symbionts at the same time[5]. Most of the work in that field has been initiated with only two of the partners: alders and ectomycorrhizal fungi[11] or alders and actinorhizal bacteria[8].

Table 3. Single stem volume index in pure and mixed plots, in cubic centimeters

Age	1 yr	2 yrs	3 yrs
Poplar "UNAL"			
Pure plots	576	1 641	5 419
Mixed plots	669	1 999	6 090
Level of significance of the differences	NS	5p. 100	NS
Black alder			
Pure plots	434	983	2 041
Mixed plots	477	1 044	1 963
Level of significance of the differences	NS	NS	NS

Controlled inoculations of alder seedlings with *Frankia* were traditionally made with crushed nodules[12,13]. This method is easy and efficient. Its disadvantage is lack of information about the number and type of strains applied. It is now preferred to inoculate with controlled strains of the endophyte. The preparation and inoculation technique applied in our experiment offers great ease from the user's standpoint.

It is well known that without the endophyte or nitrogen the alders do not develop properly. But this case does not interest a tree breeder for a practical point of view. It is not possible to deal with large quantities of seed lots and seedlings in *Frankia*-free conditions, and *Frankia*-free soils are very rare. So what we are looking for is a method to infect young seedlings as fast as possible with a concentrated inoculum obtained from efficient strains.

Studies of alder-*Frankia* interaction have been realized with *Alnus rubra, A. glutinosa, A. cordata, A. incana* and *A. crispa*[3,8,13]. The interaction values are not significant for height growth at age 21 weeks when seed lots of *A. rubra,* and corresponding nodules are collected within a small geographic range (48 km radius from Mary's Peak, Oregon, USA). When applying controlled strains of *Frankia*, it is possible to classify them for their spore (+) and spore (−) ability, for their infectiveness and for their efficiency. Maynard[8], using two spore (+) strains, was able to differentiate them for their infectiveness and efficiency, but he was not able to differentiate three spore (−) strains. Three of the conclusions of his work seem important: (1) The source of endophyte has a strong influence on both inoculation and plant growth. (2) Nodulation alone is a poor predictor of the overall symbiotic efficiency of a particular endophyte. (3) Most of the variation in results was accounted for by the main effects of host and inoculum and only a small part was accounted for by their interaction.

Our studies have two characteristics which seem new enough to be mentioned. (1) We deliberately inoculated a *Frankia* strain on a substrate which was not isolated from natural infection. Our aim was to find whether a higher concentration of inoculum at plantlet root level would increase their height growth and possibly interact with their genotypes. (2) Observations have been made and interaction noticed for the 1981 trial both after 1 and 2 vegetation periods without any new inoculation before transplanting.

In both our trials a significant alder-inoculation interaction was found at age 1 year. But it was due to different reasons: (1) Positive effect of the inoculation on *A. cordata,* on the one hand. (2) Depressive effect of the inoculation on 2 of the 3 *A. rubra* provenances on the second hand. Those two reasons are not necessarily in contradiction and they show that the strain we used reacted differently from one experiment to the other, according to the influence of different combinations of several factors. The common factors of the two trials were: the same *Frankia* strain, the same type of sowing substrate, one common provenance for *A. glutinosa* (provenance 3 of Table 2) and for *A. cordata* (provenance 2 of Table 2). Other factors were different: year effect, site in nursery and probably the inoculum concentration which was lower in 1981 than in 1982 (but this was not evaluated precisely).

Although the *Frankia* strain $ARbN4_b$ was isolated from *Alnus rubra* root nodules, results of our 1982 trial show that it is depressive on certain *Alnus rubra* provenances ($- 15\%$ and $- 26\%$ for provenance 1 and 3 respectively). As there is no evidence for competition between native strains and this one, because it would probably have affected the other alders as well, we can state that cross-inoculation studies must not be restricted to the species level but need a more accurate sampling within each species. This statement also concerns the endophyte as recalled earlier in the literature survey. And this means that a comprehensive cross-inoculation programme in nursery conditions would soon become gigantic and unrealistic.

On the other hand, the results indicate that the native population of *Frankia* in the nursery is concentrated enough to nodulate alder. Can this be considered sufficient for practical purposes? More time is needed to answer that question and information will be available when our 1981 trial which was transferred in a forest site in spring 1983 gives information.

B. Poplar-alder association

The use of interplanted alders for increasing the yield of a lumber producing species like poplars, in western Europe, is probably very old. Results may often be disappointing if one only considers the poplar growth and not the total production including alders. The reason is perhaps connected to the poplar-alder ratio per unit area and by the spacing. As a matter of fact poplars are normally planted at a 7×7 m spacing, or more, when veneer quality is looked for. At such a spacing the action of alders may be very slow.

Trials associating alders as an N_2-fixing species to another tree species give interesting results.

Miller and Murray[9] describe the long term effect of off-site red alders interplanted in 1933 within a 4-year-old Douglas-fir plantation in southwestern Washington. Insufficient available nitrogen after repeated forest fires limited tree growth. By age 48, Douglas-fir volume per hectare in the mixed stand averaged $219 \, m^3$ compared to $205 \, m^3$ in the pure stand. And total standing volume per hectare was $395 \, m^3$ if alders were included.

More recently Miller and Murray[10] have studied different ways to add nitrogen to Douglas-fir forests. They consider that alders interplanted within a young Douglas-fir stand might be a good biological solution. If alders have to be planted at the same time as Douglas-firs, the former may overgrow the latter as shown by a Nelder plot established by M. Newton in Oregon (pers. comm.). In that case it might be suggested to use a slower growing N_2-fixing species like Sitka alder (*Alnus sinuata*)[1].

When alders and poplars are associated, the poplars are seldom overgrown, even if close spacings are used. The poplar-alder ratio seems important. In their trial associating black alder and a poplar clonal mix, Hansen and Dawson[6] show that the most efficient value on poplar height growth is 66% alders. In the same way, Courrier and Garbaye[2] notice that an increased volume production is obtained for the black cottonwood "Fritzi Pauley" when associated with an increased number of black alders per unit area. In southern Washington a superiority of 1.6 was obtained in mixed black cottonwood-red alder closely spaced 2-year-old plots compared to pure black cottonwood plots[4]. But this superiority does not exist any more in the same trial at age 8 after 4 harvests. The reason is a high mortality of alder stumps (80%) compared to poplar stumps (40%) (DeBell pers. comm.).

In our trial, we never got any superiority of mixed plots compared to pure plots. This means that the nitrogen fixation of the alders has not yet been profitable to poplars. This slow response may be due to the rather wide spacing compared to other people's trial and/or to the slow nitrogen return to the soil after a slow decomposition of alder leaves and root nodules.

Future developments

A. Alder-Frankia *interaction*

Although our 1983 trial concerns 2 *Frankia* strains instead of 1 in the trials reported here, it is now clear that it is impossible to increase the number of strains and alder genotypes indefinitely in nursery conditions. This means that the cross-inoculation of alders with *Frankia* will have to be initiated in laboratory conditions using the nodulation speed as an infectivity and efficiency reference even if it is not totally reliable[8]. Performant combinations should then be confirmed in green-houses, nurseries and forest sites. Given this testing scheme one will have to choose between two strategies. (1) Select highly efficient alder-*Frankia* combinations; (2) Select *Frankia* strains for their general efficiency on a large number of alder genotypes.

And in both cases one will have to make sure that these strains resist environmental stresses and variations including competition with other soil strains.

On the pure production point of view it is possible that strategy 1 gives higher yields, but on the practical point of view, since a large number of alder genotypes will be needed for reforestation, strategy 2 seems more realistic.

B. Poplar-alder associations

These are a particular case of mixed stands associating N_2-fixing species and other species. The trial reported here was the first of the series established with different production goals: lumber or in-dustrial wood or both. The site qualities vary from very rich to very poor and the soil texture from clay to sand. The N_2-fixing species are several alders or black locust, the associated species are conifers (Con-torta pine) or broadleaf trees (aspens, poplars, mazzard). The charac-teristics assessed will be single tree growth and volume per unit area, leaf-N in the associated species and soil-N.

End note

This paper may seem a bit disappointing as far as positive results are concerned, but we deliberately decided to write it because we thought that other forest scientists would appreciate to know that some experiments even when laid out in suitable conditions do not produce results which fit perfectly with the hypotheses before hand. We hope this will be helpful to someone.

Acknowledgements This research was partly supported by the French Energy Agency (Agence pour la Maîtrise de l'Energie). The authors thank Hervé Duval, Dominique Piermant and Patrick Poursat for their technical cooperation.

References

1 Binkley D 1981 Nodule biomass and acetylene reduction rates of red alder and Sitka alder on Vancouver Island B.C. Can. J. For. Res. 11, 281–286.

2 Courrier G and Garbaye J 1981 A propos de la sylviculture des peuplements mélangés. Revue Forest. Franç. 33, 289–292.

3 Dawson J O and Sun S H 1981 The effect of *Frankia* isolates from *Comptonia peregrina* and *Alnus crispa* on the growth of *Alnus glutinosa, A. cordata* and *A. incana* clones. Can. J. For. Res. 11, 758–762.

4 DeBell D S, Strand R F and Reukema D L 1977 Short rotation production of red alder. Some options for future forest management. *In* Utilization and Management of Alder. Proceedings of Symposium. Publ. Pacific Northwest Forest and Range Experiment Station. Forest Service USDA. Portland Oregon USA 231–244.

5 Hall R B and Maynard C A 1979 Considerations in the genetic improvement of alder. *In* Symbiotic Nitrogen Fixation in the Management of Temperate Forests. Eds. J C Gordon, C T Wheeler and D A Perry. Forest Research Laboratory. Oregon State University. Corvallis (Or) USA, 222–344.

6 Hansen E A and Dawson J O 1982 Effect of *Alnus glutinosa* on hybrid *Populus* height growth in a short-rotation intensively cultured plantation. Forest Sci. 28, 49–59.

7 Lalonde M and Calvert H E 1979 Production of *Frankia* hyphae and spores as an infective inoculant for *Alnus* species. *In* Symbiotic Nitrogen Fixation in the Management of Temperate Forests. Eds. J C Gordon, C T Wheeler and D A Perry. Forest Research Laboratory Oregon State University Corvallis (Or). USA 95–110.

8 Maynard C A 1980 Host-symbiont interactions among *Frankia* strains and *Alnus* open-pollinated families. PhD Dissertation. Iowa State University, Ames, Iowa, USA 93p.

9 Miller R E and Murray M D 1977 The effect of red alder on growth of Douglas-fir. *In* Utilization and Management of Alder. Proceedings of symposium. Publ. Pacific Northwest Forest and Range Experiment Station. Forest Service USDA. Portland. Oregon USA, 283–306.

10 Miller R E and Murray M D 1979 Fertilizer versus red alder for adding nitrogen to Douglas-fir forests of the Pacific Northwest. *In* Symbiotic Nitrogen Fixation in the Management of Temperate Forests. Eds. J C Gordon, C T Wheeler and D A Perry. Forest Research Laboratory. Oregon State University. Corvallis (Or) USA. 356–373.

11 Molina R 1981 Ectomycorrhizal specificity in the genus *Alnus*. Can. J. Bot. 59, 325–334.

12 Monaco P A, Ching T M and Ching K K 1981 Variation of *Alnus rubra* for nitrogen fixation capacity and biomass production. Silvae Genetica 30, 46–50.

13 Monaco P A, Ching T M and Ching K K 1982 Host-endophyte effects on biomass production and nitrogen assimilation in *Alnus rubra* actinorhizal symbiosis. Bot. Gaz. 143, 298–303.

14 Searle S R 1971 Linear models John Wiley and Sons Inc., NY, London, Sydney, Toronto 532 p.

Plant and Soil 78, 245–258 (1984).
© 1984 Martinus Nijhoff/Dr W. Junk Publishers, The Hague.

Production, decomposition, and nitrogen dynamics of *Myrica gale* litter

CHRISTA R. SCHWINTZER
Department of Botany and Plant Pathology, University of Maine, Orono, ME 04469, USA

Key words Actinorhizal plants Litter decomposition Litter deposition
Nitrogen dynamics Nitrogen-fixing plants Peatland Wetland

Summary *Myrica gale* litter deposition and decomposition were studied in a central Massachusetts peatland to determine the amount of N made available to the ecosystem by these processes. Leaf litter added 114–140 g biomass m^{-2} annually and contained 2.12–2.59 g N m^{-2} returning about 70% as much N to the ecosystem as was fixed annually by *Myrica gale*. During the first five years of decomposition, the leaf litter lost only 40% of its initial biomass and released only 10% of its initial N content. About 60% of its original N mass was still present when the litter reached the permanently waterlogged zone, and thus was effectively lost to the vegetation. The low decomposition rate was due primarily to the chemical content of the litter because similarly low rates were observed in an upland forest where the native litter decayed rapidly. The initial lignin content (40%) of *M. gale* litter may be largely responsible for its slow decomposition in spite of its relatively high (1.69%) initial N content. *M. gale* litter decayed substantially more slowly and had a much higher initial lignin content than the litter of other woody N$_2$-fixing plants which have been examined.

Introduction

Myrica gale L. is an actinorhizal shrub capable of fixing substantial amounts of nitrogen[34] and is commonly found in open minerotrophic peatlands and along the shores of lakes and streams in northern North America and Europe. Several aspects of nitrogen fixation by *M. gale* have been studied in a peatland in central Massachusetts. Here *M. gale* is the primary agent of biological N$_2$ fixation and contributes 3.0–3.7 g N m^{-2} yr^{-1} while all other agents combined fix only 0.1 g N m^{-2} yr^{-1} [35]. *M. gale* has an annual net production of approximately 550 g m^{-2}. This contains 8.6 g N m^{-2} yr^{-1}, approximately 43% of which is supplied by N$_2$ fixation[36]. The seasonal patterns of nitrogenase activity, energy use in N$_2$ fixation, nodule growth, and morphology of the *Frankia* sp. endophyte have also been described[34,37,38].

Nitrogen fixed by *M. gale* and other nitrogen fixing plants is transferred to the rest of the ecosystem primarily through leaf litter. Additional amounts are transferred through other litter and root turnover. The availability of the N contained in these plant parts to the peatland vegetation depends on the rate of N release, the amounts of N retained in refractory organic fractions, and the amounts of N reaching the permanently waterlogged horizons.

In the present study I estimate the availability of the N_2 fixed by *M. gale* to the rest of the ecosystem. This was done by measurement of N input to the previously studied *M. gale*-dominated peatland via litter deposition and subsequent N release by litter decomposition.

Study area

The primary study site is located in a weakly minerotrophic open peatland known as Tom Swamp in the Harvard University forest at Petersham, Massachusetts. Litter deposition and decomposition were measured in a 0.7 ha area designated "Open Mat" in an earlier study[34]. The vegetation is dominated by *M. gale* (mean cover 53%) and *Chamaedaphne calyculata* (12%). The ground layer consists primarily of *Sphagnum* spp. (35%) and leaf and woody litter.

Litter decomposition was also measured at the edge of nearby Harvard Pond ("Pond Side") and in an upland forest. The vegetation at the Pond Side site consists of a narrow band of low shrubs including *M. gale* at the lake shore and a band of tall shrubs just behind. The adjacent forest is dominated by *Acer rubrum* and *Quercus rubra*. The soils are composed of a shallow layer of well decomposed organic matter overlying mineral soils consisting primarily of sand and gravel. The "Upland Forest" site is located on a gentle slope near the Harvard Forest Administration Building approximately 3 km NE of the Open Mat site. The site is occupied by a second growth forest more than 40 years old which is dominated by *Fraxinus americana* and *Acer saccharum*. The soils are derived from glacial till and have a mull humus layer. Further details concerning the Open Mat and Pond Side sites as well as information on other studies in Tom Swamp are given by Schwintzer[34].

Nomenclature of the vascular plants follows that of Fernald[15].

Methods

Litter deposition

Litter was collected in 30 traps randomly located along three widely spaced transects within the Open Mat site. Each trap consisted of a polyethylene tub 14.5 cm high with an upper diameter of 19 cm. Holes were drilled in the bottom for drainage and litter was collected on a fiberglass screen (1 X 2 mm mesh size) resting on a polyethylene support 2.5 cm above the tub bottom. Use of such small litter traps was necessitated by the high density of shrub stems (mean density of *M. gale* stems 69 stems m^{-2}) and the low mean height (63 cm) of the canopy[36]. The fiberglass screens minimized leaching losses from the litter by facilitating rapid drainage after rains. They also kept the litter above all standing water levels except the winter extremes. Litter traps were emptied biweekly during the main period of litter fall from late August through early December and monthly from early May through late August. The collected litter was separated into the four categories shown in Table 1. Bud scales were

Table 1. Biomass and N-content of litter deposited in a *Myrica gale*-dominated peatland in central Massachusetts

Litter component	Biomass, g m⁻² 1979[a]	1980	N, g m⁻² 1980	Mean % N[b]
Myrica gale leaves	133 (140)	114	2.12	1.85
Myrica gale twigs[c]	13 (27)	16	0.30	1.78
Other leaves	28 (52)	40	0.49	1.26
Other twigs[c]	1 (2)	5	0.06	1.42
Total	175 (219)	175	2.97	

[a] Measured values for Aug. 1.–Dec. 31, 1979; values in parenthesis were corrected for litter deposition prior to Aug. 1 based on the seasonal distribution of litter deposition in 1980 (Fig. 1).
[b] Annual mean % N calculated from N concentration and 100°C dry weight of individual litter collections
[c] includes catkins, fruits and seeds

included with leaves while male catkins, seeds and fruits were included with twigs. All materials were dried to constant weight at 60°C in a forced air oven to minimize loss of volatile nitrogen containing compounds, weighed, and the weights were corrected to weight at 100°C. Methods for determining N content of subsamples are given below. Because of bulk, *M. gale* leaves were subsampled three times and then ground. All other materials were ground in their entirety and then subsampled three times before digestion.

Litter decomposition
    Fiberglass litter bags containing approximately 5 g fresh leaf litter dried to constant weight at 60°C and precisely weighed were placed in three groups at each of the three sites. The litter bags were 18 × 30 cm and had a mesh size of 1 × 2 mm. The bag size was chosen so that the leaves were spread out in a manner comparable to unconfined *M. gale* leaf litter. The mesh size was chosen to prevent loss of leaf fragments while permitting most soil animals access to the leaves. The advantages and disadvantages of the litter bag technique have been thoroughly discussed by Lousier and Parkinson[21]. The leaf litter consisted of yellow and brown leaves with complete abscission layers gathered directly from *M. gale* bushes at the Open Mat site in late October. After precise weighing the leaves and coded plastic tags were placed loosely in the litter bags and allowed to hydrate over damp paper towels until they lost all brittleness before closing the bags. When the bags were placed in the field, the current year's litter was removed and the bags were placed as flat as possible on the ground. Then the current year's litter was replaced over the bags. Thus the bags at the Open Mat rested on a mat of living *Sphagnum* spp. and were about 40% covered with litter. At the Pond Side and upland Forest sites the bags rested on soil and were completely covered. The bags were put in place on 7 November 1978 and were thoroughly soaked by a 6.4 mm rain that fell that night.
    On each collection date one bag from each group at each site was collected and the amount of litter and water covering the bag as well as presence of animals, fungal rhizomorphs and roots in the bags was noted. The contents were emptied into a bowl of deionized water and animals and roots were removed. The leaves were then gently rinsed to remove mineral soil particles, dried at 60°C and weighed.
    In addition, five samples of *M. gale* leaf litter were obtained on 23 July 1982 from a depth of 20 cm below the surface in peat hollows at the Open Mat site. Twenty cm is the approximate depth of the permanently waterlogged horizon. Pits were dug to a depth of 20 cm and lined with rigid polyethylene sleeves to exclude surface materials. Approximately 100 cm³ of peat were taken and pieces of leaves recognizable as *M. gale* were removed and processed in the same manner as the leaves in the litter bags. Many leaves were still intact and easily recognizable.

Total N and C were determined in random samples of litter from each bag. Materials dried at 60°C were ground to a homogeneous mixture and then further subsampled and weights corrected to weight at 100°C[4]. Total N was determined by Kjeldahl digestion using selenium as the catalyst and analyzing the digest directly for ammonia by colorimetry using procedures modified slightly from those of Bergersen[6]. Carbon was determined by wet oxidation followed by titration as described by Allen *et al.*[4].

Annual decomposition constants, k, were calculated to permit comparison with decay rates reported in the literature. They were derived from the exponential decay formula[20, 30] as follows:

$$k = \ln (X/X_0)/t$$

where X_0 is the dry weight initially present, X is the dry weight remaining at the end of the period of measurement, and t is time in years.

Results

Litter deposition

Approximately 20% more litter was produced in 1979 than in 1980 but the proportions of the various litter fractions were similar in the two years (Table 1). *M. gale* leaves constituted by far the largest fraction of the litter comprising approximately 65% of the total. Other leaves made up another 23%.

The seasonal pattern of litter deposition is shown in Fig. 1. *M. gale* leaf litter showed a strong seasonal pattern with 95% of the annual leaf fall occurring between August and December and having a strong peak in October. Other leaves also showed a peak in October but their deposition was more evenly distributed through the year with 54% falling between December and August. The general patterns of litter deposition were similar in 1979 and 1980 but the peak of litter deposition occurred two weeks earlier in 1979.

The nitrogen content of *M. gale* leaf and twig litter was substantially higher than that of other litter throughout the year (Tables 1 and 2). Nitrogen contents showed only weak seasonal patterns with relatively high N contents occurring in June and July and lower ones during the remainder of the year.

Litter decomposition

As the leaves decomposed they became increasingly fragile and fragmented. The condition of the leaves varied with the site. At the Open Mat site most leaves remained intact throughout the 4.7 yr period. At the Pond Side site most leaves remained intact for the first three years but showed variable fragmentation during the last 1.7 yr depending on microsite. In the Upland Forest site most leaves remained intact for the first two years but showed increasing fragmentation and loss of

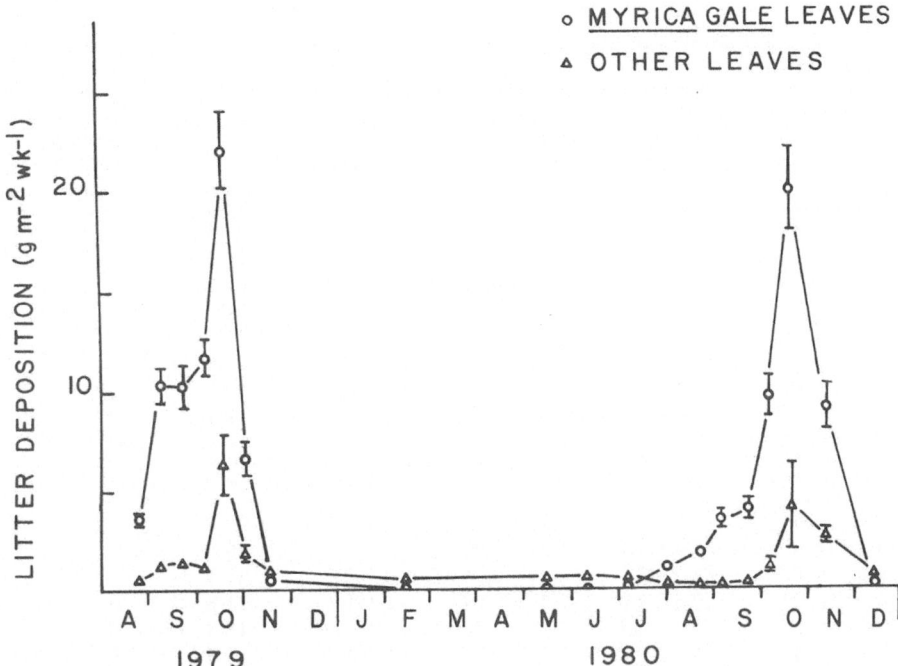

Fig. 1. Weekly litter deposition in a *Myrica gale*-dominated peatland in central Massachusetts in 1979 and 1980. Values are $\bar{X} \pm SE$; N = 30.

Table 2. Seasonal variation in N content of litter deposited in a *Myrica gale*-dominated peatland in central Massachusetts in 1980. Values are $\bar{X} \pm SE$; N = 3

Deposition period	*Myrica gale* leaves % N[b]	Deposition period	*Myrica gale* twigs[a] % N[b]	Other leaves % N[b]	Other twigs[a] % N[b]
July 1–July 25	2.13 ± 0.02	June 2–July 25	2.07 ± 0.02	1.77 ± 0.02	1.83 ± 0.00
July 25–Aug. 21	1.94 ± 0.04	July 25–Aug. 21	1.54 ± 0.01		
Aug. 21–Sept. 4	1.74 ± 0.01				
Sept. 4–Sept. 19	1.83 ± 0.08	Aug. 21–Sept. 19	1.87 ± 0.02	1.58 ± 0.02	1.52 ± 0.02
Sept. 19–Oct. 6	1.81 ± 0.04				
Oct. 6–Oct. 21	1.84 ± 0.08	Sept. 19–Oct. 21	1.68 ± 0.01	1.06 ± 0.01	1.45 ± 0.01
Oct. 21–Nov. 3	1.87 ± 0.01				
		Oct. 21–Dec. 9	1.77 ± 0.01	1.18 ± 0.01	1.17 ± 0.01

[a] includes catkins and seeds
[b] based on 100°C dry weight.

epidermal and mesophyll tissues in subsequent years. Leaves collected after 4.7 yr were not only very fragile and extensively fragmented but enriched in midribs and other vascular tissue as well.

Several characteristics of *M. gale* leaf litter decomposition at three sites are given in Table 3. The rate of decomposition was similarly low at all three sites during the first three years with roughly 60% as much biomass remaining after three years as was present originally. However, in the subsequent 1.7 yr the rate of biomass loss was substantially greater in the Upland Forest site than at the other two sites.

Table 3. Decomposition of *Myrica gale* leaf litter in litter bags in three environments. Values are $\bar{X} \pm SE; N = 3$

Date	Biomass remaining %	N concentration[a] %	N mass remaining[b] %	C:N	Decomposition constant, k[c] year[-1]
A. Open Mat					
Nov. 7, 78	100.0	1.69 ± 0.02	100.0	28.9	
April 18, 79	89.6 ± 0.1	1.78 ± 0.01	94.4	27.6	
Sept. 24, 79	83.7 ± 0.3	1.93 ± 0.06	95.6	24.7	
Nov. 30, 79	79.7 ± 0.9	2.07 ± 0.01	97.6	23.5	0.23
Nov. 14, 80	73.3 ± 0.1	2.21 ± 0.03	95.9	21.8	0.16
Nov. 23, 81	65.4 ± 0.2	2.35 ± 0.07	90.5	19.7	0.14
July 30, 83[d]	60.5	2.51	89.9	19.8	0.11
Deep litter[e]		3.18 ± 0.08		15.5	
B. Pond Side					
Nov. 7, 78	100.0	1.69 ± 0.02	100.0	28.9	
April 18, 79	92.2 ± 0.1	1.82 ± 0.05	99.3	26.3	
Sept. 24, 79	81.2 ± 2.0	2.19 ± 0.04	105.2	22.2	
Nov. 30, 79	77.3 ± 2.0	2.20 ± 0.07	100.6	21.8	0.26
Nov. 14, 80	70.2 ± 5.7	2.55 ± 0.20	105.9	18.2	0.18
Nov. 23, 81	62.6 ± 0.8	2.71 ± 0.20	100.4	16.6	0.16
July 30, 83	51.8 ± 1.0	2.62 ± 0.18	80.3	18.7	0.14
C. Upland Forest					
Nov. 7, 78	100.0	1.69 ± 0.02	100.0	28.9	
April 18, 79	92.8 ± 0.2	1.77 ± 0.01	97.2	27.7	
Sept. 24, 79	86.2 ± 0.9	2.17 ± 0.04	110.7	22.5	
Nov. 30, 79	78.4 ± 1.8	2.23 ± 0.02	103.5	21.5	0.24
Nov. 14. 80	68.3 ± 1.3	2.35 ± 0.05	95.0	19.4	0.19
Nov. 23, 81	57.7 ± 2.8	2.66 ± 0.07	90.8	16.6	0.18
July 30, 83	31.2 ± 8.6	2.67 ± 0.04	49.3	17.3	0.25

[a] % 100°C dry weight
[b] remaining N as percent of the original N mass
[c] based on litterbag weight loss and calculated according to Olson[30]. Values are for 1, 2, 3 and 4.7 yr respectively.
[d] N = 1
[e] litter taken 20 cm below the surface in peat hollows.

It was also greater than it had been previously at this site. This increased rate of loss was caused in part by loss of leaf fragments (see above) during the collection of the litter bags and processing of their contents. Consequently the true rate of loss due to decomposition during the last 1.7 yr at this site is uncertain. Fractional weight loss was greatest during the first year and substantially lower during subsequent years at all three sites with the exception of the high weight loss in the Upland Forest during the fourth plus fifth years. This is reflected in the decreasing values for k, the annual decomposition constant (Table 3). For example, at the Open Mat it declined from 0.23 for the first year to 0.11 for the 4.7 yr period. When calculated separately for the second through fifth years it was even lower with values of 0.08 (second year), and 0.11 (third year), and 0.05 (fourth plus fifth year).

The nitrogen dynamics of the decaying litter were also similar at the three sites during the first three years except that there was a net release of nitrogen during the first three years in the Open Mat and Upland Forest sites but not at the Pond Side. The N concentration in the residue increased steadily throughout the three-year period with slightly lower increases at the Pond Side and Upland Forest sites than in the Open Mat Site. Concurrently the C:N ratio declined from 28.9, its initial value, to 19.7 at the Open Mat site and 16.6 at the Pond Side and Upland Forest sites. During the first year there was a small net loss of N at all three sites during the cold months between November 1978 and April 1979. This was followed by net increases in the N mass during the warm months from May through November 1979. During the second and third years there was net mineralization at the Open Mat and Upland Forest sites resulting in release of approximately 9% of the original N mass by the end of the third year. But there was no net release of N at the Pond Side.

Conditions inside the litterbags were probably similar to those in unconfined litter except in the Upland Forest where the litter bag mesh size restricted access of larger earthworms. At the Open Mat the litter bags were covered by standing water from late fall through early spring. The amount of litter and moss growth covering the bags gradually increased from zero in April 1979 to 100% by November 1981. Fungal rhizomorphs were first seen in the bags in November 1979 and fine roots were present in all bags in 1980, 1981, and 1983. At the Pond Side the litter bags were covered by standing water during periods of exceptionally high water in late fall and winter. Litter cover over the bags was complete throughout the five year period in one group and in another it varied from 0–50%. Ectomycorrhizal roots and fungal rhizomorphs were first seen in the bags in November 1979 and became more numerous with time. In the Upland Forest the litter bags were completely covered with litter except during the late summer of the first year when most of the previous year's litter had disappeared from the surface of the mull soil. Earthworm castings and small earthworms (2–12/bag) were first seen in the bags in June 1979 and fungal rhizomorphs in September 1979. Both are continuously present after that except in July 1973 when conditions were very dry and earthworms were absent from the bags.

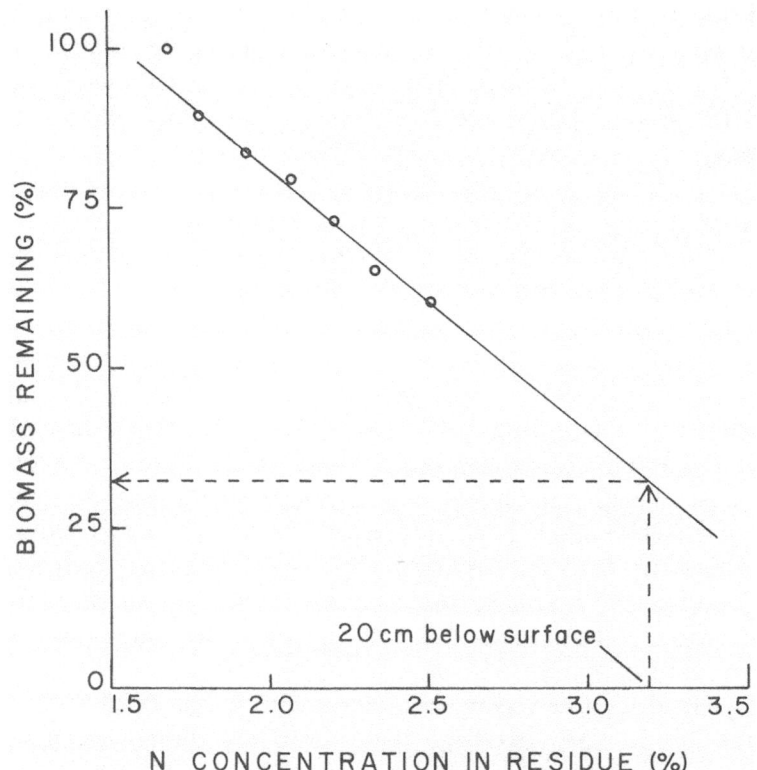

N CONCENTRATION IN RESIDUE (%)

Fig. 2. The relationship between the percentage of the original biomass remaining and the N concentration in the residual material for *Myrica gale* leaf litter decomposition in the open peatland during the 4.7 year period. The first point in the data was deleted from the regression analysis because the initial decrease in both weight and nitrogen was largely due to leaching not decomposition.

Discussion

Year to year differences in leaf litter production and the time of peak litter deposition were seen in both *M. gale* and other leaf litter. This is consistent with the observations of others and underscores the importance of measuring these parameters in more than one year to obtain representative values[12]. The litter trap method used here probably gave a reliable measure of the amount of leaf litter produced in this open peatland. The mean annual *M. gale* leaf litter production of $140\,\mathrm{g\,m^{-2}}$ in 1979 compares favorably to an estimate of $148\,\mathrm{g\,m^{-2}}$ obtained independently from leaf production. This estimate was obtained by correcting leaf production in 1979[36] for loss of weight due to translocation (20%) before litter fall[12].

Most of the N made available to the peatland via *M. gale* litter fall and its subsequent decomposition comes from leaves. Annual aboveground productivity of stems was about the same as that of leaves in 1979 but leaves have substantially higher N concentrations[36] and decompose more rapidly (*e.g.* Chamie and Richardson[11]) resulting in greater release of N before they reach the permamently waterlogged horizons of the peat where little further decomposition occurs[28]. Fruits and catkins also have relatively high N concentrations but represent only a small fraction of the total productivity[36].

The N content of *M. gale* leaves deposited in the peatland was about 37% lower than that of living leaves in late summer. The 37% reduction was calculated from 2.36% N concentration in late summer leaves[36], 1.85% N concentration in litter leaves (Table 1) and a 20% weight loss due to translocation[12]. Most of the lost N was exported to stems and rhizomes prior to leaf abscission[40] and the remainder was made available to the peatland system via leaching.

M. gale leaves deposited in the peatland contained about 2.59 g $N m^{-2}$ in 1979 (calculated from leaf litter biomass in 1979 and mean N concentration in 1980) and 2.12 g $N m^{-2}$ in 1980. This represents about 70% of the N_2 fixed in 1979 (3.72 g $N m^{-2}$ fixed) and 1980 (3.00 g $N m^{-2}$ fixed)[36]. Additional N was added in twig, flower and seed litter; in large stem litter; and in root turnover. Large stem litter and root turnover were not measured. Fine root turnover is important in forest ecosystems[18,23,32,33] and might also release significant amounts of N to the peatland system. However, in the only reported measurement of N release within the soil by an actinorhizal plant only small amounts of N were obtained by spruce (*Picea excelsa*) grown in association with alder (*Alnus glutinosa*)[43].

During the first 4.7 years of leaf litter decomposition in the peatland only 10% of the initial N mass was released (Table 3) suggesting that much of the initial N may still be present when the decomposing leaves reach the permanently waterlogged peat horizons and become effectively lost to the peatland vegetation. Indeed *M. gale* leaves still contained approximately 60% of their original N mass at a depth of 20 cm below the surface in peat hollows, the approximate depth of the permanently waterlogged zone. The N mass remaining at this depth was estimated as follows. The percentage of the original biomass remaining in the peatland was plotted as a function of the N concentration in the residue in the manner of Aber and Melillo[1]. As can be seen in Fig. 2, there is a strong inverse linear relationship ($r = 0.99, P < 0.001$) between these parameters for the full 4.7 year period. The resulting regression equation was used to estimate the fraction of biomass remaining (32.6%) in litter at the 20 cm depth from its N concentration (3.18%; Table 3). The N mass remaining was then calculated and expressed as a percentage (61%) of the original N mass. Thus less than half of the original leaf N mass was released before the leaf litter reached the permanently waterlogged horizon.

The N dynamics of the decomposing *M. gale* leaf litter (Table 3) were consistent with N dynamics observed in other ecosystems. Increases in the absolute N mass, often to values greater than the initial N mass, followed by a decrease have been observed repeatedly[14]. Increases in N concentration in the residue and a concomitant decrease in the C:N ratio have also been reported by many workers[1,3,14]. The initial release of N during the first winter was greatest in the open peatland which was flooded during most of this time and was probably due to leaching of soluble compounds[10,13,29]. The subsequent increase in the absolute N mass was probably due to fungal and bacterial immobilization of N with the added N coming from external sources including bulk precipitation, insect frass and import by fungi[8,14,17].

The relatively slow rate of *M. gale* leaf litter decomposition found at the Open Mat is consistent with decomposition rates observed in other northern peatlands (44°N to 75°N). The first year annual decomposition constant k (0.23) falls near the median of the first year k's for the northern peatlands summarized by Brinson, *et al.*[10]. Comparisons of decomposition rates beyond the first year are not possible because few studies continued for two or more years. Litter decomposition in northern peatlands is generally much slower than in other wetlands[10].

M. gale leaf litter decomposition rates do not fit the exponential decay model[20,30] very well. A perfect fit would require k to be the same each year but k was much larger at the Open Mat site during the first year (0.23) than during the second (0.08), third (0.11) and fourth plus fifth years (0.05). Similar patterns were also seen at the Pond Side and during the first three years in the Upland Forest. Larger first year k's have also been observed in terrestrial ecosystems[14,16,22]. The adequacy of the exponential decay function for describing the complex processes involved in litter decay has been questioned previously on the variety of grounds (*e.g.* Minderman[27], Anderson[5], Lousier and Parkinson[21]). In addition characteristics of several decay models commonly used to examine decomposition data have been compared by Wieder and Lang[44].

Litter decomposition is controlled by temperature, moisture, aeration, nutrient status of the substrate, presence of detritivores, and chemical composition of the litter. The rate of *M. gale* litter decomposition is surprisingly low given its relatively low initial C:N ratio. The low rate is probably caused primarily by the chemical composition

Table 4. Comparison of decomposition parameters of leaf litter from woody N_2 fixing species

Species	Biomass lost in first year %	Initial N %	Initial C:N[a]	Initial lignin %	Reference
Alnus glutinosa	> 70–90	3.06	16.3		9
		2.57	(19.5)		26
Alnus rubra	55	1.52	31.5	21	14
		2.1	(23.8)	10	2
Ceanothus spp.		0.9	(55.6)	10	2
Myrica gale	20–23	1.69	28.9	40[b]	Present study
Robinia pseudoacacia	40–41	1.67	24.2	25	19
		1.55	(32.3)	26	2

[a] Values in parantheses were calculated from the reported N concentrations assuming a C content of 50%. Ash-free values of approximately 50% C are characteristic of leaves.
[b] Determined by K. Cromack using methods slightly modified from those of Van Soest[42].

of the litter, in particular its very high initial lignin content (40%). Lignin decomposes slowly and has been identified as a major factor controlling litter decomposition in several different ecosystems[3,16,24,25,31]. Possible environmental causes of the low decomposition rates can be ruled out as a primary factor because equally low rates were observed during the first three years in the Upland Forest, where the native litter decays rapidly.

The *M. gale* leaf litter decayed substantially more slowly and has a much higher initial lignin content than the litter of other woody N_2 fixing species which have been examined (Table 4). Comparison of all species in Table 4 shows that they vary widely with respect to decomposition rate and chemical composition. For example, decomposition rates range from 20–23% biomass lost in one year by *M. gale* to more than 70–90% by *Alnus glutinosa*. Initial lignin contents range from 10% in *Ceanothus* sp. to 40% in *M. gale*. Consequently, N_2 fixing species do not share a common pattern of litter decomposition and each species must be examined individually.

Slow decomposition of litter leads to organic matter accumulation on upland as well as wetland sites. Consequently, presence of *M. gale* enhances organic matter and N accumulation in all ecosystems in which its litter is retained on the site. Interestingly organic matter and N accretion have also been reported under two other N_2 fixing species, namely *Alnus rubra*[41] and *Ceanothus velutinus*[7].

The high lignin content of *M. gale* leaves probably has one or more adaptive advantages because the large carbon cost of producing this lignin must be accompanied by some benefit to be evolutionarily retained. The high lignin content probably has two functions. First it may protect the living leaves against herbivory by greatly increasing the nondigestible fraction of the leaves and by interfering with enzymatic degradation of other plant constituents[3]. Second the high lignin content may give *M gale* an advantage over potential non-nitrogen fixing competitors by limiting the nitrogen enrichment of the site. Typically actinorhizal plants are found in early successional communities on nitrogen-poor sites and are gradually replaced by more nitrogen demanding species as nitrogen accumulates on the site[39]. By retarding nitrogen-release from *M. gale* leaf litter, its high lignin content may delay or even prevent this competitive displacement of *M. gale*.

Acknowledgements I thank S. A. Lancelle, L. D. Disney, S. L. Burpee, and A. Ostrofsky for technical assistance, K. Cromack for performing the lignin analysis, and J. D. Tjepkema for helpful discussions and comments on the manuscript. The data was collected while I held a research fellowship at Harvard University, Harvard Forest, Petersham, Massachusetts. This work was supported by NFS Grant No. DEB 81-06952.

References

1   Aber J D and Melillo J M 1980 Litter decomposition: measuring relative contributions of organic matter and nitrogen to forest soils. Can. J. Bot. 58, 416–421.
2   Aber J D and Melillo J M 1982 Nitrogen immobilization in decaying hardwood leaf litter as a function of initial nitrogen and lignin content. Can. J. Bot. 60, 2263–2269.
3   Alexander M 1977 Introduction of Soil Microbiology, 2nd ed. John Wiley and Sons, New York, 467 pp.
4   Allen S E, Grimshaw H M, Parkinson J A, Quarmby C and Roberts J D 1976 Chemical analysis. pp 411–466. *In* Methods in Plant Ecology. Ed. S B Chapman. Blackwell Scientific, Oxford.
5   Anderson J M 1973 The breakdown and decomposition of sweet chestnut (*Castanea sativa* Mill.) and beech (*Fagus sylvatica* L.) leaf litter in two deciduous woodland soils. I. Breakdown, leaching and decomposition. Oecologia 12, 251–274.
6   Bergersen F J 1980 Measurement of nitrogen fixation by direct means. pp 65–110. *In* Methods for Evaluating Biological Nitrogen Fixation. Ed. F J Bergersen. John Wiley and Sons, Chichester.
7   Binkley D, Cromack K Jr and Frederiksen R L 1982 Nitrogen accretion and availability in some snowbrush ecosystems. Forest Sci. 28, 720–724.
8   Bocock K L 1963 Changes in the amount of nitrogen in decomposing leaf litter of sessile oak (*Quercus petraea*). J. Ecol. 51, 555–566.
9   Bocock K L 1964 Changes in the amount of dry matter, nitrogen, carbon and energy in decomposing woodland leaf litter in relation to the activities of the soil fauna. J. Ecol. 52, 273–284.
10 Brinson M M, Lugo A E and Brown, S 1981 Primary productivity, decomposition and consumer activity in freshwater wetlands. Annu. Rev. Ecol. Sys. 12, 123–161.
11 Chamie J P M and Richardson C J 1978 Decomposition in northern wetlands. pp 115–130. *In* Freshwater Wetlands Ecological Processes and Management Potential. Ed. R E Good, D F Whigman and R L Simpson. Academic Press, New York.
12 Chapman S B 1976 Production ecology and nutrient budgets. pp 157–228. Methods in Plant Ecology. Ed. S B Chapman. Blackwell Scientific Publications, Oxford.
13 Day F P Jr 1983 Effects of flooding on leaf litter decomposition in microcosms. Oecologia 56, 180–184.

14 Edmonds R L 1980 Litter decomposition and nutrient release in Douglas-fir, red alder, western hemlock, and Pacific silver fir ecosystems in western Washington. Can. J. For. Res. 10, 327–337.

15 Fernald M L 1950 Gray's Manual of Botany, 8th ed. American Book Co. New York, 1632 p.

16 Fogel R and Cromack K Jr 1977 Effect of habitat and substrate quality on Douglas fir litter decomposition in western Oregon. Can. J. Bot. 55, 1632–1640.

17 Gosz J R, Likens G E and Bormann F H 1973 Nutrient release from decomposing leaf and branch litter in the Hubbard Brook Forest, New Hampshire. Ecol. Monogr. 43, 173–191.

18 Harris W F, Kinerson R S Jr. Edwards N T 1977 Comparison of below-ground biomass of natural deciduous forest and loblolly pine plantations. Pedbiologia 17, 369–381.

19 Hirschfeld J R 1982 Forest floor decomposition and nitrogen status in a northern hardwood forest ecosystem, supporting a nitrogen fixing tree species (*Robinia pseudoacacia* L.). Masters Thesis, University of Massachusetts, Amherst. 62 pp.

20 Jenny H, Gessel S P and Bingham F T 1949 Comparative study of decomposition of organic matter in temperate and tropical regions. Soil Sci. 68, 419–432.

21 Lousier J D and Parkinson D 1976 Litter decomposition in a cool temperate deciduous forest. Can. J. Bot. 54, 419–438.

22 MacLean D A and Wein R W 1978 Weight loss and nutrient changes in decomposing litter and forest floor material in New Brunswick forest stands. Can. J. Bot. 56, 2730–2749.

23 McClaugherty C A, Aber J D and Melillo J M 1982 The role of fine roots in the organic matter and nitrogen budgets of two forested ecosystems. Ecology 63, 1481–1490.

24 Meentemeyer V 1978 Macroclimate and lignin control of litter decomposition rates. Ecology 59, 465–472.

25 Melillo J M, Aber J D and Muratore J F 1982 Nitrogen and lignin control of hardwood leaf litter decomposition dynamics. Ecology 63, 621–626.

26 Mikola P 1958 Liberation of nitrogen from alder leaf litter. Acta For. Fen. 67, 1–10.

27 Minderman G 1968 Addition, decomposition and accumulation of organic matter in forests. J. Ecol. 56, 355–363.

28 Moore P D and Bellamy D J 1974 Peatlands. Springer-Verlag New York, Inc., New York, 221 pp.

29 Nykvist N 1963 Leaching and decomposition of water-soluble organic substances from different types of leaf and needle litter. Stud. For. Suec. 3, 1–31.

30 Olson J S 1963 Energy storage and the balance of producers and consumers in ecological systems. Ecology 44, 322–331.

31 Pandey U and Singh J S 1982 Leaf-litter decomposition in an oak-conifer forest in Himalaya: the effects of climate and chemical composition. Forestry 55, 47–59.

32 Persson H 1978 Root dynamics in a young Scots pine stand in central Sweden. Oikos 30, 508–519.

33 Santantonio D, Hermann R K and Overton W S 1977 Root biomass studies in forest ecosystems. Pedobiol. 17, 1–31.

34 Schwintzer C R 1979 Nitrogen fixation by *Myrica gale* root nodules in a Massachusetts wetland. Oecologia 43, 283–294.

35 Schwintzer C R 1983 Nonsymbiotic and symbiotic nitrogen fixation in a weakly minerotrophic peatland. Am. J. Bot. 70, 1071–1078.

36 Schwintzer C R 1983 Primary productivity and nitrogen, carbon and biomass distribution in a dense *Myrica gale* stand. Can. J. Bot. 61, 2943–2948.

37 Schwintzer C R, Berry A M and Disney L D 1982 Seasonal patterns of root nodule growth, endophyte morphology, nitrogenase activity, and shoot devleopment in *Myrica gale*. Can. J. Bot. 60, 746–757.

38 Schwintzer C R and Tjepkema J D 1983 Seasonal patterns of energy use, respiration and nitrogenase activity in root nodules of *Myrica gale*. Can. J. Bot. 61, 2937–2942.

39   Silvester W B 1977 Dinitrogen fixation by plant associations excluding legumes. pp. 141 – 190. *In* A Treatise on Dinitrogen Fixation. Section IV. Ed. R W F Hardy and A H Gibson. John Wiley and Sons, New York.

40   Sprent J I, Scott R and Perry K M 1978 The nitrogen economy of *Myrica gale* in the field. J. Ecol. 66, 657 – 668.

41   Tarrant R F and Miller R E 1963 Accumulation of organic matter and soil nitrogen beneath a plantation of red alder and Douglas-fir. Soil Sci. Soc. Am. Proc. 27, 231 – 234.

42   Van Soest P J 1963 Use of detergents in the analysis of fiberous feeds. II. A rapid method for the determination of fiber and lignin. J. Assoc. Off. Anal. Chem. 49, 546 – 551.

43   Virtanen A I 1957 Investigations on nitrogen fixation by alder. II. Associated culture of spruce and inoculated alder without combined nitrogen. Physiol. Plant. 10, 164 – 169.

44   Wieder R K and Lang G E 1982 A critique of the analytical methods used in examining decomposition data obtained from litter bags. Ecology 63, 1636 – 1642.